Computer Supported C

159
Experiment

185
User Profile

Page 96 ☆

HCI book
Page 145
Xerox system

Direct & indirect
spare vs place
chap 3 (P35)

Springer

London
Berlin
Heidelberg
New York
Barcelona
Hong Kong
Milan
Paris
Santa Clara
Singapore
Tokyo

Also in this series

Alan J. Munro, Kristina Höök and
David Benyon (Eds)

Social Navigation of Information Space

 Springer

Alan J. Munro, PhD, MA
Department of Computing, Napier University, Canal Court,
42 Craiglockhart Avenue, Edinburgh, EH14 1LT, UK

Kristina Höök, PhD, PhLic, MSc
Swedish Institute of Computer Science, Box 1263, S-164 28 Kista, Sweden

David Benyon, PhD
Department of Computing, Napier University, Canal Court,
42 Craiglockhart Avenue, Edinburgh, EH14 1LT, UK

Series Editors
Dan Diaper, PhD, MBCS
School of Design, Engineering and Computing, Bournemouth University,
Talbot Campus, Fern Barrow, Poole, Dorset BH12 5BB, UK

Colston Sanger
Shottersley Research Limited, Little Shottersley, Farnham Lane
Haslemere, Surrey GU27 1HA, UK

ISBN 1-85233-090-2 Springer-Verlag London Berlin Heidelberg

British Library Cataloguing in Publication Data
Social navigation of information space. – (computer supported cooperative work)
1. Information networks – Social aspects – Congresses 2. Computers and civilization –
Congresses 3. Human-computer interaction – Congresses
I. Munro, Alan J. II. Höök, Kristina III. Benyon, David
303.4'833
ISBN 1852330902

Library of Congress Cataloging-in-Publication Data
Social navigation of information space / Alan J. Munro, Kristina Höök and David Benyon (eds.).
    p.   cm. - - (Computer supported cooperative work)
    Includes bibliographical references.
    ISBN 1-85233-090-2 (alk paper)
    1. Computers and civilization. 2. Information technology – Social aspects. 3. User interfaces
    (Computer systems) I. Munro, Alan J., 1965– . II. Höök, Kristina. III. Benyon, David. IV
    Series.
QA76.9.C66S625   1999
303.48'34–dc21   99-25431

Typesetting: Gray Publishing, Tunbridge Wells, Kent
Printed and bound at The Cromwell Press, Trowbridge, Wiltshire
34/3830-543210 Printed on acid-free paper SPIN 10696950

Preface

In March 1998 an international group of 30 researchers attended a workshop on Roslagens Pärla, a small island in the Swedish archipelago. The topic of the workshop was "personal and social navigation" of information space. Although the researchers came from a wide variety of backgrounds – computer science, human–computer interaction, social science, psychology, information retrieval, computer supported co-operative work – they shared an interest in exploring new ways of thinking about the relationships between people, technologies and information. In particular they were interested in the notion of *social navigation*: how to develop and enrich the experience of dealing with information within the electronic "worlds" provided by computing and communication technologies.

The idea for the workshop arose from discussions within the group working on a research project called PERSONA, a collaboration between the Swedish Institute of Computer Science (SICS) and Napier University, Edinburgh. David Benyon from Napier and Kristina Höök from SICS had developed the project after their experiences working with intelligent user interfaces. Attempts to get the computer to make sensible inferences about what people wanted to do, so that it could tailor the provision of information to their needs, had not been successful (Höök and Svensson, Ch. 13). The dream of having intelligent interface "agents" was still a long way from becoming a reality because the techniques used to build agents and other "intelligent" user interfaces ignored a fundamental feature of people: people are social beings. The question for the workshop to consider was how to bring the social into information provision.

When people need information, they will often turn to other people rather than use more formalised information artefacts. When navigating cities, people tend to ask other people for advice rather than study maps; when trying to find information about pharmaceuticals medical doctors tend to ask other doctors for advice; if your child has red spots you might phone your mother or talk to a friend for an opinion. Even when we are not directly looking for information we use a wide range of cues, both from features of the environment and from the behaviour of other people, to manage our activities. Alan Munro observed the ways that people followed crowds or simply sat around at a venue when deciding which shows and street events to attend at the Edinburgh Festival. We might be influenced to pick up a book because it appears well thumbed, we walk into a sunny courtyard because it looks attractive or we might decide to see a film because our friends enjoyed it. Not only do we find our ways through spaces from talking to or following the trails of crowds of people, we also evaluate the things we find in these spaces through understanding them in a social context. We put them in a framework of relevance.

During a break at the workshop a group of participants were standing in the winter sunshine. They were considering how the results of the workshop could be more widely disseminated and about where the ideas raised during discussions might ultimately lead. A line of footprints led across the snow and into the woods. As a metaphor both for social navigation and for the unknown destination of this work, it seemed perfect: footprints in the snow.

This book is the outcome of the Roslagens Pärla workshop. In addition to papers from many of the workshop participants we invited contributions from others working in the area. The result is 14 chapters that explore various aspects of social navigation. There is no doubt that the book benefits from the long and interesting discussions from all those who contributed to and attended the workshop: Kathy Buckner, Nils Dahlbäck, Andreas Dieberger, Mattias Forsberg, Leif Gustavson, Phillip Jeffery, Richard Harper, Michael Klemme, Ralf Kuhnert, Paul Maglio, Teenie Matlock, Luc Mertens, David Modjeska, Per Persson, Bas Raijmakers, Paul Rankin, Harald Selke, Robert Spence, Alistair Sutcliffe, John Waterworth, Romain Zeiliger and Shumin Zhai. The workshop was co-financed by the Intelligent Information Interfaces Network of Excellence i[3]net and SICS. The work was done as part of the PERSONA project – one of the i[3] projects – funded by the European Union through the ESPRIT programme. Discussions with all the members of the PERSONA project, particularly Nils Dahlbäck, Mattias Forsberg, Leif Gustavson, Rod McCall, Catriona Macaulay, Marie Sjölinder, Martin Svensson, have significantly contributed to the ideas.

Contents

List of contributors

David Benyon
Department of Computing, HCI Group, Napier University, Canal Court,
42 Craiglockhart Avenue, Edinburgh EH14 1LT, UK
d.benyon@dcs.napier.ac.uk

Monika Buscher
Lancaster University, Department of Sociology, Lancaster LA1 4YL, UK
m.buscher@lancaster.ac.uk

Matthew Chalmers
Department of Computing Science, University of Glasgow, Glasgow, UK
matthew@dcs.gla.ac.uk

Andreas Dieberger
Emory University – Information Technology Division, 540 Asbury Circle,
Atlanta, GA 30322, USA
andreas.dieberger@acm.org

Paul Dourish
Xerox PARC, 3333 Coyote Hill Road, Palo Alto, CA 94304, USA
Dourish@parc.xerox.com

Jo Herstad
University of Oslo, P.O. Box 1080 Blindern, N-1080 Oslo, Norway
Jo.Herstad@oslo.online.no

R.H.R. Harper
Digital World Research Centre, University of Surrey, Guildford GU2 5XH, UK
R.Harper@surrey.ac.uk

Kristina Höök
HUMLE, SICS, Box 1263, S-164 29 Kista, Sweden
kia@sics.se

John Hughes
Lancaster University, Department of Sociology, Lancaster LA1 4YL, UK
j.hughes@lancaster.ac.uk

Phillip Jeffrey

Department of Computing and Information Science, University of Guelph, Guelph, Ontario, Canada N1G 2W1
Phillip.Jeffrey@acm.org

Paul P. Maglio

IBM Almaden Research Center, 650 Harry Road, NWED-B2, San Jose, CA 95120, USA
pmaglio@almaden.ibm.com

Gloria Mark

GMD-FIT, German National Research Center for Information Technology, 53754 Sankt Augustin, Germany
Gloria.Mark@gmd.de

Teenie Matlock

Cognitive Psychology Program, University of California, Santa Cruz, CA 95064, USA
tmatlock@cats.ucsc.edu

Andrew McGrath

BT Laboratories, Martlesham Heath, Suffolk, UK
andrew.mcgrath@bt.com

Alan J. Munro

Department of Computing, HCI Group, Napier University, Canal Court, 42 Craiglockhart Avenue, Edinburgh EH14 1LT, UK
a.munro@dcs.napier.ac.uk

Per Persson

Swedish Institute of Computer Science (SICS), Box 1263, S-164 28 Kista, Sweden
perp@sics.se
(Also affiliated with Department of Cinema Studies, Stockholm University, Sweden.)

Odd-Wiking Rahlff

SINTEF Telecom and Informatics, P.O. Box 124 Blindern, N-0314 Oslo, Norway
owr@informatics.sintef.no

Paul Rankin

Philips Research, Redhill, UK
rankin@prl.research.philips.com

Rolf Kenneth Rolfsen

SINTEF Telecom and Informatics, P.O. Box 124 Blindern, N-0314 Oslo, Norway
rkr@informatics.sintef.no

[9], in particular, has been very influential in identifying five features of urban environments – landmarks, nodes, edges, paths and districts – and their importance in the design of geographical space. Other influential work from the area of architecture is that of Alexander [10] who proposed a number of "patterns" for architectural features.

This work is not without criticism, however. The crucial thing missing from the traditional geographies is the failure to appreciate how environments are *conceived* by people as opposed to simply *perceived* by people. People play a role in producing the space, through their activities and practice [11]. So the social aspects of exploration and wayfinding need to be taken into consideration. Also important are the *ecological* aspects of navigation: how information is distributed throughout the people and artefacts in an environment and how that information is picked up and used by people as they navigate. The exploratory, browsing behaviour of people needs to be part of our understanding of navigation (Rankin and Spence, Ch. 10). It is one of the marks of social navigation and of the chapters in this book that the debate about how best to view navigational activities continues.

If we see the user as within the information space as opposed to outside it the job of the designer becomes one of providing graphical, auditory and haptic cues so that users will be better placed to understand, use and navigate within it; we need to "design for possible user experiences" (Waterworth, Ch. 8).

1.3.2 From Space to Place

The recognition of the importance of the social construction of space leads to the important distinction between *space* and *place* [12]; a theme picked up by several authors in this book (particularly Dourish, Chalmers, Dieberger, Jeffery and Mark, Buscher and Hughes, and Waterworth). Places are seen by a number of authors to be the *settings in which people interact*. People turn spaces into "places" where social interactions are encouraged and which are visible through the configuration of the space and how people conceive of the various interactions in it. Once again architecture is a significant influence on these ideas. Whyte [13] in particular, is interesting; the photographic studies of various different configurations of space, showing in a very detailed way what configurations of space "work" and what ones do not, what ones lead to higher crime levels, what configurations lead to sociability. One of his main findings is that what attracts people is people; something designers of information space should consider.

We must though be careful in our analysis. Particular physical configurations are associated with social situations. The consulting room is, of course, quite different from a lounge, and a pub. However, we have to bear in mind that these are *reflexively* designed: it is the architect's knowledge of what goes on in these places which conditions how they were built. However, it is not true that these places are not mutable (Dieberger, Ch. 3). Whilst certain places encourage and afford behaviours, people can change the use of that place.

To what extent lessons from architecture and the social construction of spaces can be translated into the design of information spaces is something explored in several chapters in this book. The difference between information space and information place, however, is a important feature of social navigation.

1.3.3 From Decontextualised to Social Views of Information

Just as the "space" concept of information space is important, so too is the "information" concept. Chalmers (Ch. 4) aims to provide solid theoretical underpinnings for social navigation, meditating on its relationship to "conventional" approaches to HCI and informatics. He points to Wittgenstein's concept of a "language game" as a better way of understanding information. That is, language and its relation to the world it denotes is not a fixed thing, but something which is negotiated over and over in interactions between people. Information is a social artefact and we should design information systems from this viewpoint: the relevance of a piece of information is determined from its usage.

Harper (Ch. 5) also looks at "information" as used by "information workers" who use it daily. In this setting, it is not just accuracy which is important. In fact, he shows in the context of these peoples' work, how sometimes "accuracy" is not as important as knowing where the information comes from. These people need to place information in a "framework of relevance" for the institution. Information which may be inaccurate and or biased can be more useful than more accurate and "factual" information. It can be just as important to know where information comes from, and who uses it, as for it be "factual".

There is a danger of taking an "objectivist" account of what information is; of seeing it simply as something in a database which has to be retrieved. The concept of social navigation goes beyond mining some information from large data sets like web pages. It is only when we look closer at information that as an entity it becomes strange. Through the careful study of various types of information work in a number of domains, such as Harper's work, McGrath and Munro's (Ch. 14) discussion of the nature of information, Chalmers' discussion of language games, we come toward a more problematic view of information: it is problematic, viewed as an objective, decontextualised entity.

1.3.4 Modalities, Affordances, and the Physics of Spaces

A number of writers such as Jeffery and Mark, Buscher and Hughes, Spence and Rankin, and McGrath and Munro discuss various types of virtual worlds in which people can see each other and perhaps informational objects of some type as well. In these worlds, people communicate in a number of ways, textually, visually, through audio channels and through "virtual contact". These different modes of interaction and the different representations of people and artefacts may have a significant impact on experiences within the information space. Modalities afforded by wearable computers offer other possibilities (Rahlff *et al.*, Ch. 12).

Some, for example [14, 15], suggest that textual or audio communication, because it lacks the cues of the real world, will be "depersonalised" and "antagonistic" compared to co-present communication. Whereas others put more emphasis on places and their influence on our behaviour, and the ways in which different virtual situations can engender quite different ways of relating no matter what the modality might be. One newsgroup may require a strong constitution of its members because of the constant and rolling "flame-wars", whereas another will see such things rarely, if at all. When we regard the "effects of modality" in the encounter, we do so if at all in a way which acknowledges the strong influence of social situation.

The physical laws that the spaces run under may well determine affordances of the space in ways which are more complex than we might think (Buscher and Hughes, Ch. 6). The ways in which media worlds work have an affect on the ways in which we have to communicate. In the media space work, for example, we began to see how people would inhabit these spaces and orient to these various affordances. In the real world, for example, we might find it easy to get information quickly to a number of colleagues by shouting it, and in certain environments, for example, a City dealing room this is entirely permissible. The shouted "outloud" is an efficient way to communicate with a number of people simultaneously [17]. This can only be done given that there are certain physical laws which work to afford this. However, in many virtual worlds one only hears who is close to one. In contrast, anything said in a chat room is said to all, unless one explicitly messages someone.

The physical laws of spaces allow and afford certain ways of relating with others. A number of different elements of the navigational space work together. It may be that certain types of direct or indirect social navigation are particularly hard, that other types are afforded as to be effortless. Our creation of functionality in the space can help members, or conversely create systemic and recurring issues which members have to deal with. Just now, for example, it is difficult to know more about the other users of a web page than what is given by a counter of "hits". On Internet relay chat systems, one can get an idea of the popularity of a channel, but know far less about whether people are actually attending to the channel as well as being online. We can see therefore how modality of the interface is bound up with representation, how we can visualise others. All these elements are bound up with the social, for example, when we access the channel, the "physical" laws of that space mean that we have the equivalent of everyone shouting their interaction to everyone else. The only way to form a sub-group is direct messaging or explicitly inviting someone to another, subchannel, and this may be regarded as overfamiliar and accountable (cf. [20]).

1.4 Social Navigation and Methodology

Human–computer interaction is undergoing a major change. The canonical situation is no more the one-user–one-computer scenario. We have moved on to a situation where we are beginning to recognise the fundamental effects computer applications and communication technologies have on organisations, work, society and the social communications between people. Social navigation might be seen as part of this shift. It may be seen as an attempt to bind together fundamental human social behaviour with applications with which we exist online and to populate the "dead" web spaces.

A fundamental question which stems from social navigation concerns the methods we use in order to do research in this area, to inform our concepts. Within this book we find a diversity of answers. A number of chapters use ethnographic methods of different theoretical flavours, others more "traditional", cognitivist approaches, others from recent theoretical developments in psychology and cognitive science such as experientialism.

The fact that there are a wide variety of approaches and analyses is unsurprising. Social navigation does not have a single, underlying theoretical framework and we are all learning more about the applicability of different methods in different circumstances. This book is the outcome of an organic process, from the workshop, where a number of the book chapters saw light as papers presented there, to the book. The chapters represent in some ways the networks of researchers which have helped build this field.

It is true to an extent that a number of these researchers have similar concerns: ethnographic methodology and a shared philosophical acknowledgement of (the later) Wittgenstein (though exactly how his writings are interpreted does vary). There is often an influence from the work of the psychologist James Gibson [18] on the importance of perception to thought, of George Lakoff [19] and his approach to cognition and of the sociologist Harold Garfinkel [20]. There are also, as we mention elsewhere, shared concerns about architecture, the built environment and planning. We will discuss some of these concerns below.

1.4.1 Ethnography

Ethnography is a set of methods, including participant observation, interviews, materials gathering, and desk research that requires some, usually extended, contact with a "setting", or "fieldwork". Ethnography seeks to focus understanding and describing real human behaviour in a particular context. The type of analysis, or description, that arises from these activities may be highly detailed, as in a ethnomethodological analysis, or may be much more "broad brush".

There are three chapters, those of Jeffery and Mark, Harper and Buscher and Hughes which give reports of ethnographic studies, albeit from different sets of concerns. Harper's and Buscher and Hughes' in particular is from an ethnomethodological direction, Jeffery and Mark's from a fairly eclectic orientation, though citing a number of ethnomethodological studies. Further, other chapters reference and make note of a number of ethnographic studies, particularly from the domain of CSCW.

It is interesting to note that a number of these writers cite and discuss work by the ecological psychologist J.J. Gibson. In fact, there seems a general empathy between ecological psychology and those who utilise qualitative methods and who are sympathetic to ethnomethodology. This is no accident. We argue that there are systemic reasons as to why this is so. If we go back to the work of Barker, for example [21], and others, the methodology of ecological psychology was in a similar way largely that of naturalistic study of those in "normal" settings, quite different from the settings from the laboratory, which was felt by these psychologists to be too far divorced from the everyday conditions in which they lived. Ecological psychology's methodology of studying human behaviour is thus probably more similar to ethnography than not. But we can see other commonalties also, with Whyte's [13] studies of the city. All these are contextual and naturalistic, situated. The family of concerns is thus overlapping with ethnographic work.

1.4.2 Cognitive psychology

While there is a sympathetic relationship between an ecological view of behaviour, ethnography and social navigation, this is not the only way to approach it. Another route goes through the more traditional cognitivist viewpoint. When studying individual cognitive differences and their relationship to how well users navigate large information spaces, a correlation between spatial ability and time spent to find information can be found [22–24]. The higher the spatial ability of the individual, the faster they were found able to find information. In fact, the differences observed could be quite large, sometimes 20 to 1, i.e. it takes the slowest subject 20 times longer to find information compared to the fastest subject. This is clearly important if we want to design systems that are usable by a wide variety of people; if such psychological differences are ignored many will be effectively excluded from using the system.

The cognitive demands of different designs also need to be considered. The existing means for aiding users in finding information is based on cognitively highly demanding information retrieval techniques. Users are expected to formulate their information needs in abstract languages (as with databases or search engines) or to navigate through large information space without loosing track of where they are and were to go next (as in hypermedia systems).

One solution that has been proposed to these problems is to have the computer detect difficulties that people are having and alter its displays or functions automatically. Unfortunately, attempts to adapt to users in order to compensate for the individual differences are not succeeding (Höök and Svensson, Ch. 13).

From this perspective, social navigation offers a possibility to avoid navigation which is spatially or otherwise cognitively demanding. That is, we might not need to be burdened by having to keep track of where we are. Rather we can utilise concepts of social navigation. We can then "offload" some of the cognitive burden of *remembering where we are* on the rich fount of information available directly and indirectly from the space and the people residing and acting within it.

1.4.3 Experientialism

In the early 1980s, Lakoff and Johnson [25] offered a new vision of cognitive processes as the capacity of projecting from a given domain, the well known, to a new domain. the less known. This capacity, of applying partial mappings between domains, is usually described as *metaphor*. What Lakoff and Johnson have shown is that metaphor is not only a literary trope; it is an essential cognitive process. Lakoff and Johnson developed their ideas of metaphor into a modern approach to cognition; experientialism [19, 26]. Experientialism emphasises the role of both bodily and sociocultural experience in characterising concepts and in the human imaginative capacity for creating concepts and modes of rationality that go well beyond any "mind-free" external reality.

Traditional views of cognition see meaning coming from the association of symbols with external objects whereas experiential cognition sees meaning coming from applying "imaginative projections" to some basic concepts, these basic concepts being meaningful because of their roles in bodily experience. People possess fundamental notions of spatial organisation (called image schemas) that are embedded in, and structure, our direct experience of the world. For example we recognise a container as having an inside, outside and boundary, a link joins two things together and other things have a centre and a periphery. Thinking and reasoning involve applying the imaginative processes; linking and transforming the bodily basic-level categories and image schemas into abstract concepts. We use metaphor to link two domains and may then blend this relationship with knowledge or experience from another domain.

The concepts of experiential cognition seem highly applicable to the overall concept of navigation of information spaces, and of social navigation in particular. Indeed Maglio and Matlock (Ch. 9) use some of these concepts in analysing the way that people think about navigating the web and Waterworth (Ch. 8) uses the ideas as the basis for the design of information spaces.

1.5 The Future of Social Navigation

As put by Dourish: "The concept of 'social navigation' has come of age." As the reader will see when going through the chapters, the design idea(s) behind social navigation bears the potential of helping us to take a step forward from the prevailing interface culture of today's tool-based view to richer, more social-oriented spaces.

Since the field is new, there are of course numerous problems not yet dealt with. We do not yet understand how to design for social navigation in various different domains. The relationship between spatial metaphors and social navigation is discussed much in this book, but what about spaces that are not spatial in nature and their relationship to social navigation? Are there such spaces which we do not understand as spatial (Maglio and Matlock, Ch. 9)?

We have not proposed any means by which these aspects of navigation can be evaluated – how do we know that we have created a good navigational experience? Will it be a matter of more aesthetic or emotional factors, such as feelings of flow or having a delightful experience, as opposed to the efficiency measurements usually taken for the prevailing tool-based usability evaluations?

Since social behaviours are by necessity closely tied to the culture of a community, we need to further our understanding of the new cultures we are building through applying the idea of social navigation. Even within western culture, for example, users with a lot of experience of online worlds will probably utilise them differently from those with little experience. How will the concepts extend to non-western cultures?

While the first thing that comes to mind when we talk about information spaces are work-related, objective-information-dense, spaces, in fact, the most common usage of the Internet is for entertainment and social interactions. We are seeing more and more applications in the grey zone between tools and games: "infotainment", "edutainment", and so on. Designers of information spaces have to think about issues such the natural "flow" of their designs, branding, the use of sounds to communicate "moods", and so on, much in the way that a garden designer, or interior designer considers such things. New applications may be designed to induce empathy, or use irony to enliven the space. They may encourage users to attribute intelligence and in general anthropomorphic qualities to the system [27], they may allow for a more narrative, subjective, interactive experience [28], they may induce both negative and positive affective responses in the users, for example [29].

This is a whole new bag for the field of HCI, and so we turn to other areas where entertainment is in the focus hoping to borrow some of their principles, and learn from their successes and failures. Persson (Ch. 11) makes an analysis of mainstream cinema and how the techniques used there could contribute to our understanding of how to design information spaces which are somewhere between games and tools.

At the other end of the spectrum, technology is making advances, creating completely new spaces that move us beyond the desk-top based "peep-hole" into some created virtual space. We now talk about ubiquitous computers that disappear into our furniture, walls and machines, shared soundscapes where we participate in work places through listening in on sounds of our fellow colleagues, and various augmented worlds where the reality is overlaid with information (Rahlff *et al.*, Ch. 12).

In the end, this book is not necessarily going to successful in answering questions, or giving ready solutions to building systems which will allow "social navigation". Rather we wish that any success will be in the posing of interesting and foundational issues and in helping people see new paths to take, to follow the footprints in the snow.

References

1. Benyon, D.R. Cognitive ergonomics as navigation in information space. *Ergonomics* **41:** 153–156, 1998.
2. Dourish, P. and Chalmers, M. *Running out of Space: Models of Information Navigation.* In *HCI '94.* 1994. Glasgow.
3. CACM. Special issue on recommender systems. *Communications of the ACM* 1998.
4. Dieberger, A. Supporting social navigation on the World Wide Web. *International Journal of Human Computer Studies* **46:** 805–825, 1997.
5. Benyon, D. and Höök, K. Navigation in information spaces: supporting the individual. In *INTERACT'97.* Sydney: Chapman & Hall, 1997.
6. Norman, D. User-centred systems design: new perspectives on human–computer interaction. In *User-centered Systems Design*, D. Norman and S. Draper, editors. Hillsdale, NJ: Lawrence Erlbaum, 1986.
7. Passini, *Wayfinding in Architecture.* New York, Van Nostrand Reinhold, 1992.
8. Downs, R. and Stea, D. Cognitive representations. In *Image and Environment*, R. Downs and D. Stea, editors. Chicago, IL: Aldine, 1973, 79–86.
9. Lynch, K. *The Image of the City.* Cambridge, MA: MIT Press, 1960.
10. Alexander, C. *A Pattern Language: Towns, Buildings, Construction.* New York: Oxford University Press, 1977.
11. Lefebvre, H. *The Production of Space.* Oxford: Blackwell, 1991.
12. Harrison, S. and Dourish, P. Re-place-ing space: the roles of place and space in collaborative systems. in *CSCW '96.* Cambridge, MA: ACM, 1996.
13. Whyte, W.H. *The Social Life of Small Urban Spaces.* Washington, DC: Conservation Foundation, 1980.
14. Rutter, D.R. *Communicating by Telephone.* Oxford: Pergamon Press, 1987.
15. Sproull, L. and Kiesler, S. Reducing social context cues: electronic mail in organisational communication. In *Computer Supported Co-operative Work: a Book of Readings.* San Mateo, CA: Morgan Kaufmann, 1988.
16. Heath, C. and Luff P. Media space and communicative asymmetries: preliminary observations of video mediated interaction. *Interacting With Computers* **5:** 193–216, 1992.
17. Heath, C.C. *et al.* Unpacking collaboration: the interactional organisation of trading in a city dealing room. in *ECSCW '93.* Milan, 1993.
18. Gibson, J.J. *The Ecological Approach to Visual Perception.* Hillsdale, NJ: Lawrence Erlbaum, 1986.
19. Lakoff, G. *Women, Fire and Dangerous Things: What Categories Reveal About the Mind.* Chicago, IL : Chicago University Press, 1987.
20. Garfinkel, H. *Studies in Ethnomethodology.* Englewood Cliffs, NJ: Prentice-Hall, 1967.
21. Barker, R.G. *Ecological Psychology: Concepts and Methods for Studying the Environment of Human Behaviour.* Stanford, CA: Stanford University Press, 1968.
22. Höök, K., Dahlbäck, N. and Sjölinder, M. Individual differences and navigation in hypermedia. In *ECCE 1996: European Conference on Cognitive Ergonomics.* Spain, 1996.
23. Egan, D. Individual differences. In *Handbook of Human–Computer Interaction*, M. Helander, editor. Elsevier: New York, 1988.
24. Vincente, K.J. and Williges, R.C. Accommodating individual differences in searching a hierarchical file system. *International Journal of Man Machine Studies* **29:** 647–668, 1988.
25. Lakoff, G. and Johnson, M. *Metaphors We Live By.* Chicago, IL: University of Chicago Press, 1980.
26. Johnson, M. *The Body in the Mind. The Bodily Basis of Reason and Imagination.* Chicago, IL: University of Chicago Press, 1987.
27. Reeves, B. and Nass, C. *The Media Equation: How People Treat Computers, Television, and New media Like Real People and Places.* Cambridge, Cambridge University Press, 1996.
28. Murray, J. *Hamlet on the Holodeck: The Future of Narrative in Cyberspace.* Cambridge, MA: MIT Press, 1997.
29. Picard, R. *Affective Computing.* Cambridge, MA: MIT Press, 1997.

Chapter 2

Where the Footprints Lead: Tracking Down Other Roles for Social Navigation

Paul Dourish

Abstract

Collaborative filtering was proposed in the early 1990s as a way of managing access to large information spaces by capturing and exploiting aspects of the experiences of previous users of the same information. Social navigation is a more general form of this style of interaction, and with the widening scope of the Internet as an information provider, systems of this sort have moved rapidly from early research prototypes to deployed services in everyday use.

On the other hand, to most of the HCI community, the term "social navigation" is largely synonymous with "recommendation systems": systems that match your interests to those of others and, on that basis, provide recommendations about such things as music, books, articles and films that you might enjoy. The challenge for social navigation, as an area of research and development endeavour, is to move beyond this rather limited view of the role; and to do this, we must try to take a broader view of both our remit and our opportunities.

This chapter will revisit the original motivations, and chart something of the path that recent developments have taken. Based on reflections upon the original concerns that motivated research into social navigation, it will explore some new avenues of research. In particular, it will focus on two: the first is social navigation within the framework of "awareness" provisions in collaborative systems generally; and the second is the relationship of social navigation systems to spatial models and the ideas of "space" and "place" in collaborative settings.

By exploring these two ideas, two related goals can be achieved. The first is to draw attention to ways in which current research into social navigation can be made relevant to other areas of research endeavour; and the second is to re-motivate the idea of "social navigation" as a fundamental model for collaboration in information-seeking.

2.1 Introduction

In the early 1980s, researchers working in the area of interactive systems became increasingly interested in the topic of co-operative work. Human beings are, after all, social animals, and most activities in which we engage are conditioned by and conducted in co-ordination with other individuals. We work collectively. However, until this point, the focus of interactive systems research had largely been on a single user sitting at a desk in front of a computer screen. The field of computer-supported co-operative work (CSCW) emerged in response to this focus. It drew attention to the range of concerns that lay outside this computer–human dyad but which fundamentally affected the nature of work at the computer, such as the social setting in which the activity took place, and the role that the activity played in an individual's collaborative actions. In the years that have followed, research into CSCW has legitimised this concern and reoriented human–computer interaction (HCI) to take into account the social context in which work is conducted. At the same time, it has also had an influence on the development of everyday technologies for computer-based work; as the Internet has become a more commonplace computational phenomenon, so technologies such as workflow and groupware have moved out of the research laboratory and into everyday computational practice.

The topic of social navigation offers an opportunity to take this even further. Social navigation is one of the most direct expressions of the fundamental principle that the action of a user at a computer is driven not simply by a set of internal goals and cognitive processes, but by the social setting in which people find themselves, by the action of others, individually and collectively, and by the social nature of the work being conducted and the goals sought. In other words, social navigation provides an opportunity to take those aspects of computer-based interaction and information seeking which might still be regarded as primarily single-user activities, and to invest them with a sense of the social. In systems supporting social navigation, the social and collaborative aspects of work are not limited to the overtly co-operative activities such as meeting support, communication and shared workspace manipulation, but to activities that would seem, still, to be grounded in the model of a single user sitting at a computer. Reading online documents, navigating file systems, searching a database, finding a book to read and sorting electronic mail might all seem like fundamentally individual activities; private and encapsulated in one person's head. However, the use of social navigation techniques allows us to be able to capitalise upon the social nature of the everyday world, and so to enrich the interface with collaborative support for individual tasks.

This book presents a wide variety of current research into the role and function of social navigation. This chapter is concerned with potential new opportunities for the application of the social navigation concept in interactive systems in general, and will argue that social navigation is a more general phenomenon than current practice would suggest. As a result, the focus here will be less on current research on social navigation and more on those other areas to which it might be applied. The body of the chapter will discuss how we might apply the understanding of social navigation to address wider problems in the design, development and analysis of interactive and collaborative systems. A consideration of new opportunities in research on social navigation, however, should begin by looking at the origins and scope of the term, and so that will be the first topic.

2.2 Perspectives on Social Navigation

How should we consider social navigation? What range of phenomena and systems does the term name, and how do we come to consider them as aspects of a single concept?

Although there had been investigations of collaborative systems as far back as the NLS/Augment work of Engelbart and his colleagues in the 1960s [1] and which persisted throughout subsequent developments in data communication networks, it was only in the early 1980s that the term "computer-supported co-operative work" was coined and the field began to develop a more uniform identity and a shared base of understandings from which to build. At this point, a variety of systematic investigations began into the role of mutual cooperative behaviour through computer systems, the role of computational technologies to support coordination and collaboration, and the use of data-sharing to support concerted action.

The Tapestry system [2, 3], developed at Xerox PARC, introduced the idea of "collaborative filtering", which became one of the key ideas that has driven subsequent work on social navigation. Collaborative filtering systems provide a user with recommendations of their likely interest in data items on the basis of "interest matches" derived from previous ratings from a set of users. Tapestry operated over a database of electronic-mail messages and allowed users to exploit "votes" that previous users had entered concerning the value or interest of the message. By selecting an appropriate group of users based on overlaps in interest, Tapestry could allow an individual user to exploit collective experience.

Recommender systems, as these sort of applications have become known, have become one of the most visible forms of social navigation in everyday systems. The work of Pattie Maes' group at the MIT Media Lab [4], Will Hill and his colleagues at Bellcore [5] and others has explored the use of collaborative recommendation of books, films and music, and these technologies have been incorporated into deployed technologies such as electronic commerce systems. Recommendation systems vary along a number of different dimensions, but the basic insight is similar to that of the collaborative filtering system in that they provide users with a means to use information that others have left behind as a cue to exploring an information space. In particular, recommendation systems typically use information about individual interests (e.g. which books they like and which they dislike) to build a profile which can be matched with the profiles of other individuals. If the system can determine that two people have similar tastes in books (because there is enough overlap in their stated likes and dislikes to imply a correspondence), then it can suggest that books that one likes but the other hasn't seen are probably good ones to recommend. In electronic commerce settings (such as an online store), profiles can be developed on the basis of shopping habits, reducing the overhead in generating the profiles in the first place.

2.3 Spatial, Semantic and Social Navigation

In a short paper presented at HCI'94, Dourish and Chalmers [6] introduced the term "social navigation", which included systems of this sort. However, the focus of that brief exploration was broader, and the term "social navigation" was coined to describe a particular phenomenon, in which a user's navigation through an information space was primarily guided and structured by the activities of others within that space.

"Social" navigation was contrasted with "spatial" and "semantic" navigation. Spatial navigation relies on the structure of the space itself, often a two- or three-dimensional metaphor of some spatially organised real-world phenomenon (such as an office, street or landscape). Virtual reality systems, for example, place considerable reliance on spatial navigation, offering users a spatial organisation by which to explore an environment. "Semantic" navigation, in contrast, relies on the semantic structure of the space. A hypertext system, for example, provides "links" between semantically related items and offers a means to move from one item to another according to these semantic relationships.

Dourish and Chalmers were attempting to clarify the distinctions between the three forms to support better evaluation of the important features of navigational systems. This concern arose from the emergence of collaborative virtual reality systems or "collaborative virtual environments" (CVEs), with their strong appeal to spatial models. CVEs use a three-dimensional visualisation metaphor to organise data and interaction, and so are firmly grounded on a spatial model; by exploiting our familiarity with the spatial structure of the everyday world, they can provide smooth and intuitive interaction with large information spaces. In CVEs such as DIVE [7] and MASSIVE [8], a user may have visual access not only to data items or objects in the space but also to other individuals and to their interactions. This gives them the opportunity not only to interact with data items, but also to be able to see (and hence be influenced by) other people's interactions in the space; and researchers studying CVEs have developed a range of mechanisms for providing participants with a control over views of each other's actions [9].[1] However, the arguments that apply to the familiarity of interaction with objects in a three-dimensional space to do not apply unproblematically to interactions between individuals. Is inter-personal interaction in the three-dimensional space driven primarily by the structure of that space? Is spatial structure enabling social action, or is social action happening within a spatial structure?

---

[1]However, the restricted field of view and cumbersome mechanisms of navigation and exploration afforded by current CVEs can introduce problems for mutual orientation and reference [10].

This confusion arises because semantic and social navigation may well take place in spatially organised settings. Semantic and social navigation do not name types of systems; rather, they name *phenomena of interaction*. They may all occur in the same sort of "space". For instance, imagine browsing in a bookstore. If I pick up a new book because it is sitting on the shelf next to one I've just been examining, then I'm navigating spatially. If I pick up another book because it was referred to in a citation in the first book, then I'm navigating semantically; and if I pick up yet another because it was recommended to me by someone whose opinion I trust, then I am navigating socially. The conceptual separation between "spatial", "semantic" and "social" styles of information navigation was intended to provide terms in which these different forms of behaviour could be discussed.

This is the basic principle from which the investigations to be presented here proceed: that social navigation is an *interactive phenomenon* rather than a class of technology. In this chapter, I will take the term "navigation" as an adequate term for all information-seeking activities. The space to be navigated is the notional space of potential information items, be those book recommendations, web pages, interesting data items, other people, or whatever. Navigation does not, in itself, imply "movement" in any typical, spatial sense. What will be of concern is how the information-seeking activities can be framed socially and how this framing can assist an individual in the course of their information-seeking activity.

2.4 New Opportunities for Social Navigation

In light of these considerations, we can take a number of different perspectives on social navigation. One would be to think of social navigation as an explicit activity, in which a user calls on others to request advice or pointers, either directly or via an intermediate (perhaps an artefact like a "frequently asked question" list). Another, more commonly accepted among researchers in the field, is to think of social navigation as essentially an intrinsic element of individual interaction, one which respects that individual action is carried out within a complex of social relationships and provides a means to exploit this connectivity to other individuals to help people organise information.

In this chapter, some alternative conceptions of social navigation will be presented and explored. In the first, social navigation will be considered as an aspect of collaborative work, in which information can be shared within a group to help each group member work effectively, exploiting overlap in concerns and activities for mutual co-ordination. In the second, it will be presented as a way of moving through an information space and exploiting the activities and orientations of others in that space as a way of managing one's own spatial activity.

These last two perspectives will be the main topics for the rest of this chapter. The following section considers the notion of social navigation in the context of collaborative systems, and particularly with relation to the idea of "awareness" in collaborative workspaces. The subsequent section will consider the relationship between social navigation and the conceptions of "space" prevalent in collaborative and interactive systems. By taking these two particular perspectives I hope to be able, first, to cast some light on the fundamental nature of social navigation broadly construed and, second, to open up new avenues for the use and design of social navigation facilities in interactive tools.

2.5 Social Navigation and Collaborative Awareness

The idea of social navigation is firmly based on the fact that information about others and about others' activities can be beneficial to an individual in the conduct of their activity. In collaborative systems, this idea is clearly strongly related to the topic of the mutual "awareness" between collaborators of each other's activity. In this section, I will pursue this relationship in more detail and use the perspective of social navigation to highlight opportunities for a broader conception of collaborative awareness.

2.5.1 Awareness in Collaboration

The topic of "awareness" has become a significant focus of research activity in the CSCW domain. The principal idea behind this notion is that the activity of others within a group setting can be a critical resource for the individual in managing their own work [17]. The term "awareness" has come to refer to the ways in which systems can present this information and so provide to participants in a collaborative activity an "awareness" of the action and activity of others.

This feature of collaborative activity is clearly not restricted to electronically mediated settings. Excellent examples can be found, for instance, in the work of Heath and his colleagues studying a variety of collaborative settings such as transit system control rooms [11] and dealing rooms [12]. In these settings, a common observation is that the detail of coordination between individuals is tied to a "peripheral monitoring" by which they maintain an understanding of the state of each other's activities, so as to be able to organise their own behaviour in correspondence. At the same time, individuals also choose to organise their work so that it can be successfully monitored; that is, they display the state of their own activity.

Consider an example. In their study of a London Underground control room, Heath and Luff [11] presented a detailed analysis of an interaction between two individuals in the control room. One, the Line Controller (Controller), has responsibility for the movement of trains on the line, while the other, the Divisional Information Assistant (DIA), has responsibility for information provided to passengers about the service. Health and Luff observed that the DIA's actions can be seen to be tied to the precise detail of what the Controller is doing, to the extent that, on overhearing the Controller discussing a problem with a driver over the in-cab intercom, the DIA can tell that this problem is going to result in a service disruption and is *already beginning* to announce to passengers the consequences of this disruption *before* the Controller has informed him of the nature of the problem. This sort of very tightly coupled interaction depends on a detailed understanding, on the part of the DIA, of the Controller's activity and the consequences it might hold. Conversely, when "reworking" the schedule (to accommodate changes in the service resulting from local disruptions), the Controller will talk through the changes he is introducing in such a way that he can be overheard by the DIA precisely to enable this sort of interaction.

In other words, what we find in studies of naturalistic collaborative work is that the twin mechanisms of peripheral monitoring and explicit "display" provide collaborators with the means by which to couple their work. These observations have motivated the development of technological approaches that provide for the framing of individual activity within the context of the group.

2.5.2 Designing for Awareness

The developers of CSCW systems have drawn from these observations the importance of awareness information for coordinating collaborative activity and so they have developed a range of mechanisms for providing awareness information in collaborative technologies, with considerable success in a variety of domains [13–16]. However, the emphasis in these technical developments has largely been on the first of the two mechanism found in real-world settings, that is, on the peripheral monitoring. Systems have provided a variety of means for what Dourish and Bellotti [17] referred to as "shared feedback", by which collaborators can see the results of each others' actions. For instance, the SASSE system [18] provided a variety of ways by which users could see each other's work, including "radar views" and distortion effects on the display of the textual workspace. Gutwin et al. [19] provide an overview of a variety of techniques for visualising the activity of others.

These systems have provided for ways to see the work of others, and so have supported the "peripheral monitoring" aspect of the awareness-based coordination of activities that Heath and Luff have so vividly demonstrated in their studies. Relatively little work, however, has gone into the area of the explicit production of activity information for coordination purposes. In fact, in general, this question of how information is to be produced for awareness purposes – corresponding, for example, with Heath and Luff's observations of the mumbling Line Controller – has remained relatively unexplored so far.

Certainly, it raises some hard problems. It immediately leads to a set of concerns about what kinds of uses people expect the information to be put to, not to mention the sorts of privacy and protection issues raised, in the context of ubiquitous computing environments, by Bellotti and Sellen [20]. However, some potential solutions to these problems lie in the ways in which similar issues have been tackled in research into social navigation.

2.5.3 Taking the Social Navigation Perspective

The fundamental relationship between social navigation and collaborative awareness technologies is simple, and lies in their mutual concern with the means by which an individual can exploit the information of others. What is interesting is that so many social navigation systems, and particularly the recommendation systems that have become so popular, are based not simply on the information of others, but information about the activity of others, which is fundamental to collaborative awareness. So I can capitalise on information about what books other people have bought, what web pages they have visited, what newsgroup articles they have read, and so forth. The reason this tends to be the case is that activity information is easy to capture at low overhead to the user; it is much more convenient for my system to record which books you have bought from an online store than it is to present you with a list of 300 books and ask you to rate them each on a scale from one to ten. However, despite this pragmatic explanation, the relationship between the two concerns is still suggestive.

It is particularly suggestive because it opens up opportunities for further cross-fertilisation. In particular, there are a number of features from social navigation systems that could fruitfully applied in systems providing collaborative awareness.

Aggregation

One interesting feature of social navigation systems in comparison to traditional collaborative awareness technologies is that of aggregation. By this, I mean that social navigation systems might bring together information from a number of individuals at the same moment, presenting either information about a plurality of users or about a fictional composite user.

Consider the case of the book recommendation system again. The sorts of observations we might expect from such a system are ones like, "People who liked *The Thought Gang* also liked *Mr. Vertigo*, *My Idea of Fun* and *Towing Jehovah*". What is significant about this, from the perspective of collaborative awareness systems, it is that the information is not about a single individual, but rather, reports a trend which is based on observation from a variety of individuals. In all likelihood, there are some individuals who purchased all four books, but there may not be; what is presented is an aggregate observation based on data gathered from a number of individuals.

In collaborative awareness systems, however, the focus has normally been on presenting information from different individuals separately. Most displays tend to present me with the information about the current activity of Joe, Katie and Brian independently, presenting a fairly direct and faithful rendition of their current activity. Awareness systems generally do not provide any sort of interpolation or generalisation of information. And yet, at the same time, this introduces serious problems about scalability. Since there is no generalisation, and only individual reports, problems arise over the number of individual reports that can be accommodated, and how they can be presented together. The social navigation approach offers an alternative that could be of considerable benefit in trying to address this problem.

Presentation

A related issue concerns the terms in which information is delivered and presented. Partially, perhaps, as a result of the potential aggregation of information from different sources, social navigation systems have had to consider the possibility that the terms in which information about the activity of others is presented to a user might not be the same terms in which it was gathered. In the case of inherently aggregated information, such as votes, this is inevitable; the gathering of information is in terms of positive or negative votes about some item, whereas the presentation of the information is in terms of the mass of votes for or against; a quantitative measure, and so of quite a different sort. However, there are other ways in which the form of the information is altered for presentation – web pages rather than links, books rather than purchases, or whatever. In other words, the designers of social navigation systems, since they are focused primarily on the notion of helping the end-user, rather than simply presenting the information, are open to the fact that the form in which the information is most usefully presented might not be the same as that in which it was gathered; while the designers of collaborative awareness systems have often been concerned with verisimilitude as a primary design criterion, showing the movement of mouse cursors and text edit points as accurately as possible, or giving direct views of the activity of others (as in video-based systems such as Portholes [21]). Exploiting, again, the clear foundational relationship between social navigation and collaborative awareness opens this up to some question, and points the way towards new approaches to awareness that consider more how the information is to be exploited, and hence how it might best be transformed and presented.

Decoupling

One final feature of social navigation systems relevant here is that they are both synchronous and asynchronous. Most awareness systems, on the other hand, tend to be synchronous only, displaying the user information about other collaborating individuals in real-time, but giving them less information about what the others may have done in the past when the user was disconnected or engaged in another task. Although there has been some work on asynchronous means for conveying awareness information, such as Hill et al.'s "edit wear" and "read wear" [22], collaborative awareness systems have largely focused on synchronous solutions. The value of social navigation systems as an asynchronous mechanism for sharing activity information and for capitalising on such shared information is, surely, ample indication of the value to be derived from asynchronous collaborative awareness.

The problem here is one of coupling. In the typical, synchronous, collaborative case, the presentation of information about a user's activity is directly coupled to the activity itself. As surely as one user's keystrokes result in characters being entered into a text buffer, they also result in indications of typing activity on another user's screen, and so on. There are not, typically, mechanisms for decoupling the awareness information (although some awareness "widgets" are a start in that direction [23]). In the same way that technologies for social navigation aggregate and re-represent information about individual actions for presentation as indices of group trends, they also introduce a decoupling that extends the usefulness of this information beyond the synchronous case.

2.5.4 Summary

What this exploration suggests then is that the concerns of collaboration awareness and the phenomenon of social navigation are closely related. They both rely on being able to present information about the activity of individuals in a way that allows other individuals to capitalise on it for the management of their own activities. What is more, by looking at collaborative awareness through the lens of social navigation, we have been able to identify a number of aspects of system design – aggregation, presentation and decoupling – that could help address current problems in the development of awareness technologies when we think of awareness as being a form of social navigation and information management.

The phenomenon of social navigation can also be applied to other areas of concern in collaborative systems. In particular, we will now go on to explore it in another context. Taking the term "navigation" more seriously, we can consider social navigation in relation to recent trends investigating the notions of "space" and "place" in CSCW systems.

2.6 Social Navigation and Models of Space

The very term "social *navigation*" invokes a spatially organised world, a world of paths, proximity and wayfaring. Spatial metaphors are one of the most widespread conventions in interactive system design, and are also carried across to the design of collaborative systems. However, in a collaborative setting we are more directly faced with a need to pay attention to the social phenomena of space, and the social construction of meaning in spatial settings. The distinction at work here is that between "space" and "place", and the relationship between the two. Collaborative systems design has traditionally paid little attention to this relationship, although it has begun to receive some more attention lately.

In this section, we shall consider this relationship, starting from the position, again, that social navigation is a phenomenon of interaction. This section will briefly recap on its consequences for the design of collaborative systems, and then consider how social navigation fits into the picture when the notion of space has been reformulated.

2.6.1 On "Space" and "Place"

Collaborative systems are frequently organised around computational spaces of some form. We encounter systems employing media spaces [24], shared workspaces [25], argumentation spaces [26], etc. The idea of "space" is widespread because it is so fundamental to our everyday experience. Our conception of the world is fundamentally spatial; our own three-dimensional embodiments in the world are the most fundamental part of our everyday experience. The use of space as an organising metaphor for interaction (and, indeed, for many other things besides [27]) is a natural one.

On the other hand, and just as we found earlier in considering navigation in shared virtual reality systems, the idea of "space" is one that deserves some consideration. Spatial settings are conveniently familiar and all-encompassing, but just what role is spatiality playing when it is adopted as an interactional metaphor? Further, what role does it play when the interaction is collaborative or social in nature?

Interface developers have only been in the space business for a couple of decades; perhaps it's not surprising that our view is relatively unsophisticated. So, in trying to look at these questions, one place to turn is to the Built Environment (architecture, urban design, and so on), where issues of space, of interaction and of design have been combined for thousands of years. Drawing on architectural and social theorising, I have argued, elsewhere, for a reconsideration of the notion of "space", and in particular, a reassessment of the relationship of "space" to "place" [28]. The argument presented in that paper was rooted in the philosophy behind the design of "media space" technologies. We observed, in particular, that media spaces had been designed around an understanding of the relationship between the structure of the environment and emergent understandings of the action that takes place there. It is no coincidence that the original developers of media space technologies had backgrounds in architecture. The technology was dubbed media "space" precisely because its design took an emergent view of the relationship between space and place.

What is this relationship? At its most primitive, it is the relationship between structural and social aspects of the designed environment. *Space* refers to the three-dimensional structure of the world, and the configurations of light, air and material that create lots, buildings, rooms, conference centres, churches, theatres, casinos, shopping malls, offices, nooks, parks, and the various other familiar configurations of spatial setting with which we are familiar in the everyday world. Alongside this world of spatial settings is a world of places. While spaces take their sense from configurations of brick, mortar, wood and glass, places take their sense from configurations of social actions. Places provide what we call *appropriate behavioural framing*; on the basis of patterns of social action and accountability, places engender a set of patterned social responses. Spaces provide physical constraints and affordances, based on things like the fact that it is easier to go downhill than up, that people cannot walk through walls, and that light passes through glass. In parallel, places provide social constraints and affordances, based on things like the fact that western society frowns on public nudity, courts and churches are places for more dignified affairs than nightclubs, and that joyful exuberance is an acceptable response to sporting events but not to conference presentations.

Until recently this relationship has largely gone unexamined in the CSCW literature. On the other hand, the emergence of collaborative virtual reality systems and observations of the natural and compelling nature of interaction in spatial settings has become more prevalent. This has resulted in a variety of recent explorations of collaborative and interaction systems from the perspective of architectural and urban design, including the work of Erickson [29], Fitzpatrick et al. [30], and Benyon and Höök [31], as well as the contributions of Dieberger and Chalmers in this book (Chs 3 and 4).

Space and place are fundamentally intertwined, of course; they each influence and condition the other. However, they are distinguishable. The design of the built environment is precisely about the artful manipulation of the relationship between space and place, between physical structure and social action (and, of course, about the history of the relationship between the two).

In his book *City: Rediscovering the Center* [32], William Whyte presented a detailed study of the everyday elements of urban life. His concern was with the functioning of urban spaces, in particular the densely populated daytime downtown areas. On the basis of photographic and statistical studies, he built up an image of spaces that "work" and spaces that "don't"; ones which succeed or fail at creating a sense of place. He documented where people like to stop and have conversations, or eat their lunch; which spots attract crowds and which are deserted. Throughout his discussion of patterns of activity on streets, at plazas and in parks, he emphasised the idea that "what attracts people is people." The role of space is to frame human action.

Space and Place in Collaboration

What can we learn from this? What would such a reconsideration say about collaborative systems?

The functioning of virtual spaces and physical spaces are clearly very different (it is all too easy to forget that the relationship is metaphorical), but we can find evidence for similar relationships between space and place in a virtual setting. Harrison and Dourish present a number of examples from the research literature of virtual spaces working or not working. For instance, experiments in the use of video technology to link public spaces for the purposes of informal communication have had very mixed results. While some, such as the PARC/Portland link reported on by Bly et al. [24] have been extremely successful and popular with participants, others such as the virtual window discussed by Fish et al. [33] have languished largely unused.

We can speculate about the reasons, and the potential consequences for design; clearly, the set of issues surrounding the deployment of these sorts of technologies are extremely complex. On the other hand, one clear indication is that the immediate social context is important. The issue is how the virtual space that the technology creates comes to be peopled and inhabited, how people come to have an understanding of what it does for them and how they should act. Harrison and Dourish point particularly towards a notion of *appropriation* – the extent to which the technology lends itself to be taken over by the participants and turned to their own uses, so that they can structure the space around their own needs and activities and make the technology their own. In other words, what we find to be important is not the technology itself, not the "space" that it creates, but the *creative peopling* of that space to turn it into a place, where people do things.

The other significant consequence for the development of collaborative systems echoes an earlier observation. It was noted earlier in contrasting social navigation with spatial and semantic forms that social navigation can take place both in and out of spatial settings. It may be that social navigation is affected in a spatially organised environment, but it may also happen in non-spatial settings. Analogously, at a conceptual level we may find "places" that are non-spatial at heart (or, at least, ones whose spatial nature has nothing to do with their functioning as places). One simple example is an electronic discussion group such as a mailing list, web forum or USENET news group. These have many features that we associate with "place"; they have a set of behavioural norms and expectations that frame the activity of individuals and against which activity can be considered. However, they do not offer what we would think of as "spatial" features; there is no up and down, no near and far, none of the "technology" of the everyday world. They are space-less places. Space might be a convenient, compelling and familiar metaphor for the development of collaborative systems, but it is not fundamental, nor necessary to their social functioning.

This reflects the fact that what places (rather than spaces) offer is what we called appropriate behavioural framing. Dieberger (Ch. 3) presents interesting evidence for the ways in which people have mutually held understandings of what he calls the "social connotations" of places; social connotations are the consequences of behavioural framing, and they operate in both real world and virtual environments. Discussions of appropriate postings to newsgroups or other forms of discussion forum reflect the same concerns in non-spatial settings.

So these two concerns – appropriation and appropriate behavioural framing – help to set up an initial framework for considering how the ideas of space and place apply to collaborative technologies in general. The next question to ask is how they help explain a relationship between space, place and social navigation.

Social Navigation from the Perspective of Spatial Settings

When we think about it from the perspective of spatial settings, we immediately encounter some interesting features of the term "social navigation" itself. Indeed, we are immediately confronted with two diametrically opposed positions, from which "social navigation" is either an oxymoron or a tautology.

Social navigation as an oxymoron. The first position is that "social navigation" is an oxymoron. After all, navigation is a phenomenon fundamentally rooted in the physical world, an arrangement of, and orientation to, the elements of the physical world. From this perspective, navigation is a purely physical phenomenon, and not a social one.

There would seem to be at least two responses to this. The first is that, although navigation in the everyday world might be a purely physical phenomenon, there is no physical instantiation of the virtual information spaces we are concerned with when applying "social navigation" in electronic settings, and so the support of social features is a more pressing requirement in virtual environments. The second is that the purely physical construal of everyday navigation is a flawed one, and that, in everyday settings and the physical world, our navigation and wayfaring are actually highly socially conditioned phenomena. This response leads us to a second position.

Social navigation as a tautology. The second position is that "social navigation" is a tautology. From this position, all navigation behaviour is social action and fundamentally conditioned by social forces. We can follow up this observation in a number of different ways.

One example is to take "navigation" itself, as a practical matter. Hutchins has developed this position in considerable depth in a number of investigations [34, 35]. His particular concern has been the practice of navigation on board large ships. He has produced a detailed analysis of the practice of shipboard navigation, from the perspective of distributed cognition, observing how the measurements and calculations which comprise navigation on ships are distributed over a group of individuals working together. Navigation is not so much performed as achieved through a complex of interlocking individual activities.

The tools of navigation are, themselves, socially constructed. Maps, for example, which might seem to be objective representations, faithfully detailing the structure of the environment, are rife with social and political concerns. Who draws the maps? What ends do they serve? Who decides what constitutes a "permanent feature", what is or not named, what is or is not mapped and recorded? What is the reality of the straight black lines that delimit one region from another? Wood [36] presents a wide-ranging analysis of the power of maps, arguing that maps construct the world rather than representing it.

The other tool of navigation, of course, is the world itself, but even that is not straightforward. Lynch [37], for example, discusses the "imageability" of cities. Not all places are equally navigable, and the actual practice of navigation depends on specific features of the environment. Chalmers et al. [38] has explored how imageability features could be incorporated into virtual information spaces.

However, what those features might be, what constitutes navigability and how navigation takes place – none of these are human constants. A number of social scientists, including Suchman [39] and Hutchins [40] have explored the metaphor of the differences between western and Polynesian navigational practices, which have allowed Polynesian sailors to navigate, without instruments, long voyages inconceivable to western observers.

2.6.2 Place for Social Navigation

The distinction between "place" and "space" provides a frame for analysing the structure of, and action within, computational environments for collaboration. How does this frame relate to social navigation? There are at least two ways we can see a relationship between these ideas. The first is the question of how a space becomes populated; and the second is how social activity might transform it.

If it is a sense of "appropriate behavioural framing" that distinguishes, conceptually, between space and place, then the most important factor is the presence and activity of people. In other words, the fundamental benefit we gain from social navigation is that it is a way of populating the information space. Social navigation, after all, hinges on two fundamental features; first, the presence of multiple individuals within some space, and second, the communication of aspects of their activity to each other (which we called "awareness" in the earlier discussion).

By seeing something of the activity of others, I can gain an understanding of the behaviour of individuals in a space, and so gain a sense of the style of appropriate action. When the space is populated, it becomes invested with a sense of appropriateness. That sense of appropriateness is, of course, a phenomenon which emerges from the activity of the individuals themselves; it is subject to change over time, and continually evolves around the contents of the space, the proclivities of the people there, the affordances it offers, and so on. It becomes a place, with a set of understood behaviours, norms and expected practices.

The power of social navigation, or the foundations on which social navigation is built, is to give the space meaning. That meaning comes from the collective sense of the action that takes place within the space, which itself is captured, made manifest and communicated by social navigation technologies. Since a user can see something of what has happened in the space before, then they can gain a sense of the history of the space and hence gain a sense of a set of spatially oriented practices. The book recommendation system that an online bookstore provides gives someone a sense of what other texts are read by people who are interested in a topic that they are investigating, which in turn gives them a sense of which authors are prominent and respected in the area (and by extension which ones are not), which books are regarded as definitive, and even, perhaps, of what other people find interesting with regard to this topic, other things they like, and so forth. The space of books has become populated by other people, and an individual's activities within the space are guided by a sense of how that space is currently configured and inhabited. Recently, others such as Benford et al. [41] have begin to investigate the opportunities for considering the World Wide Web as a space to be populated, so that the space itself begins to reflect something of the actions that take place within it.

The other side of the coin is the impact of the user's action on the space. In other words, if a place is a setting for action, how is the setting influenced or changed by the action that takes place there? One form of this impact is, again, the domain of workspace awareness technologies described earlier; for example, the "edit wear" approach is a way of leaving a mark in the space as a sort of "computational erosion".

One particularly interesting place to look at the way in which a place can not only frame action but also serve to represent and communicate it is in multi-user dimensions (MUDs), persistent text-based virtual reality environments [42]. Most modern MUDs are programmable, allowing users to create new objects and behaviours that can be left in the space, cloned, copied, given to other users, used generally, and so forth. One facet of this is that a common way of learning the programming language that supports the development of new MUD objects is to examine and change other objects. In this way, programmed objects become units of exchange in what MacLean et al. [43], in a different context, have called a "tailoring culture". The MUD provides the mechanisms that support the development of this culture by allowing customisations to be made explicit and shared. What is particularly interesting about this setting, however, is that the MUD is also a place of social interaction and behaviour. This has two effects of the development of programmable artefacts in MUDs. The first is that the "objects" that are created are not simply artefacts, but can also be new modes of behaviour (new capabilities for players achieved through the creation of new "verbs" in the available lexicon) and new responsive places for interaction to happen (new "rooms" with special abilities). The second is that the patterns of social interaction in the environment set a context against which the development of these new programmable artefacts is set. For example, Cherny [44] presents an analysis of conversational behaviour in a MUD and documents the emergence of specific conversational patterns that are part of the established culture of that specific environment. One interesting feature is the way that, although these patterns of conversational behaviour emerge independently, they consequently became embodied in the user-configurable *technology* of the environment, and, hence, became available and inspectable to the participants in that environment. In other words, the place itself changed to reflect the commonly occurring patterns of activity within that space, and in such a way that aspects of those behaviours could be "read off" the space.

This ability to reflect patterns of activity such that they can be "read off" the setting in which they occur is, clearly, a generalised form of the approach taken by social navigation systems. The perspective of "space" and "place" gives us a means to understand the role that the setting, and in particular, the setting as a populated site of social interaction, plays in making these connections.

2.7 Conclusions: New Opportunities for Social Navigation

The concept of "social navigation" has come of age. Since early observations about the role that could be played in information systems by information about the activities of others, and how, through that channel, individual use of information systems could be enriched more directly with social practice, these forms of systems have become familiar, accepted and even commonplace. In particular, with the rapid spread of the Internet and networked information systems, the use of social navigation, principally but not exclusively in the form of recommendation systems, has become integral to how we consider asynchronous interpersonal interaction and the management of socially organised behaviour in virtual information environments.

It is time, then, to step back and consider just what social navigation *is*, how it works, and what it means. It is important to recognise that social navigation is not a sort of technology, but rather is a *phenomenon of interaction*. In particular, it is fruitful to consider how it relates to other perspectives on collaborative computational practice. In this chapter, I have been concerned, firstly, with these sorts of reflections and, secondly, with what we can learn from them for the design of new technologies.

This chapter has focused, in particular, on two current perspectives in collaborative systems. One is the perspective of collaborative awareness. The notion of collaborative awareness as a fundamental concern in the development of (especially synchronous) CSCW tools has a longer history than social navigation, but shares a great deal in common with it. What it particularly interesting is how the two areas of research have taken different perspectives on the same fundamental issue of providing a means for individuals to discover and exploit information about others. The techniques which have been developed in social navigation systems, in addition, hold promise for tackling some of the problems with current approaches to awareness. We might hope, in the end, to see collaborative awareness and social navigation seen as aspects of the same phenomenon, and a unified design approach emerge.

The second perspective that has been explored here is that of the conceptual roles played by notions of "space" and "place" in collaborative settings, and how each of these influences the behaviour of individuals and groups in collaborative settings. One interesting feature here is that social navigation is built upon the same foundations that motivate a "place"-centric perspective on collaborative systems, one oriented around "peopled" spaces and a sense of "appropriate behavioural framing" that emerges from the visibility of social conduct within a space. As such, then, the lessons learned from investigations of social navigation have an important role to play in explorations and further development of this place-centric view.

Along the way, various underlying concepts have emerged as features of the landscape. Social navigation systems support *aggregation*, *transformation* and *decoupling* of information, allowing them to present awareness information in terms of trends rather than specific actions. Populated social places offer *appropriation* and *appropriate behavioural framing*, distinguishing them from simple spaces, which are characterised in terms of their dimensionality.

What emerges from this is a picture of a reflexively populated information space. By "populated", I mean that it contains not just information, but also people who are acting on that information, and who can see the effects of each others' actions and exploit that information in managing their own activities. By "reflexively" populated, I mean that not only does the structure of the space have an impact on the action of the users, but that the action of the users can also have an impact on the space. The space is malleable, adjustable to reflect patterns of action and the needs of the users who inhabit it. It supports the forms of appropriation encountered in the discussions of media space design and similar environments. The lesson of those experiences is that simply populating a space is not sufficient, but that appropriation and malleability are equally important. Similarly, the lesson of social navigation is that the social element of information seeking are critical resources in the development of collaborative systems, whether those are based on physical real-world metaphors or chart new virtual spaces. The explorations in this chapter have suggested that these lessons have some bearing on each other, and that social navigation has a role to play in the broader design of interactive and collaborative systems. At the same time, considering the role that social navigation can play in other sorts of collaborative setting has prompted reflections on underlying principles that may, in the future, help to frame a theoretical account of the mechanisms supporting socially supported information seeking.

Acknowledgements

Consideration of these issues over the years has been hugely enriched by stimulating discussions with a variety of colleagues, particularly Victoria Bellotti, Matthew Chalmers, Tom Erickson, Bill Gaver, Steve Harrison and Christian Heath.

References

1. Engelbart, D. and English, W. A research center for augmenting human intellect. *Proceedings of the Fall Joint Computer Conference* (San Francisco, CA). Reston, VA: AFIPS, 1968, pp. 393–410.
2. Goldberg, D, Nichols, D., Oki, B. and Terry, D. Using collaborative filtering to weave an information tapestry. *Communications of the ACM* **35:** 61–70, 1992.
3. Terry, D. A tour through tapestry. *Proceedings of the ACM Conference on Organisational Computing Systems COOCS'93.* New York: ACM, 1993, pp. 21–30.
4. Shardanand, U. and Maes, P. Social information filtering: algorithms for automating "word of mouth". *Proceedings of the ACM Conference on Human Factors in Computing Systems CHI'95.* New York: ACM, 1995, pp. 210–217.
5. Hill, W., Stead, L., Rosenstein, M. and Furnas, G. Recommending and evaluating choices in a virtual community of use. *Proceedings of the ACM Conference on Human Factors in Computing Systems CHI'95.* New York: ACM, 1995, pp. 194–201.
6. Dourish, P. and Chalmers, M. Running out of space: models of information navigation. Short paper presented at HCI'94, Glasgow, 1994.

7. Carlsson, C. and Hagsand, O. DIVE: a platform for multi-user virtual environments. *Computers and Graphics* 17: 663–669, 1993.
8. Greenhalgh, C. and Benford, S. Virtual reality tele-conferencing: implementation and experience. *Proceedings of the European Conference on Computer-supported Cooperative Work ECSCW'95.* Dordrecht: Kluwer, 1995, pp. 165–180.
9. Benford, S., Bowers, J., Fahlen, L, Greenhalgh, C., and Snowdon, D. User embodiment in collaborative virtual environments. *Proceedings of the ACM Conference Human Factors in Computing Systems CHI'95.* New York: ACM, 1995, pp. 242–249.
10. Hindmarsh, J., Fraser, M., Heath, C., Benford, S. and Greenhalgh, C. Fragmented interaction: establishing mutual orientation in virtual environments. *Proceedings of the ACM Conference on Computer-supported Cooperative Work CSCW'98* (Seattle, WA). New York: ACM, 1998, pp. 217–226.
11. Heath, C. and Luff, P. Collaboration and control: crisis management and multimedia technology in London Underground line control rooms. *Computer Supported Cooperative Work* 1: 69–94, 1992.
12. Heath, C., Jirotka, M., Luff, P. and Hindmarsh, J. Unpacking collaboration: the interactional organisation of trading in a city dealing room. *Computer Supported Cooperative Work* 3: 147–165, 1994.
13. Tang, J., Isaacs, E. and Rua, M. Supporting distributed groups with a montage of lightweight connections. *Proceedings of the ACM Conference on Computer-supported Cooperative Work CSCW'94.* New York: ACM, 1994, pp. 23–34.
14. Donath, J. Visual who: animating the affinities and activities of an electronic community. *Proceedings of the ACM International Multimedia Conference.* New York: ACM, 1995, pp. 99–107.
15. Fuchs, L., Pankoke-Babatz, U. and Prinz, W. Supporting cooperative awareness with local event mechanisms: the GroupDesk system. *Proceedings of the European Conference on Computer-supported Cooperative Work ECSCW'95.* Dordrecht: Kluwer, 1995, pp. 247–262.
16. Palfreyman, K. and Rodden, T. A protocol for user awareness on the world wide web. *Proceedings of the ACM Conference Computer Supported Cooperative Work CSCW'96.* New York: ACM, 1996, pp. 130–139.
17. Dourish, P. and Bellotti, V. Awareness and coordination in shared workspaces. *Proceedings of the ACM Conference on Computer Supported Cooperative Work CSCW'92.* New York: ACM, 1992, pp. 107–114.
18. Baecker, R., Nastos, D., Posner, I. and Mawby. K. The user-centered iterative design of collaborative writing systems. *Proceedings of INTERCHI'93.* New York: ACM, 1993, pp. 399–405.
19. Gutwin, C., Greenberg, S. and Roseman, M. Workspace awareness in real-time distributed groupware: framework, widgets and evaluation. In *People and Computers XI: Proceedings of HCI'96,* Sasse, Cunningham and Winder, editors. London: Springer, 1996, pp. 281–298.
20. Bellotti, V. and Sellen, A. Design for privacy in ubiquitous computing environments. *Proceedings of the European Conference on Computer-supported Cooperative Work ECSCW'93.* Dordrecht: Kluwer, 1993.
21. Dourish, P. and Bly, S. Portholes: supporting awareness in a distributed work group. *Proceedings of the ACM Conference on Human Factors in Computing Systems CHI'92.* New York: ACM, 1992, pp. 541–547.
22. Hill, W., Hollan, J., Wroblewski, D. and McCandless, T. Edit wear and read wear. *Proceedings of the ACM Conference on Human Factors in Computing Systems CHI'92.* New York: ACM, 1992, pp. 3–9.
23. Ackerman, M. and Starr, B. Social activity indicators: interface components for CSCW systems. *Proceedings of the ACM Symposium on User Interface Software and Technology UIST'95.* New York: ACM, 1995, pp. 159–168.
24. Bly, S., Harrison, S. and Irwin, S. Media spaces: bringing people together in a video, audio and computing environment. *Communications of the ACM* 36: 28–47, 1993.
25. Ishii, H. TeamWorkStation: towards a seamless shared workspace. *Proceedings of the ACM Conference on Computer-supported Cooperative Work CSCW'90.* New York: ACM, 1990, pp. 13–26.
26. Streitz, N., Hannemann, J. and Thuring, M. From ideas and arguments to hyperdocuments: travelling through activity spaces. *Proceedings of the ACM Conference on Hypertext.* New York: ACM, 1989, pp. 343–364.
27. Lakoff, G. and Johnson, M. *Metaphors We Live By.* Chicago, IL: University of Chicago Press, 1980.

28. Harrison, S. and Dourish, P. Re-place-ing space: the roles of space and place in collaborative systems. *Proceedings of the ACM Conference on Computer-supported Cooperative Work CSCW'96.* New York: ACM, 1996, pp. 67–76.

29. Erickson, T. From interface to interplace: the spatial environment and a medium for interaction. *Proceedings of COSIT'93* (Elba), pp. 391–405, 1993.

30. Fitzpatrick, G., Kaplan, S. and Mansfield, T. Physical spaces, virtual places and social worlds: a study of work in the virtual. *Proceedings of the ACM Conference on Computer-supported Cooperative Work CSCW'96* (Boston, MA). New York: ACM, 1996, pp. 334–343.

31. Benyon, D. and Hook, K. Navigating in information spaces: supporting the individual. *Proceedings of INTERACT'97.* London: Chapman and Hall, 1997, pp. 39–46.

32. Whyte, W. *City: Rediscovering the Center.* New York: Doubleday, 1988.

33. Fish, R., Kraut, R. and Chalfonte, B. The VideoWindow system in informal communication. *Proceedings of the ACM Conference on Computer-supported Cooperative Work CSCW'90.* New York: ACM, 1990, pp. 1–11.

34. Hutchins, E. The technology of team navigation. In *Intellectual Teamwork: The Social and Technological Foundations of Cooperative Work*, Galagher, Kraut and Egido, editors. Hillsdale, NJ: Erlbaum, 1990, pp. 191–221.

35. Hutchins, E. *Cognition in the Wild.* Cambridge, MA: MIT Press, 1995.

36. Wood, D. *The Power of Maps.* New York: Guilford Press, 1992.

37. Lynch, K. *The Image of the City.* Cambridge, MA: MIT Press, 1960.

38. Chalmers, M., Ingram, R. and Pfranger, C. Adding imageability features to information displays. *Proceedings of the ACM Symposium on User Interface Software and Technology UIST'96, ,* New York: ACM, 1996, pp. 33–39.

39. Suchman, L. *Plans and Situated Actions: The Problem of Human–Machine Communication.* Cambridge: Cambridge University Press, 1987.

40. Hutchins, E. Understand micronesian navigation. In *Mental Models*, Gentner and Stevens, editors. Hillsdale, NJ: Erlbaum, 1983.

41. Benford, S., Snowdon, D., Brown, C., Reynard, G. and Ingram, R. The populated web: browsing, searching and inhabiting the WWW using collaborative virtual environments. *Proceedings of INTERACT'97.* London: Chapman and Hall, 1997, pp. 539–546.

42. Dourish, P. (editor). Interaction and collaboration in muds. *Computer Supported Cooperative Work* [special issue] 7(1/2), 1998.

43. MacLean, A., Carter, K., Moran, T. and Lovstrand, L. User-tailorable systems: pressing the issues with buttons. *Proceedings of the ACM Conference on Human Factors in Computing Systems CHI'90.* New York: ACM, 1990.

44. Cherny, L. The MUD register: conversational modes of action in a text-based virtual reality. PhD dissertation, Palo Alto, CA: Stanford University, 1995.

Chapter **3**

Social Connotations of Space in the Design for Virtual Communities and Social Navigation

Andreas Dieberger

Abstract

Future information systems will be populated information spaces. Users of these systems will be aware of the activities of others, and what information they find useful or not. They will be able to point out and share information easily and even guide each other. These systems therefore will be social spaces. People associate social connotations with various types of spaces. These social connotations raise expectations about appropriate behaviour, privacy, trust, and private or public space in these virtual communities. This chapter discusses the relationship between social navigation and spatial metaphors and how social connotations of spatial metaphors can influence social activities in virtual space. We discuss a number of open issues in virtual communities in general and social navigation in particular, that are related to social connotations. We also present a pilot study that indicates that social connotations are perceived differently in real and virtual spaces.

3.1 Introduction

Although many people may access an information system at the same time, most systems maintain the illusion of a dedicated resource and the only indication of a large number of users simultaneously accessing a system might be an unusually slow response time. Humans are social animals, but our social skills are mostly unused in today's information systems. Admittedly there are virtual communities and chat systems, but these are only small pockets of social activity in a sea of antisocial information spaces.

Navigation is a social and frequently a collaborative process [1]. A goal of social navigation is to utilise information about other people's behaviour for our own navigational decisions. Future information systems will be populated spaces where there is awareness of other people's activities, where people rate and annotate information, and give guidance either directly or through intermediaries.

The success of chat rooms and web bulletin boards shows that people feel a need to interact and to share thoughts and information. Social navigation processes are very common in everyday life. The number of cars parked in front of a restaurant is an indication for its popularity as is the length of a waiting line before a theatre. We frequently base the choice of a movie or restaurant on friends' recommendations or on articles written by well-known critics. Few people would choose a dentist based solely on the listings in a doctor's directory if they can get a personal recommendation from somebody they know. The same is true for buying cars, selecting books and sometimes even meeting people. All of these processes are variants of what we call social navigation.

Social navigation is not identical with recommender systems. As Dourish points out in Ch. 2, social navigation is a much broader concept. Dieberger describes several simple forms of social navigation that evolved on the Internet by themselves, even without tools that supported them [2]. It is time that we realise the need for well-supported social interaction in information spaces and move beyond these simple tools.

Social activity typically is rooted in a space. A spatial metaphor can provide a framework to structure an information space and give a spatial basis to social interaction, even if the metaphor is only hinted at. People associate "social connotations" with spaces. These social connotations trigger expectations of appropriate behaviour and the social content of a space. A space with social meaning can become a place and serve as a social metaphor for interactions in it.

The main reason for studying social connotations is that we need to better understand what expectations and social rules people infer from particular types of spaces and places. Many current computer-mediated communication (CMC) and computer-supported collaborative work (CSCW) systems employ very generic spatial metaphors, like "room" or "conference". By choosing such a nondescript metaphor the system misses the opportunity to convey information on appropriate behaviour, social norms, and on what to expect in the system. The space we choose for a system communicates a message on privacy, trust, the use of private and public spaces and so on to the user. In current CMC systems these issues are often dealt with using technical features like encryption or passwords. In physical spaces such issues can often be solved through a shared understanding of what is appropriate behaviour and what is not [3]. We advocate that a thoughtful design of CMC and CSCW systems can create a similar shared understanding of what is appropriate in virtual interaction spaces without imposing hard restrictions on users.

In this chapter, social navigation is first introduced in a more detailed form, it is then contrasted with recommender systems, and examples of social navigation tools on the web are given. We discuss the relationship between spatial metaphors and social connotations and how social metaphors can shape "places".

We then focus on open issues for social navigation systems, such as how to address concerns about privacy, trust and how to find expertise in a community. We will describe how social connotations can set expectations for a shared space.

In the last section a pilot study on how social connotations are perceived in real and virtual spaces is presented.

3.2 Social Navigation   *Definition*

Social navigation encompasses all activities where two or more users collaborate directly or indirectly in a navigational task. We use the term "navigation" in a very general sense. It may encompass finding information, deciding on the usefulness of information based on other peoples' recommendations, or deciding whether to join a group of people for a chat. Dourish and Chalmers introduced the term "social navigation" in their discussion on spatial, semantic and social navigation [4].

Early forms of social navigation evolved out of the web's lack of perceivable structure. People began creating structure in the form of pointer pages or favourite link pages [5]. These web pages represented personalised information spaces that were shared freely on the Internet. Web sites with useful pointer lists, frequently asked questions, and other helpful information evolved into navigation landmarks and simple social navigation tools [2]. Other early examples were e-mail or news group messages with pointers to web pages. Dieberger has described several such early examples of social navigation [2]. Since then our tools have improved tremendously and we can post all kinds of documents on the web, embed links to them in e-mails and word-processing documents. It is even a well-established process (albeit not an easy one) to point out web pages in spoken conversations, television and radio commercials. Systems like Tribal Voice's Powwow (http://www.powwow.com/) even allow users to give guided tours through web space and thus support "direct social navigation". Despite these advances we consider these tools rudimentary, because they are far from making social navigation as natural a way to interact as it is in the physical world.

Over the past few years there is growing interest in systems that support choices based on the experiences of other people. These "recommender systems" are also instances of social navigation. For many people the terms "social navigation" and "recommender systems" are even seen as synonymous. However, we agree with Dourish that social navigation can be a fundamental model of collaboration in information seeking, which is a broader concept than making choices based on other people's preferences and recommendations.

A good overview of recommender systems can be found in a special issue of *Communications of the ACM* [6]. Among the systems discussed is Phoaks (People Helping One Another Know Stuff), a system that generates recommendations by mining newsnet messages for recommendations through a "frequency of mention" [7]. The Referral Web [8] uses a representation of a social network to find expertise in an organisation. Recommender systems typically base their recommendations on interest profiles of other people. An exception is content-based recommendation like in the Fab system [9]. It points out items that are similar to documents that a user liked in the past.

While recommender system often base their recommendations on the profiles of single users, indirect social navigation systems combine information of a large number of users and thus can avoid many scaling and privacy issues (see below and also Dourish's chapter, Ch. 2). Social navigation can also serve as a mechanism to provide awareness about the aggregated activities of a user community. Many CSCW systems strive to achieve a feeling of mutual awareness among their users, because this is generally regarded an important ingredient to collaborative work [10–13].

3.2.1 Direct social navigation

Direct social navigation typically is a synchronous process and involves direct communication between two or more individuals. An example of direct social navigation is asking for directions in a virtual environment or guiding each other. Such a direct exchange may be instantaneous, like in a chat or MUD system when both participants are present at the same time.[1] Using e-mail lists or newsgroups such an exchange can be asynchronous and the reply to a question might be seen by more than one person. Design issues in supporting direct social navigation are how to realise turn taking, how to find somebody who can give directions, how to approach somebody for help, etc. We will address several of these issues in more detail below.

3.2.2 Indirect social navigation

Indirect social navigation is asynchronous and often provides information metaphorically. A typical example for indirect social navigation is a bulletin board system that points out the most active discussions and topics by placing the image of a hot pepper next to it. What sets indirect social navigation apart from direct social navigation is that it is a by-product of our activities – we do not have to actively create indirect social navigation information.

This author is working on a modified collaborative web server that places a footprint symbol next to links leading to recently accessed pages.[2] Depending on the number of accesses within the last 24 hours these footprints are shown in different colours and thus give an indication on the amount of activity. Other markers point out pages that have not been accessed for a very long time. The system also visualises the age of pages using various markers and maintains lists of recently accessed and modified pages. The markers create an accurate feeling for what areas are visited frequently and where there are active discussions. The

---

[1]MUD stands for multi-user dungeon or dimension; they are text-based virtual environments often used for adventure-type games.

[2]This collaborative web server is a modified version of a Swiki "Pluggable web server" written in the Squeak dialect of Smalltalk by Mark Guzdial of Georgia Tech. For more information see: http://www.cc.gatech.edu/fac/mark.guzdial/squeak/pws/.

markers not only guide users to active discussions, but also increase the awareness of other people's activities in the information space.

The footprint symbol is an application of the "read wear" concept introduced by Hill and Hollan [14]. Read wear enriches information environments by visualising which parts of it have been used most frequently. An example for such a history-enriched object is a reference book in a library that falls open at a frequently needed chapter. It also might show dog ears, highlighter marks and annotations, all of which tell us a story about the usage history of this information object [15]. Other examples for the application of the read wear principle are the Juggler and the Vortex system described in [2] or bulletin board systems indicating especially active discussions.

Wexelblat and Maes' Footprints system goes a step further by visualising the usage history of an information space as commonly used paths [16]. This is useful both for social navigation as well as for identifying clusters of pages in a site and areas that are difficult to reach.

The classification into direct and indirect social navigation is not always sufficient. Social filtering is an indirect (asynchronous) form of communication that can help in navigation – see for example [17]. Social filtering often requires a conscious effort, and therefore an act of information sharing. This act involves issues of trust, a feeling of ownership and a resulting desire to contribute to a community.

An important aspect of social navigation is its close relationship to a spatial framework. People associate social meaning with various types of spaces. The simple fact that an information space is presented as a meeting room, rather than a café, can influence social behaviour and therefore change navigation behaviour in this space.

3.2.3 Why spatial metaphors?

Humans are social animals operating in space. Rooting activities in a spatial reference frame makes people feel more comfortable, even if this reference frame is only hinted at. People often refer to spatial imagery to describe activities in information spaces. Good examples are the trajectory metaphors observed by Maglio and Matlock (Ch. 9). Web navigation obviously is conceived in terms of a cognitive map similar to a cognitive map of a physical space, in terms of landmarks and routes (see Ch. 9).

Many information systems feature metaphorical spaces. Harrison and Dourish distinguish between spatial, semantic and social navigation [4, 18], see also the chapter by Dourish (Ch. 2). Spatial navigation relies on the structure of space itself and typically is based on two- or three-dimensional metaphors of space. Semantic navigation instead relies on a semantic space, very much like one would see in hypertext. Social navigation on the other hand is primarily guided by the activities of others and thus is navigation based on a social space. Semantic and social navigation are not systems per se, but rather "phenomena of interaction" and can take place in the context of a spatial setting (see Ch. 2).

We argue that social navigation benefits from a spatial framework, even if this framework is weak. A common example is a web chat room: all users in the chat room are seen as having the same distance from each other. Although some systems allow users' avatars to change position, this position does not influence how people can communicate. Talking to a person at the other side of the room is as easy as chatting with a person you are touching. In a chat room all users are talking at the same time in the same loud voice, whereas in a real world situation a person's voice is less perceivable the farther one moves away from the person. Similarly, private communication (whispering) works independently of location in a chat room, which clearly is a stretch of the metaphor.[1]

Despite these gross inconsistencies a chat room is a functional social space and carries meaning for the user. People even react sensitively to other people's perceived positions and move their avatar away if somebody else suggests a movement that would bring them too close. This behaviour is similar to what would occur in an encounter in physical space [20], see also Jeffrey and Mark (Ch. 7).

3.2.4 Space vs Place

Harrison and Dourish point out that we should not focus on spatial metaphors, but rather talk about "places" [18]. Places are "spaces invested with under-standings of behavioural appropriateness, cultural expectations, and so forth". We do live in space, but we act in place.

People perceive spatiality in information systems even from a simple description and infer assumptions about social processes in this space or place. These expectations are shaped from previous experiences in the real world or virtual spaces. In that respect these expectations could be called "social perceived affordances". Perceived affordances can signal a user that certain activities are supported by an object on a display. For example a displayed button can be perceived as affording to be clicked [21, 40]. We believe that these expectations go beyond being simply conventions, because users can infer expectations about spaces without having learnt conventions about them. Social perceived affordances tell users whether a space affords loud or silent conversations, giving a party, asking questions, etc. Social perceived affordances can tell users whether a space affords loud or silent conversations, giving a party, asking questions, etc.

Instead of the term "social perceived affordances" we will use the term "social connotations". Social connotations stem from a shared understanding of a space, based on experiences with similar real world spaces. Social connotations can change over time (see below). The social connotations of the place can be changed temporarily if, for example, a library is used for a reception after hours,

---

[1]Features that stretch a metaphor beyond its breaking point are "magic features" [19], because if they occurred in a physical space we probably would speak of magic. Magic features are sometimes used to deliberately break a metaphor to provide shortcuts and special functionality that does not fit into the overall user interface metaphor.

where the main goal of people is to communicate with each other, and not to read.

In a spatial metaphor, the limits of the metaphor are quite strict. Given a space that functions like physical space, a user cannot walk through a wall (without hurting herself). Social connotations impose weaker restrictions on behaviour than spatial constraints. It may be considered rude to yell at somebody in a library, but it is impossible to reliably prevent people from doing so. Social connotations are thus weak "rules or guidelines of behaviour". There might be no immediate penalty for not following them, but one might be expelled from a community for repeated violations of a shared "code of conduct".

3.3 Open Issues in Social Navigation

In this section we will discuss a number of open issues in social navigation. Several of them are relevant mainly for direct social navigation systems, other issues concern only indirect social navigation. Social connotations are of relevance for any type of virtual community and therefore both for direct and indirect social navigation. For most of these issues there are no clear-cut solutions but every system and community has to find a reasonable way to handle them. How well the community succeeds in this task can well decide whether users accept a system or not.

3.3.1 Why Should People Help Each Other?

Direct social navigation sounds almost too good to be true: we live in an information space where everybody has plenty of time to help each other out. Everybody is friendly and helpful. In reality this will be seldom the case. In competitive or time-pressed situations users probably won't bother to point out information or give a guided tour. One could therefore assume that direct social navigation is rare.

Experience shows that people are more than happy to provide help and guidance, but that it is not possible to rely on (immediate) assistance. In the real world a limited amount of information is often given for free (asking somebody for the way in a foreign city), but a larger amount of information has to be paid for (a guidebook or guided tour through town, for instance).

"... in many situations behaviour that is reasonable and justifiable for the individual leads to a poorer outcome for all. Such situations are termed social dilemmas" [22]. A famous example of such a social dilemma is the "tragedy of the commons" [23]. Hardin describes a group of herders having open access to a common parcel of land on which they can let their cows graze. It is in each herder's interest to put as many cows on the land as possible, but this might damage the commons such that the overall benefit for the group of herders would be eliminated.

Another common social dilemma is the "free-rider problem". Ostrom describes this problem such that "Whenever one person cannot be excluded from the benefits that other people provide, each person is motivated not to contribute to the joint effort, but to free-ride on the efforts of others. If all participants choose to free-ride the collective benefit will not be produced" [24]. Ostrom describes a variety of communities to determine what features of each contributed to the success or failure in managing collective goods.

In virtual communities we can observe both of these social dilemmas. A common good that has to be managed reasonably is bandwidth, or attention span. If one participant uses all the bandwidth in a community by spamming or posting a lot of biased information the overall community goal cannot be achieved. Such behaviour can even permanently damage a community if there are no regulatory mechanisms in place to cope with such a disturbance [22]. An example is the alt.hypertext newsgroup, that was a popular place for hypertext researchers in the past but has been practically abandoned because of excessive spamming and unsolicited advertising posted by newcomers and outsiders.

Likewise, a virtual community consisting only of consumers will not be successful either. Virtual communities tend to find a balance between consuming and contributing such that the overall common goal is achieved. People tend to understand that even a small contribution can make a difference. Terveen and Hill found that the majority of participants in newsgroups post only one or two messages over half a year and that only a few participants post a large number of messages [25].

Besides self-regulatory processes virtual communities can also be based on economical processes such that people trade information that is especially valuable. An example for such an economy is implemented in the Java developers' connection web site (http://developer.javasoft.com/). DukeDollars are a virtual currency that is used only inside the Java developers' connection. New members get an initial amount of 10 units. They can earn DukeDollars by visiting the site every week, which encourages people to visit regularly. They can also earn DukeDollars by answering other people's questions in the bulletin boards. The amount they receive depends on how useful the receiver rates their information. DukeDollars are spent when asking questions: the more DukeDollars somebody is willing to spend, the more likely he or she is to find somebody who provides a detailed solution very quickly. We mentioned above that one cannot rely on getting an immediate free answer to a question in a virtual community. The DukeDollar concept tries to tackle this issue, by allowing users to signal urgency through the amount of virtual money they are willing to spend.

An interesting aspect of indirect social navigation systems is that nobody is only a consumer. Social navigation information is collected automatically by the system when it is used. As there is no effort involved, people tend not to notice that they are contributing in very small increments to a shared resource that all community members benefit from. Indirect social navigation processes ideally occur permanently and are never disabled. Of course this permanent monitoring of a community's activities raises issues of privacy.

3.3.2 Privacy

Providing recommendations and read wear are forms of annotation. People sometimes are very picky about who can access their annotations. At other times people don't seem to care at all if anybody can access even sensitive information they have added to a document. In a study on how students annotated textbooks, Marshall reports she even found social security numbers and bank statements in books that were sold back to a university bookstore [15]. Annotations are typically created for private use, but are rarely removed from a (paper) document before this document is circulated among colleagues or sold back to the bookstore.

It appears that users of the World Wide Web do not really care too much about privacy – they feel safe because of the sheer number of people on the Internet and because they think "nobody cares about my data anyway". On the other hand we do know that as computers get more powerful it is easily possible to scan large amounts of e-mail or newsgroup messages for key phrases and potentially harmful information. The Internet is not a safe haven for sensitive information.

Systems collecting information on our activities put us in a similar situation as in a media space where a camera is constantly pointed at us. After a certain time most people forget about the camera and live blissfully unaware of the fact that they are broadcasting live. An example is the "Jennicam" (http://www.jennicam. org/index.html) which is a web camera installed in a student's room. Jenny reports on her page that she lives a normal life completely ignoring the camera and the potentially thousands of people watching her day and night, even while she is intimate with her boyfriend.

People tend to forget if somebody can see then, when they cannot see the other person. Bellotti reports [26] that people forgot they were still broadcasting live if a conversation partner switched off their cameras. In indirect social navigation systems we have a similar situation, because there is no apparent observer. The system collects data about our activities, no matter what we do and makes this information available to all users in an aggregated form. Because of this fact there is a certain risk that people underestimate the privacy issues of social navigation systems.

Care must be taken that a system cannot reveal private information about a user. A possible solution is to show read wear only after a sufficient amount of data from a larger number of users has been collected. This guarantees both privacy and unbiased information. If a user consciously wants to give recommendations or openly talk about his or her interests using direct social navigation that is fine, but the system cannot make the decision what to publish and what to keep secret. Privacy issues as well as issues of when awareness of other users becomes an intrusion are discussed by Hudson and Smith [10].

3.3.3 Finding Expertise

An important goal of recommender systems is locating expertise to solve a problem. Examples are Answer Garden [27] or Answer Garden 2 [28] that are designed to facilitate informal flow and capture of information on expertise and thus build an organisational memory. Referral Web [8] helps finding an expert by assuming topical expertise among a number of co-authors and identifies experts by their participation in co-author relationships. Phoaks (People Helping One Another Know Stuff) [7] is a collaborative filtering system that allows identifying who contributed information. Identifying contributors is an important feature, because it allows the user to make judgements about a contributor's expertise.

Less ambitious than these systems, a few virtual communities on the web allow users to identify favourite community members. For example, on the Motley Fool (http://www.fool.com/) a community to exchange investment information, people can identify other users as "favourite fools" such that all postings of these "experts" are highlighted. It is even possible to get a summary of all postings by a "favourite fool" over a period of time, no matter what board they were posted to. It is also possible to ignore people whose contributions were considered useless in the past. While ignoring spammers and obnoxious community members is a common feature in virtual communities, highlighting contributions of people who are considered experts is relatively rare.

A problem most of these systems do not address is that finding expertise involves more complex issues than just tracking down an expert. McDonald and Ackerman report [29] that people often choose not to ask an expert for assistance for political reasons. For example it might be considered embarrassing if a work group cannot solve a problem internally. Sometimes, when experts are perceived as having a "bad attitude" asking them is reserved for the most critical problems.

People who are most approachable or who give most recommendations are not necessarily the ones who are the most appropriate people to ask. The number of recommendations by a user does not determine the usefulness of these recommendations. Whittaker et al. report [30] that 3% of the users of a newsgroup account for over 25% of the messages. Similar ratios are likely to occur in direct social navigation. The number of postings does not necessarily indicate the quality of these contributions.

Identifying a user as expert is especially important in direct social navigation situations. In indirect social navigation information is averaged over a larger number of people and it is impossible to track down the contributors anyway. An interesting aspect of indirect social navigation is whether information (for example, read wear) caused by expert users should have a different weight than that of ordinary users.

A special case of social navigation occurs when the provider and consumer of social navigation information are one and the same. This type of (social?) navigation can occur in information spaces used by only one person. Read wear of this kind leaves a track record on past activities and can be useful to ease access to commonly used documents. An example for such a use of read wear occurred in the Vortex system [2]. Frequently used information items migrated to the top of a list of resources and were shown in a more prominent font. A snapshot of this list of resources provides information not only on the resources, but also on their frequency of use by the owner of the information space. The advantage of this situation is that the receiver of the social information can have complete trust in the information (see below). This type of social navigation information could be also used in an educational setting: imagine an information space containing resources, with the instructor's read wear on that space. It would provide information on the resources relative usefulness and frequency of access based on the activities of the instructor. Trust is not an issue in this case, because students typically can trust the instructor's recommendations.

3.3.4 Trust

While a recommendation from an expert might be more valuable, because one can be sure the information is correct, it may be harder to get that recommendation. People might therefore look for alternative sources of recommendations. But how can one know whose recommendations to trust, and whether indirect social navigation information is relevant for the task at hand?

A possible solution for issues of trust is to have people vote on the usefulness of other people's contributions. MessageWorld [31] was an early system that forced people to rate every message they saw. Experiences in this system showed that any additional effort required to provide a rating is prohibitive. The Java developers' connection also requires rating of messages, but the rationale there is to determine what amount of DukeDollars the author of the message is to receive. As these messages are in direct response to a question the reader has sent out, the motivation for providing the rating is very different.

Many of these issues are eliminated, when the social navigation system supports a closely-knit group of users with similar interests – for example a work group in a company. If a person knows all the people in the system she knows approximately what the expertise of each colleague is, and whose recommendations to trust. For this reason we think that social navigation can be especially useful in well-defined communities of users with related interests.

If all users in the community have common goals, it is safe to assume that no recommendation will be deliberately misleading. Similarly, information from special sources can be considered trustworthy and correct. An example would be a "knowledge concierge" for a work group or a company. Knowledge concierges have the additional advantage that the social cost of asking is very low, as it is their job to answer questions.

Nielsen reports that the number of outbound hypertext links on a web site increases the credibility on a web page (see Nielsen's Alertbox of 1 October 1997 at http://www.useit.com/alertbox/9710a.html). Terveen and Hill describe a study where links between web sites indicate emerging collaboration between web authors [25]. These studies indicate that the amount of references to outside material is an important factor in the perception of material and that referring to other people in a recommendation can also raise the level of trust in the material.

3.3.5 "Yesterday's News" – Ageing of Social Navigation Information

All meta information has a time component. A recommendation from last week may indeed be last week's news and therefore irrelevant for today's work. Different types of information age differently. While the population statistics for a country are useful for a long time, information about a sudden rise in the stock market may be useless after only half an hour.

When we receive a recommendation in a direct, live conversation, the age of the recommendation is known and we can check on the age of any underlying information. Indirect social navigation tools, however, aggregate information over time. How to decide over what time periods to aggregate information? Wexelblat defines six dimensions of history, one of them being the rate of change: "the rate of accretion and the rate of decay will vary in any history-right interface" [16]. This rate of ageing became a very bothersome problem in the Juggler system: read wear cues sometimes pointed newcomers to information items that were outdated. By accessing these items users increased the read wear again and further extended the life span of the outdated information or even caused newer, more relevant information not to be seen [2, 32].

We partly solved this issue by defining a decay function on the read wear. A similar problem in the modified Swiki server mentioned above was avoided by pointing out only read wear within the past 24 hours, which is a reasonably short amount of time. Information ages at different rates. Some areas on the Swiki server get visited frequently, whereas others are often empty for days. Each of these areas requires a different rate of ageing on read wear, footprints and recommendations – whatever social navigation cues might be used.

Related to the issue of age of recommendations is a change of interests. Users' interest profiles not only change over time, but they change between projects and contexts. When browsing the web at home I might be interested in seeing information on the stock market or cheap airfares to the Caribbean. While working I would like to see information related to my current project, or maybe a project I'm supposed to work on next week. A possibility to cope with such different contexts could be virtual workspaces like they were used in the Rooms project [33]. While in a context (Room) that contains the tools and documents for a project, I will see recommendations and read wear according to these interests. When changing to a different work context, the weights in my interest profile change as do recommendations and social navigation information.

3.3.6 Change in Virtual Communities

Social connotations may give a good initial indication of what social interactions are appropriate in a space, but it is a "shared understanding of a space's meaning" that distinguishes space from place and this shared understanding can change over time.

A virtual or real community is a living organism and appropriate behaviour changes with what people are present, and with the context of their activities.

This effect is very visible in chat rooms or MUD systems over the course of a day. People tend to log into these systems early in the morning and in the evening. Depending on the time people from all over the world come online and offline the behaviour and style of conversations in the community changes. It is a remarkable experience to stay online in a familiar virtual community for over 24 hours and experience this constant change. The change is perceived especially intense in a system based on a strong spatial metaphor, for example, a MUD community. MUD environments typically feature elaborate spatial environments with villages, buildings, a pub, monsters and treasures to hunt for etc. [34, 35]. Such a familiar environment can change in an almost frightening way when there is not a single person around that you know. To make things worse, these people behave as if they had always lived in this space, they express ownership, but they talk differently, maybe have different rituals and behaviours and a different common history.

A radical change in the social connotations of a space can also occur when a space is used for a different activity than usual. An example is the space of a classroom that is used for a student party. In such a situation all assumptions about communication patterns and appropriate behaviour are completely upturned. This shows that social connotations are influenced not only by space, but also by the people and activities of the space.

The degree to which participants are able to influence the character of a virtual community might be one of the key ingredients to a successful one. Harrison and Dourish write [18]:

> "It is only over time, and with active participation and appropriation, that a sense of place begins to permeate these systems. The users must forge the sense of place; it cannot be inherent in the system itself. Space is the opportunity, and place is the understood reality."

Change in the social connotations thus is a key ingredient to make a space into a place.

3.3.7 Social Connotations and Communications Patterns

Social connotations may influence dominant communication patterns in a space. In a lecture hall, one can assume a very specific communication pattern between one teacher and a group of students. Communication occurs mainly from the teacher to the student and is discouraged between students. The communication pattern changes entirely when the class engages in a group discussion.

A meeting room shows a different pattern. There often is a focus person, but turn taking and discussion among the other participants occurs frequently. The chairperson might intervene if a discussion moves off track, but – depending on the meeting style – the communication pattern is likely to be a network.

In a café the pattern resembles islands of communication with little communication in between, whereas at a party similar islands form, but with constant movement between the islands. In a sports bar people cluster along a bar to watch a game on a big-screen television. In this space we encounter yet another communication pattern: attention is focused on the television, and communication is open along a line perpendicular to the television. However communication is easiest with the people close to you because of the general noise level. Communicating with a person further down the bar might involve changing one's position.

This short list of places shows what a wide variety of communications patterns might evolve if virtual communities managed to move beyond the concept of a plain chat room with its lack of any communication pattern.

3.3.8 Perceiving Social Connotations from the Outside

In our discussion we generally assumed a user being inside a virtual community. One of the most promising uses of social connotations might be for users outside virtual communities. In the physical world it is typically possible to assess the character of a place already from the outside. Amy Bruckman (personal communication) describes the example of a fancy restaurant or a biker bar. In both cases it is possible to see even from the other side of the street what type of environment to expect. Similar visibility of social connotations for virtual communities is much harder to achieve, maybe because many virtual communities do not define what comprises appropriate and inappropriate behaviour as clearly as real world spaces do.

Instead virtual communities rely on a code of conduct sheet that often becomes available only after you have joined a community. Communities sometimes refer to rating codes and short descriptions. Communities change though, sometimes even for short times and it is unlikely that a community's description correctly mirrors the character of a community. We consider it one of the biggest challenges for virtual communities to become visible from the outside.

3.4 Social Connotations in Real and Virtual Spaces

Architects tend to know how to design physical spaces to convey certain social connotations although, according to Whyte, it is surprising how often architects manage to do it wrong [36]. Metaphors map behaviour from a known source domain to a target domain. We know that certain spatial properties of spatial metaphors map well from physical space to the virtual space. However, we do not really know how well social connotations carry over to the virtual domain. To use more specific spatial metaphors in groupware systems we need to achieve a better understanding of how social connotations differ in real and virtual spaces. Only with this understanding can spatial metaphors help us convey a desired social metaphor to the user.

In information systems we generally talk about metaphorical spaces, whereas most of our knowledge about social spaces is based on physical spaces and places. Metaphors provide a mapping from a known source domain into an unknown target domain [37]. These mappings are not necessarily perfect mappings. Erickson describes an example of problems caused by an imperfect metaphorical mapping: the message of a voice mail system conveyed the impression that a co-worker was taking a message, creating the assumption that the message would be delivered the moment the receiver walks into his office. The reality is that the owner of a voice mailbox has to initiate action to retrieve messages, which can cause significant delays in the delivery of the message [38]. Social connotations of spaces and social metaphors also rely on a mapping from a source domain into a target domain. Virtual space can be perceived quite differently than real, physical space, especially in information spaces that are purely metaphorical as in chat rooms or MUD systems.

We describe a pilot study that tries to shed light on how social connotations are perceived in real and virtual spaces. We used textual descriptions of spaces because many virtual communities are textual. The study focused on questions that compared participants' expectations on appropriate behaviour for asking for help, on the number of people in a space, on privacy issues and so on. The study was conducted using two web questionnaires in February and October 1998.

The study consisted of 10 situations (hotel lobby, mall, café, cubicle maze, conference reception, etc.). The situations were set up such that participants would need to ask other people for help. We asked participants about appropriate and inappropriate behaviour, if they perceived a space as private or public, if they expected to know people and so on. We further studied people's assumptions on the presence of surveillance cameras in a space. This question aimed at expectations on logging of conversations and privacy issues.

The first run of the questionnaire used situations that were described as real space scenarios. The participants came from Internet discussion forums for human factors professionals who – we assumed – would host a wide variety of people, not only technically oriented (group A). In a second run the study setup was slightly modified to put the situations in the context of a virtual space. That second questionnaire was sent to experienced MUD users (group B). MUD users spend a lot of time in text based virtual environments and tend to be more technology oriented. For information on MUDs and textual virtual environments see references [34, 39]. It is a safe assumption that group B's results are representative for many designers of (textual) virtual communities.

Our first hypothesis was that group B would perceive social connotations differently than group A. Our second hypothesis was that these differences depended on the type of space.

One hundred and one participants participated in the first run, 73 of which completed all 10 situations. In the second run we had 50 participants, 26 of which completed all situations. In our analysis we looked at the percentage of participants perceiving a particular social connotation, and then subtracted the results of group B from those of group A. Negative values thus indicate that group B perceived a social connotation stronger than group A. We considered only differences above 30% in individual situations to be significant.

Results of web questionnaires must be taken with a grain of salt, because the participants are self-selected. Yet, we believe that the results of this pilot study are useful to point the direction to further research.

Hypothesis 1

Reported perceptions of certain social connotations were virtually identical in both groups, for example the perception of a space as *public, informal,* or if there were *rules of behaviour.* Other connotations showed significant differences, for example real spaces were consistently perceived as *safer* than virtual spaces (mean = 0.2, standard deviation (SD) = 0.16).

Most group B participants assumed the presence of *surveillance cameras.* We interpret this as expectation that conversations are logged. The average over all situations was not significant (mean = 0.1, SD = 0.224), but in some situations the difference was very strong: *reception, café* and *discussion* (0.4) and *bus station* (0.3).

A less surprising result was that group B assumed virtual spaces to be less populated than real spaces. We used categories of 1, 2, 3–5, 6–9, 10–30, 31–100, 100+ people in a space. Virtual spaces were at least one class below the real spaces, for example 10–30 people instead of 31–100, with library stacks and an apartment building as the only exceptions, which were in the same class.

Hypothesis 2

In several cases the perception of social connotations differed strongly depending on the situation. *Anonymity* was stronger in the real space in: *mall, café, bus station* and *concert or sports event* (0.3), however it was perceived much stronger in a *conference reception* in virtual space (−0.4).

Group A assumed to know some people in a *discussion* (0.5) and the *cubicle maze* (0.3), however group B perceived this stronger in a *concert* or *sports event* (−0.3).

3.4.1 Discussion

The results indicate that social connotations are not always identical in real spaces and spatial metaphors and that the differences may indeed depend on the type of spatial metaphor. Of particular interest is the difference in the perception of anonymity. A virtual conference reception, virtual parties and similar events are frequent uses for virtual communities. According to our pilot study people perceive such an event to be especially anonymous, which is exactly the opposite of the desired effect. We were also quite surprised to see that real spaces were perceived as safer than virtual spaces. People often justify visiting virtual communities because of their anonymity and safety relative to meeting strangers for real.

The more or less consistent difference in the number of people in real and virtual spaces is explainable from the simple fact that virtual communities do not support crowds well. In a large room a party of 30–40 people is nothing unusual. Thirty people in a chat room, all talking at the same time, tends to be a disaster. As we discussed above the reason for this problem is that the perceived distance between people does not have any influence on their communication in many virtual communities. In a chat room, participants either whisper into somebody else's ear or they yell at the top of their lungs. Therefore, most chat rooms either limit the number of people in a room or the majority of users ignores the public talk and resorts to "private communication" with one or two people using whispering or personal messages.

The results of our pilot study indicate that there is a clear need for more research into the differences of social connotations in real and virtual spaces. The influence social connotations can have especially on early phases of virtual communities make this a very important field of future research.

3.5 Summary

We motivated the study of social connotations through their connections with spatial metaphors in information spaces. Future information systems will be social spaces where people are aware of each other's activities and where it will be much easier to share and recommend useful information, or to even guide each other. These systems will be social spaces.

We described the concept of social navigation and argued that social navigation is not identical with recommender systems. Social navigation rather is a fundamental model of collaboration in information seeking. We distinguish direct and indirect social navigation processes: direct social navigation involves direct communication between two or more participants, indirect social navigation information is a by-product of people's activities in an information space. This distinction seems almost too simple to capture the wide variety of social navigation observable. However, it provides a useful distinction between processes that actively involve a limited number of participants, and processes that generate information on the activities of entire communities, which happen as invisible processes that users often are unaware of.

The concept of space, even if only hinted at, is an important foundation for social activities. People associate meaning with spaces and develop a shared understanding of appropriate behaviour, about private and public space, how to ask questions and so forth. We called these assumptions social connotations. Social connotations and user's activities in a space can turn it into a space invested with meaning – a place.

We argue that spatial metaphors in CMC and CSCW systems can be used as social metaphors that encourage and discourage certain behaviour in virtual communities. What is presently achieved using technical protocols, like password protection and other forms of access control might be achievable through shared understanding of social appropriate behaviour – as is the case in many communities in the physical world.

An important use for social connotations is to indicate the character of a place already from the outside. Most current virtual communities are unsuccessful at giving people a clear idea of their character before they get actually immersed in the community. This obviously is an important issue with communities featuring adult content. While this specific case can be solved using rating systems, there are many cases where the situation is not that clear-cut and where people entering a community are left unaware of what to expect.

In the section on open issues of social navigation we discuss how the social dilemmas of the "tragedy of the commons" and the "free-rider problem" affect virtual communities and especially direct social navigation. An advantage of indirect social navigation is that these dilemmas do not apply.

We identified and discussed a number of other open issues in direct and indirect social navigation systems, like changing social connotations, the rate of ageing of social navigation information, how to find expertise and so forth. Most of these issues are directly related to the perception of social connotations. This supports our claim that groupware systems need to consider the use of more elaborate spatial metaphors than "room" or "conference" to communicate social connotations and social metaphors.

What makes this choice more difficult is that social connotations can be perceived differently in physical and virtual spaces. We report on a pilot study that indicates that social connotations sometimes are perceived very differently in the physical space and in spatial metaphors and that these differences can depend on the type of space used. While only a pilot study these results clearly indicate that additional research into social connotations is needed.

Acknowledgements

I'd like to thank Andrew Frank for making me aware of social connotations of space, especially of their implications for communication patterns in a place. I'm also indebted to Elisabeth Churchill, Paul Dourish, Thomas Erickson, Andrew Frank, Kristina Höök, Werner Kuhn, and Teenie Matlock for valuable discussions on social connotations and their input on the web study. Furthermore, I'd like to thank David Benyon, Phillip Jeffrey, Teenie Matlock, and Alan Munro for their comments on drafts of this chapter. Last but not least I'd like to express thanks to Alan Cattier. Without his ongoing support of my research activities neither the web study nor this chapter would exist.

References

1. Hutchins, E. *Cognition in the Wild*. Cambridge, MA: MIT Press, 1995.
2. Dieberger, A. Supporting social navigation on the World Wide Web. *International Journal of Human–Computer Studies* **46**: 805–825, 1997.

3. Catedra, M. "Through the door": a view of space from an anthropological perspective. In *Cognitive and Linguistic Aspects of Geographic Space*, D.M. Mark and A.U. Frank, editors. Dordrecht: Kluwer, 1991, pp. 53–63.

4. Dourish, P. and Chalmers, M. Running out of space: models of information navigation [short paper]. *Proceedings of HCI'94*. British Computer Society, 1994.

5. Dieberger, A. Providing spatial navigation for the World Wide Web. In *Spatial Information Theory – Proceedings of COSIT'95*, A.U. Frank, and W. Kuhn, editors. LNCS 988. London: Springer, 1995, pp. 93–106.

6. Resnick, P. and Varian, H.R. Recommender systems. *Communications of the ACM* **40:** 56–58, 1997.

7. Terveen, L. et al. PHOAKS: a system for sharing recommendations. *Communications of the ACM* **40:** 59–62, 1997.

8. Kautz, H., Selman, B. and Shah, M. Referral Web: combining social networks and collaborative filtering. *Communications of the ACM* **40:** 63–65, 1997.

9. Balabanovic, M. and Shoham, Y. Fab: content-based, collaborative recommendation. *Communications of the ACM* **40:** 66–72, 1997.

10. Hudson, S. and Smith, I. Techniques for addressing fundamental privacy and disruption tradeoffs in awareness support systems. *Proceedings of CSCW'96*, Boston, MA. New York: ACM, 1996, pp. 248–257.

11. Gutwin, C. and Greenberg, S. Design for individuals, design for groups: tradeoffs between power and workspace awareness. *Proceedings of CSCW'98*, Seattle, WA. New York: ACM, 1998, pp. 207–216.

12. Gutwin, C. and Greenberg, S. Effects of awareness support on groupware usability. *Proceedings of CHI'98*, Los Angeles, CA. New York: ACM, 1998, pp. 511–518.

13. Karsenty, A. Easing interaction through user-awareness, *Proceedings of Intelligent User Interfaces '97*, Orlando, FL. New York: ACM, 1997, pp. 225–228.

14. Hill, W.C. and Hollan, J.D. Edit wear and read wear. *Proceedings of CHI'92*, Monterey, CA. New York: ACM, 1992, pp. 3–9.

15. Marshall, C.C. Toward an ecology of hypertext annotation. *Proceedings of Hypertext'98*, Pittsburgh, PA. New York: ACM Press, 1998, pp. 40–49.

16. Wexelblat, A. History-rich tools for social navigation. *Proceedings of CHI'98 - Conference Summary*, Los Angeles, CA. New York: ACM, 1998, pp. 359–360.

17. Konstan, J.A. et al. GroupLens: applying collaborative filtering to usenet news. *Communications of the ACM* **40:** 77–87, 1997.

18. Harrison, S. and Dourish, P. Re-Place-ing space: the roles of place and space in collaborative systems. *Proceedings of CSCW'96*, Boston, MA. New York: ACM, 1996, pp. 67–76.

19. Dieberger, A. On magic features in (spatial) metaphors. *SigLink Newsletter* 4(3): 8–10, 1995.

20. Jeffrey, P. and Mark, G. Constructing social spaces in virtual environments: a study of navigation and interaction. *Workshop on Personalized and Social Navigation, SICS Technical Report 98:01*. Stockholm, 1998, pp. 24–38.

21. Norman, D.A. *The Psychology of Everyday Things*. New York: Basic Books, 1988.

22. Kollock, P. and Smith, M. Managing the virtual commons: cooperation and conflict in computer communities. In *Computer-mediated Communication. Linguistic, Social and Cross-cultural Perspectives*, S.C. Herring, editor. Philadelphia, PA: John Benjamins, 1996, pp. 109–128.

23. Hardin, G. The tragedy of the commons. In *Managing the Commons*, G. Hardin and J. Baden, editors. San Francisco, CA: WH Freeman, 1977, pp. 16–30.

24. Ostrom, E. *Governing the Commons: The Evolution of Institutions for Collective Action*, Cambridge: Cambridge University Press, 1991.

25. Terveen, L. and Hill, W. Evaluating emergent collaboration on the web. *Proceedings of CSCW'98*, Seattle, WA. New York: ACM, 1998, pp. 355–362.

26. Bellotti, V. Design for privacy in multimedia computing and communications environments. In *Technology and Privacy: The New Landscape*, P.E. Agre and M. Rotenberg, editors. Cambridge, MA: MIT Press, 1997, pp. 63–98.

27. Ackerman, M.S. Augmenting the organizational memory: a field study of answer garden. *Proceedings of CSCW'94*, Chapel Hills, NC. New York: ACM, 1994, pp. 243–252.

28. Ackerman, M. and McDonald, D. Answer Garden 2: merging organizational memory with collaborative help. *Proceedings of CSCW'96*, Boston, MA. New York: ACM, 1996, pp. 97–105.

29. McDonald, D.W. and Ackerman, M.S. Just talk to me: a field study of expertise location. *Proceedings of CSCW'98*, Seattle, WA. New York: ACM, 1998, pp. 315–324.

30. Whittaker, S. et al. The dynamics of mass interaction. *Proceedings of CSCW'98*, Seattle, WA. New York: ACM, 1998, pp. 257–264.

31. Rose, D.E., Borenstein, J.J. and Tiene, K. MessageWorld: A new approach to facilitating asynchronous group communication. *CIKM'95 (Conference on Information and Knowledge Management)*, Baltimore, MD, 1995, pp. 266–273.

32. Dieberger, A. Browsing the WWW by interacting with a textual virtual environment – a framework for experimenting with navigational metaphors. *Proceedings of Hypertext'96*, Washington DC, 1996, pp. 170–179.

33. Henderson, D.A. and Card, S. Rooms: the use of multiple virtual workspaces to reduce space contention in a window-based graphical user interface. *ACM Transactions on Graphics* **5:** 211–243, 1986.

34. Erickson, T. From interface to interplace: the spatial environment as a medium for interaction. *COSIT'93*, Elba, 1993, pp. 391–405.

35. Schiano, D.J. and White, S. The first noble truth of cyberspace: people are people (even when they MOO). *Proceedings of CHI'98*, Los Angeles, CA. New York: ACM, 1998, pp. 352–359.

36. Whyte, W.H. *City – Rediscovering the Center*. New York: Anchor Books, 1988.

37. Carroll, J.M., Mack, R.L. and Kellogg, W.A. Interface metaphors and user interface design. In *Handbook of Human–Computer Interaction*, Helander, editor, 1988, pp. 67–85.

38. Erickson, T. Working with interface metaphors. In *The Art of Human Computer Interface Design*, B. Laurel, editor. Addison-Wesley, 1990, pp. 65–74.

39. Bruckman, A. and Resnick, M. Virtual professional community: results from the MediaMOO project. *Convergence* **1:** 1 (1995).

40. Norman, D.A. *Affordance, convention and design interactions* **6:** 3 38–43, 1999.

Chapter **4**

Informatics, Architecture and Language

Matthew Chalmers

Abstract

Two complementary schools of thought exist with regard to the basic underlying assumptions and philosophies that guide our research in information navigation and access. As with all of human–computer interaction (HCI), and indeed most of informatics, we can place theories and design practices based in objectivity and mathematics at one end of a spectrum, and those emphasising subjectivity and language at the other. The first school of thought sees itself as part of traditional computer science, rooted in models that encompass the individual variations of users and that are often derived from experimentation and observation in controlled conditions. Mainstream information retrieval, cognitive psychology and task analysis exemplify such a philosophy. Complementary views are held by those who hold the sociological and the semiological as primary, and consider that objective categorical models are insufficient to model the complexity of human activity and ultimately of limited utility in guiding system design and development. Collaborative filtering, ecological psychology and ethnography are examples here. The techniques and systems presented in this book do not all lie towards one end of this spectrum, but instead show a variety of choices and emphases. This chapter, however, focuses on theory firmly towards the subjective and linguistic end of the spectrum: tools to let us place, compare and design techniques and systems. Such theory is noticeable by its absence in the majority of literature in this burgeoning research area. Here we try to redress the balance, aiming to build up a more abstract and general view of our work.

At the most applied level, this chapter deals with one approach to social information navigation systems, the path model [1], and describes its origin in an analogy with a theory of urban form: Hillier's space syntax [2]. More generally, we relate the use of and movement through information to use and movement in urban space. While architecture has already affected informatics in a number of areas, for example, in the pattern languages of Alexander, here we use architecture as a stepping stone between linguistics and informatics. Through these links we wish to reinforce the view that all three are instances or subfields of semiology. In so doing, we aim to make more visible the range of assumptions and models that underlie all interactive information systems. We are often unaware of the models of knowledge and information that we build on, and the possible alternatives. Here we aim to make clearer some of those buried layers –

the "archaeology of knowledge" [3] that determines many of the strengths and weaknesses of any system for information navigation.

4.1 What Underlies HCI?

The opposing schools of thought mentioned above were exemplified in a recent exchange in the pages of *SIGCHI Bulletin*, discussing appropriate metaphors for navigation and organisation of files on the desktop. The most recent contribution (at the time of writing) was that of Nardi and Bureau [4]. At issue was the importance of location-based searching over logical retrieval. Fertig, Freeman and Gelernter put forward what might be considered a traditional computer science viewpoint, suggesting that continuing research, gradually adapting and extending the analysis approaches rooted in 1960s "document retrieval", would eventually succeed. It would create better automatic tools for indexing the content of large databases and collections, made up of varied data types such as textual documents, images, sound files and so forth. It would, therefore, allow users to gain a full return on their investment in storing large amounts of information. This despite the fact that textual document retrieval has, even by its own measures of performance, not dramatically improved since its inception, and retrieval of image, video and audio content is still rudimentary. Nardi and Barreau suggested that such improvements would be valuable, but would not address what they consider to be the paramount information management problem: the volume and heterogeneity of ephemeral information that comes and goes in everyday work. They see the traditional computer science approach as giving insufficient attention to the intricate details of how information on an individual's desktop is interwoven with the rest of that person's working environment of people, institutions and cultures. Such concerns are largely absent from the practice and theory of data retrieval.

At the core of Nardi and Barreau's objections is a concern that "the alternatives offered by many developers of personal information management systems seem to view documents in the work space as a collection that can be easily characterized, ordered and retrieved based upon common characteristics, or based upon full text retrieval. These approaches ignore the complexity and variety of information in personal electronic environments. ... Schemes that automatically characterize information may not provide enough flexibility to consider the richness of these environments, and schemes that allow characterization for visual retrieval may not easily accommodate all of the desired dimensions" [4].

The same issues arise in the World Wide Web, where complexity and heterogeneity of information representation are increasingly problematic. The mix of data types restricts the coverage of traditional indexing techniques and limits the consequent power of search engines. As Tim Berners-Lee pointed out with regard to search engines for web data [5], "they are notorious in their ineffectiveness. ... A web indexer has to read a page of hypertext and try to deduce the sorts of questions for which the page might provide the answer."

Images, numerical data, audio, programs and applets: the variety of information is increasing along with the volume. Attempts are being made by various researchers to solve these problems, usually by adding some form of metadata. Metadata is data that represents the meaning or content of other data. It is often expressed in a formal vocabulary, and intended to allow programs to uniformly compare data originally from a variety of authors and sources.

Metadata may be handled automatically and with rigorous consistency inside indexers and search engines, but is often created manually. If metadata is manually created, it is open to be written, read and interpreted as each person sees fit. For example, web page authors or site managers may enter informally structured textual data such as captions or tags into a formally manipulated metadata scheme. For example, the PICS format proposed as a standard for web metadata [6] involves "labels" that describe a web page or site, for example, its suitability for children, whether it is commercial or not, and even, as Resnick and Miller suggest, "coolness". They also suggest that there might be many rating services, each of which could choose its own rating vocabulary. This opens the door to completely subjective metadata, however. For example, how objectively and uniformly with regard to its competitors will a large corporation describe its own web pages and products? How "cool" will each home page be according to its author?

Also, even if the metadata describing a site is, to give an optimistic figure, one thousandth the size of its referent, the vast size of the web would lead to gigabytes of metadata overall. Ironically, we would then need "metametadata" tools to find good rating services. We have not solved the problem, but deferred or even exacerbated it by adding to the indirection and complexity involved in acquiring useful information.

In other words, we only delay the problem of matching available information to users' interests and activities by stepping up to a metalevel. Since metadata open to individuals' use slips down to be just more data, we are back where we started.

Nardi and Barreau are among those involved in contemporary HCI theory, which is critical of modelling the mind (and hence the user) as an algorithmic processor – an approach that was until recently considered the firmest foundation on which to build interactive information systems. Typical research involved short-term controlled experiments in a laboratory-like setting, with experimental subjects introduced to new tools and techniques to be used in isolation from the other tools familiar from their everyday work, and away from their colleagues and workplace. Modern HCI theory criticises this as excessively reductionist, unrealistically examining use of tools that would normally be interleaved with other tools, as part of long-term work within a community of use. Consequently they fail to take account of the complexity and situatedness of interaction, and this approach has not offered the hoped-for practical benefit to designers. Work such as activity theory [7] takes a more realistic view of the subjectivity, dynamics and social context of individual action. It broadens consideration from just the actor and the tool being designed to the other tools used, the intended outcome (at various levels of abstraction), and the community within which activity takes place. Activity theory has, however, been better for analysis and criticism than for driving system design, which we suggest is related to its greater concentration on activities than on artefacts, i.e. on work's goals and actions rather than on the information and tools that represent and mediate work. Here we shift the balance the opposite way, focusing on information representation, categorisation and interpretation, and so becoming more directly linked to system design.

The claim that algorithms, themselves founded in mathematical logic and formal languages, form the ultimate foundation of informatics has also been denied by formal informatics itself. Wegner recently published a proof that interactive computing is an inherently more powerful computational paradigm than purely algorithmic computing [8]. He shows that the complexity and unpredictability of human interaction offers greater expressiveness and analytic power than formal algorithms. Interaction involves more than algorithms can express, that is, interaction is not reducible to algorithms, and interaction-based approaches to system design are more powerful than purely algorithmic ones. Computer science's traditional demand that one should be able to use formal languages to ground the behaviour of programs is therefore seen as inhibiting the expressiveness and power of the programs it builds. Wegner uses formal informatics to show its own limits, echoing the history of mathematics (and indeed physics) where the dream of reducing the world to a pure, clean and objective mathematical model has been shown to be an illusion.

We suggest that a belief or assumption that mathematical logic and objectivity form the ultimate basis of informatics is a naïve, reductionist stance. This is all too common even among computer science professionals; this is not based on the mathematicians' view of mathematics but on "folk mathematics", i.e. a common-sense framework for understanding mathematics, widespread among the wider population. This is analogous to "folk physics" that has a rough notion of inertia and Newtonian mechanics, and only a vague grasp of relativistic physics and its successors, and "folk psychology" the pre-scientific framework commonly used in comprehension and discussion of human behaviour [9].

In the next section, we attempt to look beyond "folk informatics" and see what we can learn from mathematics' reconsideration of its foundations this century. In effect we ask the question: even if informatics is based on mathematics, mathematics is not firmly grounded in objective logic, so what is mathematics based on?

4.1.1 From Mathematics to Language

Some programs represent models and theories which are not intended to relate to the physical universe or our lives in it, for example in some branches of abstract mathematical theory. More often, however, information representations model aspects of our lives. We reduce and select from the infinite complexity of the world in order to gain the ability to store, index, manipulate and retrieve. This involves finite models, expressed in mathematically based formal languages, that represent properties and relations between real-world objects. It has long been recognised in mathematics that such languages comprise of three types of structure: algebraic structures such as sets, structures of order such as lists, and topological structures such as graphs.

Since we use such mathematical schemes as foundations for building information representations, we should be aware of how solid and objective they are. This has of course been a concern within mathematics itself, for example in Cantor's demonstration of a paradox in set theory. The set of all sets is shown to be indeterminable because the "metaset" of all sets is itself a set, and therefore should be a member of itself along with many other sets, but this is inconsistent with our definition of what a set is. The metaset slips down from metadata to data. Similarly we are familiar with the paradox in statements such as "This sentence is false."

To those determined to find objective foundations for mathematics, the persistent lack of a non-perceptual basis for such an apparently simple and familiar mathematical framework as Euclidean geometry was a particular concern [10]. This framework consists of five axioms, but no-one had been able to show that the axiom of parallels was founded in more than (human) claims of apparent self-evidence. This, to use Everitt and Fisher's phrasing of it, postulates that given a line and a point not on that line, there is exactly one line through the point parallel to the given line. Also, Euclidean geometry is not in accord with modern physics' observations. On a cosmological scale, parallel lines do eventually meet. Even logic was under pressure, as quantum physics began to cast doubt on the law of the excluded middle whereby, given a logical statement *p*, only one of *p* and *not p* can be true. Hilbert led the formalist or foundationalist approach, trying to render mathematics genuinely consistent and independent of perception. As cited by Karatani [11], this meant that "the solid foundation of mathematics is in the consistency of its formal system: mathematics does not have to be 'true' as long as it is 'consistent,' and as long as this is the case, there is no need for further foundation".

Abandoning the claim to truth was not sufficient, however. The major blows to Hilbert's approach came from two directions. One was from within, in the form of Gödel's incompleteness theorem. This, in a self-referential manner related to that of Cantor, set up a paradox like "This sentence is false" where meta-mathematics, understood as a class, slips down into the formal system as a member of itself. Gödel proved that for any system of mathematics of significant expressive power, it was always possible to set up such a paradox or inconsistency. As later echoed by Wegner's proof, he therefore showed that the consistency of a formal language was only obtainable at the price of limited expressiveness. Gödel thus helped release mathematics from the illusion that it could consistently represent the world.

The other blow was from without, from Wittgenstein, who reinforced the view that mathematics was part of human history, and not an abstract ideal independent of it. He rejected the notion that mathematics' formal system can be solidly deduced from axioms. Proof is just another "language game" [12], involving our invention of rules, systems and notations whose truth, as with all our natural language, is determined by our own social use rather than from axiomatic deduction.

Gödel demonstrated the problem of inconsistency in the mathematical structures we use for information representation, while Wittgenstein replaced the axiomatic basis of their truth and meaning with a social, linguistic basis. Together they were instrumental in discrediting positivism, the previously dominant paradigm whereby a symbol was a "positive term", i.e. an objective absolute, based purely on a logical process of naming a thing in the real world. This concept of naming and reference, that connected mathematical symbols with things in the world, had been the core of positivists' concept of "truth". This concept was seen to be no longer justifiable. As a result, mathematics no longer claims to offer a means to consistently, truthfully and absolutely represent the world. We can choose to use it as a tool, pragmatically accepting its limitations and historical biases, but even then we should be aware that mathematics has a history of paradigm shifts and scientific revolutions, and we have no reason to believe that this historical process has ended.

Underneath mathematics we find language: systems of symbols, with subjectivity tempered by socially constructed practices of proof and experiment. The contemporary view of language is that we cannot dig further: there is "no exit from language" partly because all our consideration and modelling of the world is ultimately understood and used by means of perception and language. If natural language is formalised or abstractly represented, then when we interact with that representation, use it in the world, in our human interpretation and activity, we necessarily involve subjective perception. We may try to step up to a metalevel by means of formalism and abstraction, as with the web metadata example earlier in this chapter, but with human activity we slip back down to language. Within a controlled environment such as the computer we can manipulate such a "meta-representation" as part of a more formal, finitely defined system. In interaction, however, the informal, subjective and infinite reassert themselves.

What, then, can we take from contemporary linguistics and semiology, that will help in our theory-building? Only a few decades before Gödel and Wittgenstein revolutionised mathematics' core, a similar paradigm shift had taken place within linguistics. Wittgenstein's language games have often been identified with linguistics' new paradigm: structuralism. Structuralism [13] combined linguistics and semiology, and displaced positivism in those fields.

Saussure's view was that, unlike positivism, naming is a relative or differential process, in that the elements of a language at any given time form a structure where any element only has meaning because of its relations and differences with other elements. Again we see a contrast with a positivist view of a one-to-one relationship of naming a unique, absolute, ideal thing in the world. If naming and language was based on such absolutes, then observed temporal variations of meaning could not happen, for example, "cattle" refers to bovine animals in this century but meant "all kinds of personal property" in the Middle Ages. Also, how could one language use only one word where another language uses two or more, for example, we use "river" to cover both "riviére" and "fleuve" in French. (A "fleuve" is a large river, which may flow into the sea but the word does not refer to just an estuary or firth.) Japanese has several words for the number "one", used to suit the type of thing being enumerated. While we may use a word to signify a thing in the world, it does not refer to one absolute and abstract thing that each other language also has exactly one word for. Our meaning is derived from our use of the word's similarities and differences to the other words of our language. A word means what we use it to mean, or, to quote Wittgenstein, "the meaning of a word is its use in the language".

In Saussure's theory of natural language, the medium can be anything, including speech, written text and physical motion of the body, for example, sign language and dance. We can choose to use anything and any combination of media to communicate. It is this interpretative choice or reaction that creates significance, and so any action in any medium can be taken as significant, and hence as a symbol.

4.1.2 Representation in HCI and Information Access

Now we can see a fundamental limitation in the information navigation and access approaches that rely solely on the content of each information object in isolation, such the words inside a document, and ignore objects' use in human activity and objects' inter-relatedness. To assume that the words contained inside a document faithfully and fully describe the meaning of a document, irrespective of its use in language, is a naïvely positivist approach. Traditional content-based approaches can be seen as emphasising and operating on symbols and attributes which are contextually independent. For example, no matter who has a document and what activity they are involved in, the same set of words are contained inside the document. Of course, this specificity affords highly useful techniques such as indexing of contained words to allow quick searching, but we rely on the assumption that the context of use and the person involved are not significant. This is true when one wishes to find all documents that contain a given word, but false if one wishes to find all documents that one's colleagues find useful, or that conform to one's interpretation of a given topic. The assumption in itself is neither good nor bad; it is lack of awareness of it and its consequences that causes us problems. We should realise what assumptions our information representations are built on, and hence what they afford and what they inhibit.

That perception of a structure or representation is bound up with the perception of use became familiar to many within HCI via the ecological theory of perception of Gibson [14], later popularised by authors such as Don Norman. Gibson stresses the complementarity of perceiver and environment. The values and meanings of things in the environment arise from the perception of what those things provide or offer as potential actions or uses to the perceiver – in Gibson's terms, their affordances – and not by universally naming and categorising absolute or objective properties. He emphasises the way that a theory of meaning must avoid the philosophical muddle of assuming fixed classes of objects, each defined by its common features and then given a name. As Ludwig Wittgenstein knew, you cannot specify the necessary and sufficient features of the class of things to which a name is given. ... You do not have to classify and label things in order to perceive what they afford [14].

While many in HCI and information access have read Gibson or Norman, our field has not taken full account of the way that information representation schemes have affordances: they are also objects with characteristic strengths and weaknesses to choose in accordance with our uses, interests and abilities. Revealing the underlying models of knowledge and interpretation gives a common framework and vocabulary for comparing and analysing such schemes, as is the subject of a forthcoming paper [15].

Within HCI, Lucy Suchman [16] has been instrumental in establishing the importance of situated action: how particular concrete circumstances have a strong influence on behaviour, and how strict plans are often merely resources for more flexible, dynamic, contingent action, that is, more like maps than scripts [17]. Like Gibson, she generalises over objects in interfaces and objects in the physical world, treating them as elements of sign systems, as linguistic expressions.

The significance of a linguistic expression on some actual occasion ... lies in its relationship to circumstances that are presupposed or indicated by, but not actually captured in, the expression itself. Language takes its significance from the embedding world, even while it transforms the world into something that can be thought of and talked about [16, p. 58].

Again we see the need to look beyond the content of the expression or object, towards the co-dependence and co-evolution of human behaviour and information structure, and the influence of context and situation of use not usually represented in our information systems.

4.2 Space Syntax

Having put forward the argument that information use is semiological, we can draw on other fields or disciplines that have been accepted as semiological for some time. We focus on architecture here, among many semiological fields, as the path model of information access, discussed in a later section, was based by analogy with what we characterise as a structuralist theory of urban form, Hillier's "space syntax" theory [2]. The path model treats information objects like spatial forms, histories of information use like individuals' paths though the city, and language as the city. The notion of "language as city" was also at the centre of Wittgenstein's language games [12, p. 8].

Around the start of the twentieth century, around the same time as linguistics' and mathematics' revolutions, architecture took its most decisive steps on from the notion that buildings and cities are purely functional objects, exempt from significance to and influence from cultural and symbolic concerns [18]. The shaping and use of architectural form was then understood as being equally as semiological as the shaping and use of letter, word and document form in written texts. Word and written symbol usage corresponds to the motion and occupation of rooms and spaces in being a temporally ordered sequence of significant action.

In this section we outline space syntax, prior to a section using it to briefly characterise several traditional information access approaches and a section which lays out the path model in detail. Space syntax offers a view of city structure and development based on individuals' movement. Moving through the city may be due to a variety of motives – plans, contingencies and so on – but Hillier and his colleagues have found that patterns of use and meaning for people are correlated with statistical consistencies in peoples' paths. Such consistencies are correlated not so much with the nature of individual building forms and functions, but instead with the patterns of connectivity and visibility that make up the urban configuration.

4.2.1 Non-discursivity

As with other forms of language, and as activity theory and situated action emphasise, we often do not know how to talk about why we act as we do – why we read the city in the ways we do, or why we collectively tend to favour certain paths and routes and not others, or why the city configuration works or fails. This non-discursivity of configuration is an important issue in Hillier's theory and in HCI theory, in Gibson and Suchman. Just as we are able to speak understandably and to understand what we hear without being able to express – to make discursive – why our language is as it is, movement in the configuration of the city is generally non-discursive. Space syntax is an attempt to find consistencies of use that we can talk about as theorists and designers, i.e. to bring non-discursive aspects of architecture into the discursive.

Space provides the potential for paths of movement and view, infinite in the case of void, empty space but increasingly constrained as we introduce forms such as buildings. The city is at once a record of the functional processes that historically created it, and at the same time the strongest constraint on future development. Specific activities such as finding a particular building and assessing the most profitable location to place a specialist shop relate to the spatial form of the city through "general functionality": the ways that we as individuals find a system of spaces intelligible, and the ways we move around in it.

4.2.2 Integration

The "integration" of a configuration is a metric based on connectivity of spaces, and is a fundamental part of the vocabulary of the analytic side of the theory as a whole. We represent an urban configuration, such as the rooms of a building or the streets of a city, by a graph – an abstract representation of the configuration that simplifies but affords some useful analysis. Each node represents a separate subspace such as a room. The links between the graph nodes correspond to the

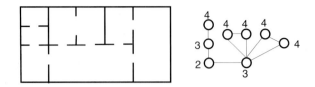

Figure 4.1 A floor plan with its connectivity graph. Each node in the graph represents one "room" and each arc between nodes represents a door connecting two rooms. Each node is labelled with its integration, showing the maximum path length to other nodes. The node of lowest integration may not be as good a movement node as the well-connected central node that bridges two rings. The top left "dead end" node is a good occupation node.

connectivity of these spaces. Figure 4.1 shows an example. The integration of a node is the maximum of the lengths of the most direct paths one could take to each other node, i.e. the maximum shortest path length to any other node. We can combine the effects of the nodes that make up a graph by summing these integrations. There is an additional normalisation factor to take account of the size of the graph, but essentially we obtain a figure that describes the graph, and hence the urban configuration. This figure, the integration of the graph, shows (among other possible interpretations) how well connected the space is, that is the degree to which moving around the space requires long routes that often connect you to only a small proportion of the nodes.

4.2.3 Visibility and Intelligibility

Visibility also plays its part. Although connectivity may make paths feasible, if the consequences of movement are not apparent – one cannot see where or to what a path may lead – then we diminish the confidence with which we make choices of movement. As we look from any point, we are aware of how our lines of visibility extend outwards and connect to spaces we could move into. An "intelligible" configuration is one in which what you see from the nodes of the system is a good guide to the global pattern of depth. More formally, Hillier treated intelligibility as the correlation between integration and the connectivity of lines of visibility.

Hillier suggests that minimising graph depth and integration support intelligibility, and are themselves best supported by combinations of occupation and movement spaces. Occupation means the use of space for activities that are at least partly and often largely static, such as conversing, reading, sleeping, cooking, and working at a laboratory bench. "Dead end" spaces with a single link are essentially occupation-only spaces, links that collectively form rings tend to help the integration of spaces, and spaces with more than two links and connect at least two rings are the best movement spaces. Different types of space afford different patterns of occupation and movement according to their patterns of branches and rings. Occupation spaces afford detailed, localised activity, while movement spaces maximise the flexibility and efficiency of motion.

Movement through the city means that we encounter buildings and people along the way. No matter how planned such activity is, we gain a by-product in the encounters along our way. Configuration affects how well we can find our goals as well as come across the situations where unforeseen actions may take place, for example, in going to one shop in the city centre, we might pass another that prompts us to take a detour. Hillier suggests that maximised integration is the best way to gain the usefulness of this by-product, and leads to positive feedback between movement and development – the "multiplier effects which are the root source of the life of cities". To use Gibson's term, maximising affordances for encounter is what makes cities good.

Architectural space limits the patterns of co-presence amongst the individuals living in and passing through an area, and therefore co-awareness and affordances for interaction. In this way the spatial becomes involved with the social. As a social fact and a social resource, spatial configuration defines the "virtual community" of an area: the pattern of natural co-presence brought about through the influence of spatial design on movement and other related aspects of space use.

Configuration does not always incrementally co-evolve with the varied movements of everyday life and work in a "natural" or "passive" way. Cities also show structures that are the result of active planning and shaping on a large scale. This may be for economic reasons, for example, a city constructing a train line to encourage public access to its centre, or it may be for political or ritualistic reasons: "to make a statement". Examples are the dominant central axis of Brasilia, and the streets and spaces that focus attention on the centres of US government in Washington, DC. Grand axes are extreme cases of how, at all scales, we can use paths and configurations as symbols and instruments.

Throughout space syntax, the function, categorisation, or content of the buildings involved are deliberately de-emphasised, and indeed it is Hillier's thesis that intelligibility and movement dominate them because all other aspects of function pass through them, and influence the urban form through them. These are strong claims, and yet Hillier presents many predictive analyses, backed up by a good match with observed behaviour in existing urban forms, without recourse to content analysis. He does not need it for his theory to work, and even apparently intangible phenomena such as house prices, burglary rates and locations of teenage and drunkard "hangouts" are quite accurately predicted by this very powerful theory which is now beginning to have wider application [19].

4.3 Information Structures Seen as Architectural Plans

Part of the strength of Hillier's presentation of his theory was the analysis of example configurations of streets in cities such as London, and his involvement in urban regeneration projects such as King's Cross. We now use space syntax to look at simple prototypical cases of traditional information access techniques, as represented by simple connectivity graphs.

We employ several criteria in discussing each technique. Integration is usually first along with perceptibility. We must use the words "perceptible" and "visible" more loosely here, as we have no analogue of the objectively measurable lines of sight. Intelligibility, describing their informal correlation, is then mentioned. Another issue related to intelligibility is scale. This is not considered at length by Hillier, perhaps due to the size and flatness of cities generally restricting the variation in the number of perceptible buildings to within a few orders of magnitude. We discuss scale here, because in information environments the number of objects varies from a handful to billions. We also look for good occupation and movement objects, and a tendency towards the minimisation of graph depth.

4.3.1 Hypermedia

Let us reuse the architectural floor plan from above, and consider the graph of connectivity to represent bidirectional links in a simple hypermedia structure, as in Fig. 4.2. We do not include the object that provides access to the hypermedia structure, for example, the desktop or command line, in this or the other figures.

In simple or "pure" hypertext, connectivity is the sole associative medium and the only perceptible nodes on a page are those which are linked to it. The scale of a hypermedia object is usually relatively small, being kept under the control of a small number of authors – often one. Integration is dependent on the author of the links, and will worsen in larger graphs unless connectivity is kept high. This not generally being the case, there is a tendency towards deep tangled networks of connections. Perceptibility tends to be poor and intelligibility tends to be low – it is easy to become "lost in hyperspace". This also contributes to the poor general functionality of hypermedia; the authoring process does not tend to produce shallow graphs, or a mixture of objects serving as good general-purpose circulation spaces and others as local "occupation" spaces.

One of the reasons connectivity stays low is to take advantage of one of hypermedia's strengths, namely that the explicit association expressed by links is directly under the subjective control of the author. The author need not rely on indirect association by, for example, using rules for categorisation or metrics of similarity to classify or measure objects' association with others. He or she doesn't have to say *why* two objects are associated, just that they are associated.

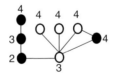

Figure 4.2 A simple hypermedia graph. Textual nodes are marked in black, while nodes of other types are in white. Arcs are bidirectional links. Integration values are unchanged from the architectural floor plan of Fig. 4.1.

4.3.2 The Web – Indexed Hypermedia

If the web was purely a large hypermedia structure then it would suffer badly from the scale-associated problems described earlier. Unlike "normal" hypermedia its authors are huge in number and there is no sense of collaborative authoring of one large design or configuration. These problems of scale have been attacked primarily by indexing, which boosts connectivity. Considering our hypermedia example above as a web site, we can see the effects of adding a search engine in Fig. 4.3. Indexing has helped the integration of the non-textual objects and some of the textual objects, but the central non-textual nodes are still distant from the textual nodes to the left. In an architectural analogy, the search engine is like a narrow corridor that connects all objects but does not help with perceptibility.

Improving connectivity with indexing tends to improve integration, especially for structures of large scale. This is dependent on the data types of objects, however: if many objects cannot be indexed then integration becomes skewed. Site authors can allow nodes to be indexed by adding captions or other metadata but then face the difficult task of guessing what indices to offer (as in the Berners-Lee quote above). Perceptibility is not improved and, since highly connected areas still tend to have low perceptibility, intelligibility is also not improved. Search engines are good circulation spaces, as we tend to pass through them for varied reasons and in varied directions. They tend to reduce overall graph depth and make other nodes more like occupation spaces. In this way they aid the general functionality of the overall configuration, and would be good places to introduce features that afforded awareness of other people if the web were to directly support the social aspects of use.

Another asymmetry arises with search engines: the most used indices are in search engines and catalogues which cover many sites, and yet site authors and administrators have direct control only over their own site. They have a tendency to ignore these "external" links and work primarily with their own "internal" links. From the user's point of view this distinction is not so important, and they

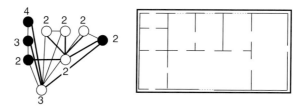

Figure 4.3 A text-based search engine, accessible from any object, has been added to the structure of Fig. 4.2. Bidirectional links connect the engine to textual objects while unidirectional links (lighter lines) show the linkage from non-textual objects. Integration values confirm that indexed objects are still distant from non-textual objects. On the right is a "floor plan" with the search engine as a narrow corridor accessible from all rooms but with "one-way" doors (in lighter lines) blocking access into non-textual rooms.

need not give greater authority to connections defined by site authors when they can use the links from external search engines.

4.3.3 Indexed/Categorised Databases

If we now do not consider inter-object links, and focus on indexing alone, we rely on search engines completely (Fig. 4.4). Generally we categorise some pre-existing corpus of data and use category-specific indices as our associative medium. We thus partition information in ways that may not be in accord with use.

An index and search engine generally aid integration on a large scale, but do little to support perceptibility between objects of a type. They tend to offer neither between objects of different types. More formally, in a heterogeneous set of objects integration values are very high as one must move via different engines. Intelligibility is therefore extremely poor. As with the web, indexing aids general functionality. Direct association such as authored, tailored hypermedia linkage is de-emphasised in order to take advantage of indexing's handling of scale. Access to manageably sized subsets and more specific association depends on subclassification schemes (classifications of objects' components) and metrics such as Boolean matching, treatment as a high-dimensional vector space, and probabilistic models. These are indirect associative media, and problems of handling non-discursivity and overly objective formalisms tend to occur.

4.3.4 Visualisation

In visualisation, the primary medium for association is the low-dimensional space that objects are positioned in. Configurationally, this is similar to an index (Fig. 4.5). Visualisation offers a means to connect heterogeneous objects and allows us to make perceptible complex interrelationships.

Integration is generally very good, and the potential for perceptibility is high due to the great deal of information that can be handled when presentation and interaction are in accord with our perceptual skills. Realising the potential for perceptibility is not a trivial matter, however. Position, colour and other

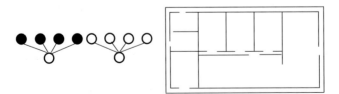

Figure 4.4 On the left we see two sets of four objects. Each set contains homogeneous objects. Shown below each set is an index for that type. On the right is a corresponding floor plan, with two groups of rooms inaccessible from each other. Connectivity amongst each set is afforded by its index, represented as a narrow corridor that offers little perceptibility.

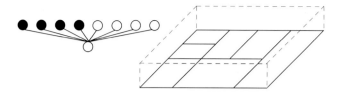

Figure 4.5 Objects of arbitrary types can be connected by placement in a visualised space. This space plays a configurational role similar to that of an index and offers the potential for maximal perceptibility. It is represented here as a space above the rooms, cf. the narrow corridor of Fig. 4.3, as it offers access to and visibility of all rooms.

graphical dimensions must convey association, and their perceptual subtleties and interrelationships – bound up with our culture and language – make this a difficult design problem. Scale is another significant problem. Features such as screen resolution, legibility of symbols, occlusion and clutter all inhibit perceptibility of object sets of the scale handled routinely by large database systems. Many visualisation systems add in a search engine in order to help with such problems. While intelligibility can be very high for smaller sets of objects, this tends to decline for larger sets.

Just as search engines aided general functionality, so with the embedding space of visualisation. The space is even more clearly a "circulation space", and positioned objects are dead-end occupation spaces in a low-depth configuration that supports a variety of paths of movement.

4.4 The Path Model: An Information Access Approach Based on Activity

Given the common semiological features of architecture and informatics, and the independence of categorisation of spaces and predictive power of Hillier's architectural theory, it would appear useful to work towards an analogous informational theory. This would be based on movement through and visibility of computer-based information. Association and similarity of information would be interwoven via consistent patterns of use. The configuration would record the functional processes that created it, and in this way constrain and guide future development of structure. The path model is a step in this direction.

What should we take as the fundamental basis, or model, of information access? Traditionally, informatics has focused on the form of the datum, with functions involving use and activity being built around the categorisation and formalisation of data representations. As Gödel showed, this approach leads to formal inconsistency – the metaclass slips into the class. Wittgenstein showed that separation between the structure and use of a language of representation ignores the social origins of meaning. We need to shift away from a view centred on formal categorisations that are separated from use, with activity peripheral.

The analytic and predictive power of Hillier's theory comes from activity being at the centre of the model. Movement and visibility in space, rather than the individual building, is at the centre of his theory. Presence at a spatial position and visual perception were quite easy to define as separate phenomena in the case of the city. With information we are forced to conflate perception of an object and presence in an object, since we have no strict equivalent of the reader being at a precise position within one and only one object. So, while movement through information space is an analogous basis for information representation, movement should be construed as the change in perception of information as one reads, writes, hears, scrolls, changes three-dimensional viewpoint, and so forth. We move among information in many ways and, whether explicitly or implicitly, in this way we build a path of associations.

In the case of a simple web browser we might construct paths by logging which pages are successively loaded, approximating the order in which they are presented to the user. Use of web tools is usually interwoven with that of others such as mail tools and editors. Also, there is little, if any, fundamental difference between objects on and off the web. Therefore to understand web use we should also log use of documents and tools on the desktop. Ideally we would look at all media that form the environment of the user, for example, where that person is, the papers and books on the desk, what music they can hear, who else is in the room, what the weather is. The list appears endless and, somewhat inconveniently, it is. We should, of course, concentrate our efforts first on the most obvious and amenable targets for logging. Nevertheless, we as system builders (or information authors) should be aware of our assumptions about which categories of information and tools – computer-based or not – will be important to our systems' users (readers).

As with the city, our reasons for movement may be myriad. The association between items of information may express pre-planned action or our behaviour may be more contingent on the situation. Rather than attempt to model the meanings in users' minds, Hillier was able to take statistical consistencies in users' paths as the social expression of meaning, and in this way bring the non-discursive into the discursive. Similarly we do not attempt to directly represent or formalise the consciousness of each user, but instead look for meaning in statistical consistencies within the collection of users' paths through information objects.

In the architectural case, uniqueness or identity of spaces was relatively straightforward and familiar, because of the way that individual spaces were defined as partitioning the total space under consideration. Here we begin by assuming that each information object has a unique name or identifier. An object might be a document, a program, an image, a person, or even a path, but its identifier is not dependent on the type or form of the object. Internal changes do not change the identifier of an object, and a copy of an object has a different identifier to the original. The metric of object similarity should not be based on identifiers or naming but on association, or linkage, in paths.

Let us consider that at any given time we model our perception of objects as a set of object identifiers. A path has an identifier, an author identifier, and a time-ordered sequence of object sets. For example:

```
{path12, Matthew,((doc77, image3),(doc77, image3, Bill),(image3, Bill))}
```

Periodically, we record the set of objects most obviously perceptible to a user, and add this set to their path. Each such record might be made either after a chosen time interval (which might lead to contiguous object sets which are equal) or when some discrete move was considered to have been made (i.e. a change in the set of perceptible objects). Other schemes are also clearly feasible, for example timestamping each set. (Current implementations rely on timestamp-ing.) More difficult to justify and to implement would be the association with each object of a probability describing relative strength of perceptibility. Perhaps some tools involved in information access will not be able to co-ordinate with others in order to define the set for a given moment in time. Instead, either concurrently or as part of a later analysis, another program combines individual logs. By means of timestamps, for example, it weaves logs together to form the sets of contemporaneously accessed objects.

A null path has no objects in it. Two paths are equivalent if they have the same author and contain the same object sets in the same order. Timestamps would offer even greater specificity. One can imagine a variety of metrics of path similarity, with each giving a maximum value for equivalent paths, and decreasing as authors, object sets and orders differ. Contiguous path segments, and paths defined by selectively filtering out particular objects from a larger path or inserting selected objects into a shorter one, could be compared similarly. This allows us to compare our recent behaviour with our own past (i.e. comparing a short segment of a path, with end "now", with the full path so as to find similar episodes in the past) and of course with the paths of others.

Two objects are similar if they consistently tend to be in similar paths. This is irrespective of the objects' heterogeneity or homogeneity of content or form, as it is their use that we focus on here. A path's start and end might be defined by the earliest and latest of all logged activity by the person involved. Alternatively we might delimit paths on noticing prolonged periods of inactivity, or abrupt changes in the membership of the set of objects worked with. One danger here is that inactivity in the logged behaviour may be due to an inability to log significant activity, rather than a lack of activity that might be logged. We should also be aware that such partitioning based on inactivity, object set consistency, etc. is pre-categorisation or pre-judgement that may not suit later use.

We emphasise that we do not assume that following paths in this way is the only way to move through information, or that a tool based on path extension would be the only one in active use on a computer at any one time. Instead here we look for a way to model the movement made via whatever ensemble of tools the user cares to use. This ensemble may or may not include a tool based on paths, but nevertheless we can consider the tool that initially offers information access, for example, the desktop on login, to be an object. Like most objects, it offers access to a number of subsequent objects, in this case mail tools, web browsers, path-based tools and so forth. In this way, our analysis can always involve a single configuration of objects.

4.4.1 Perceptibility and Intelligibility

When perception of an object A makes another object B perceptible then we say that A is connected to B. This does not mean that the ability to perceive the connection is necessarily easy or natural. As mentioned earlier, we diverge from Hillier here, whose "line of sight" visibility was simpler to measure, treated as if it was an objective, formal property. Here, it is more obvious that perceptibility may involve complex actions and interpretations understood only by the community of use (and not by us outsiders, the analysts), but nevertheless the space syntax analogy suggests we represent this by a practical, binary property: connectivity. If one could use perceptibility to define (unidirectional) connectivity, one could form graphs of connections between objects, and apply Hillier's integration to express how well connected individual objects and configurations are, and intelligibility as the correlation between visibility and integration. The degree of perceptibility for the community of users will hinder or help their everyday use, and will be affected by the degree to which their informal language of use is in accord with the formal language of analysis. As yet, we cannot go so far. Here we must rely on more intuitive notions of perceptibility. Note another aspect of movement and view that Hillier assumes but that we cannot: people can see each other within the same space, and often in connected spaces. This is a simpler assumption to accept for cities than for most information systems.

Perceptibility is an essential feature of movement, determining affordances for movement and our confidence in moving. An intelligible configuration is therefore one in which what is perceptible of the objects is a good guide to the global structure. One perceives where to go, as well as how objects fit together at the larger scale. When at a given object, it is often the case that some connected objects are more perceptible than others. An HTML page makes linked pages visible by colour, font, location on the page, a textual description and so forth. The changing background colour of the page, a position far down a very long page, or having an unhelpful description may make a link less perceptible. The page design may also reflect, or fail to reflect, where we might move in the larger-scale configuration, e.g., a clear step up to the parent directory in a graphical file browser, or another confusing step in a maze of "see also" cross-references in manual pages.

An extreme case, showing maximal connectivity and minimal perceptibility, would be a tool akin to the "go to" field in a web browser, but isolated and used only once. One could type in an arbitrary URL, and then jump to the page. All objects are connected to the tool, but none are perceptible. This is somewhat similar to the case of web search engines, which are connected to a very large number of pages but do not make those pages easily visible or perceptible.

When we look at objects and how we move around them, we can again focus on occupation and movement. Occupation means the use of objects for activities that are at least partly and often largely "static": reading a long piece of text on a page, interacting with an embedded Java application. Making a strong analogy with Hillier would suggest that "dead end" spaces may be best for occupation and that such "focused" texts or applications should be kept distinct from highly-connected objects. Connecting objects, such as well-connected ring-bridging objects, complement occupation objects. They offer flexibility and efficiency of movement and support more dynamic activities. Together with designing so as to minimise depth, implementing such distinctions should enhance the general functionality of the configuration, increasing the affordances of encounters with objects and, assuming some kind of mechanism for perceiving people in nearby information spaces, of encounters with people.

4.4.2 Adaptation and Social Effects

Some information access techniques require a start-up or training period, whereas others operate immediately. In either case, the pace of creation and evolution can be much faster than that of urban environments. Cities generally present the same visible configuration to all and, like the basic access techniques discussed above, do not adapt differently and specifically to each individual user. Given the dynamics of computers, however, one must ask how quickly information access can be tailored to individual habits and needs.

In the path model, we define a new path as we move through information, ostensively expressing our interests or activities. The path will be continually logged, adding to the configuration of paths shared by our community of users and adapting the relative path similarities. Given our current position at the end of our path we may find corresponding periods in other similar paths, and objects that consistently arise slightly further along from those periods. These can be presented to the user, i.e. made perceptible, affording future movement and triggering off adaptation in the form of further logging, reconfiguration, and presentation. Now we see what "retrieval" can mean here. Having created or accessed an information object in the course of our work, we can be presented with other "relevant" objects without dependence on their homogeneity. Furthermore, relevance is derived from the context of similar human activity, and not raw data content or "expert opinion".

It is feasible that mere presentation of possible paths for information movement might be logged, trigger reconfiguration, and spark new presentation. Setting such a "hair trigger" which did not wait for (or give weight to) user choice among possible paths of movement might offer an interesting "guided tour" if the pace were calm. If too fast, however, the tool might bolt off wildly into the informational horizon. This sprint toward infinity, a surfeit of adaptation, would be at least as bad as having no adaptation at all. In between these two extremes, controlled adaptation would move at the pace of the user, taking account of evolving activities and choices, and continually offering appropriate awareness of objects and affordances for action.

Since paths are associated with people we make manifest the community's use of information. Association and meaning are thus socially determined. The patterns of co-perceptibility in the city are strongly linked to co-awareness and the social, and similarities in the way people act in information potentially can make those people perceptible to others. Paths offer a means to find out about people's past activities and, as paths grow and adapt, people's ongoing activity. This may be useful when, for example, this enhances interpersonal communication and community. The act of writing a proposal for a research topic could trigger presentation of relevant references and authors, and also local people with experience in the topic. Note, however, that we do not have inherent symmetry as is normal in physical visibility on a street, or audibility in conversation. Like all information that identifies people and may be accessible to others, either at the time of action or at some later date, paths raise issues of privacy and invasiveness [20].

4.4.3 Implementations

So far, application of the path model has been in two areas: a URL and file recommender tool and two-dimensional visualisation of URLs and files. This work was carried out while the author was at UBS Ubilab, in Zürich, Switzerland. An earlier system, for URLs only, was described in [1], and so we offer only a brief description of the more technical details. Web usage was a convenient area for early experimentation because the URL naming system offered unambiguous references to heterogeneous objects, we use the web regularly, and browser activity is easy to log. We extended the Muffin web proxy (http://muffin.doit.org) to log URLs in a relational database. We now also log the use of local files inside the *xemacs* editor, with logging triggered by switches between editor buffers. Since we only recorded loading of URLs and switching of buffers, and not visibility on screen, a path involves not object sets (as discussed above) but a sequence of individual timestamped URLs. Each user can turn path logging on and off at will. By default, each path is potentially visible to all those who contribute paths, i.e. the set of paths is treated as a shared resource. We are also experimenting with adding the content of web pages and files to the path. (See Fig. 4.6.)

We treat the person's most recent path entries as an implicit request for recommendations. Every few minutes, the system takes the most recent path entries, i.e. the end of the path, and searches for past occurrences of each URL (or, more generally speaking, each symbol). Currently, this search can either cover one's own path or all paths within the shared path set. In the interface, a slider is used to set how long this "recent" period is, and one can select which paths (or owners) to draw from, thus allowing people to use knowledge of their colleagues to steer the recommendation process.

http://www.arosabergbahnen.ch/	18.0
http://www.arosabergbahnen.ch/Grafiken/collage.jpg	15.0
http:www.klosters.ch/images/tn_gotschna.jpg	9.0
http://www,Klosters.ch/gotschna.html	9.0
http://194.158.230.224:9090/telenet/CH/180/2.g-html	9.0
http://www.arosa.ch/main.html	9.0
http://www.arosa.ch/skiauswahl.html	9.0
http://ad.adsmart.net/src/goski/mountains^1?adtype=ac&bgcol or=F. . .	9.0
http://www.arosabergbahnen.ch/home.html	9.0
http://www.arosa.ch/	9.0

Figure 4.6 A snapshot of a URL recommendation list from our system. Starting to choose a day for a ski trip, the author accessed web pages with detailed weather reports for the mountains of Switzerland, including the telenet service of a local university. Recommendations were drawn from six sequences of past activity in three people's paths, and were mostly for web sites of ski resorts near Zürich, such as Arosa and Klosters. The numbers in the right-hand column are the tally values in ranking relevant URLs.

The system then collects the context of each past occurrence of each of the most recently used symbols – the path entries following soon after each past occurrence. Another slider sets this period's length. The system then collates these symbols, removes any which are among the set of recently used symbols (since we do not need to recommend symbols the user just used), and then presents the top ten from this ranking as a recommendation list. The system thus recommends to the user symbols that were frequently used in similar contexts but that it has not observed recently in the user's path. The people whose paths contributed to recommendations are not identified in this list. An example recommendation list is shown in Fig. 4.6, and four snapshots of a visualisation of the same example are shown in Fig. 4.7.

This example is intended to demonstrate how recommendations, such as ski information given weather pages, might not be the most obvious thing for a system to do until one considers the context of use and history behind it. The recommendations suited the author extremely well, as they were useful in his particular situation. If the winter weather was good, then a ski trip could go ahead. In this case, he got the Klosters information from his own path, and the Arosa information from another person's path. Never having been to Arosa, or to the Arosa web site, the recommendation was therefore both novel and relevant. The example also offers a contrast with tools based on content. The likely recommendation would simply have been for yet more weather pages. Lastly, the recommendations include heterogeneous data: JPEG images as well as pages of HTML. This demonstrates the ability to mix media that are usually indexed and searched by disjoint systems. Since paths involve identifiers, and similarity involves patterns of use of identifiers, it doesn't matter what the content "inside" an identifier is.

As with [1], co-occurrence statistics were used to define a similarity metric in the layout algorithm of [21], and the resultant layouts were displayed in the visualisation tool of [22]. Sections of paths cross at multiply accessed symbols. Symbols adjacent to these recurrences are brought near to each other, building associations between different periods of time. Layouts now involve a relatively small number of symbols, which tend to form "strands", each of which is a sequence of symbols collected prior to symbol tallying, i.e. one of the similar periods of past activity used in the recommendation process. Our previous visualisations involved layouts of complete paths, involving thousands of URLs. In both recommendations and in visualisations, we then used statistics of co-occurrence over entire paths. This diluted the specificity of recommendations and also involved a co-occurrence matrix of size $O(N^2)$, where N is the number of symbols in paths. We now avoid this quadratic data structure, taking advantage of database indexing facilities to generate required statistics on the fly.

Recommendations are made periodically – another slider in the interface sets how often – and take a few seconds to make. Most of the work takes place on a server machine devoted to the database. We can make a matching visualisation in under one minute on a Silicon Graphics O2.

Statistics for recommendation and visualisation thus form categorisations or abstractions over path entries that are not fixed *a priori*. Instead, recurrence statistics are made anew for each person at the time of, and using the context of, each recommendation operation. Each new entry in a path changes the pattern of symbol co-occurrences throughout the shared configuration of paths. Even if you have not accessed new information recently, your recommendations may still vary as other people's activities change their paths.

Ongoing work focuses on adding the symbols contained in web documents (such as words and URLs) to paths, in extending the interface to allow more active steering of the recommendation process, and in more specialised tools for logging and recommending Java software components. In adding content, the goal is to treat words, filenames and URLs equally within the same model. In active steering, the user could choose to add symbols to (and remove others from) the "recent" symbol set. At one end of a spectrum would be the removal of all automatically collected symbols, and the explicit specification of a query-like set of terms. At the other end would be the current implicit or passive specification.

4.5 Conclusion

In the path model and related systems, information representation has human activity at its centre. Categorisation, in the form of statistical consistencies in similar past activity, is built around this shared, adaptive, relative system of symbols. In our recommendations and visualisations, we aim to show and support user activity, and to take account of the interdependence and interweaving of information tools such as web browsers, editors and so forth. We have described how the path model was built by analogy with a theory of

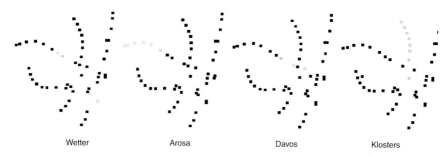

Figure 4.7 Four snapshots of an interactive "map" of URLs based on the path model, scaled down from actual screen size, and with the frame and controls of the surrounding visualisation tool (described in [22]) cropped. Following access of web pages with detailed weather (*wetter* in German) reports for the mountains of Switzerland, recommendations involved URLs for web sites of ski resorts near Zürich, such Arosa, Davos and Klosters, as well as URLs for other weather pages. Each snapshot shows the same positions, based on co-occurrence of URLs, but shows in a lighter colour those URLs that match the given search string, e.g. *arosa* in *www.arosa.ch*. The *Klosters* snapshot also shows three URLs that have "popped up" by the tool to show textual detail.

architecture, and more generally tried to make clearer the similarities between the fields that can profitably be used in both system design and theory. These similarities exist even while there are substantial differences. For example, mutability: bits are easier to move around than bricks. This difference is diminishing, however, as some information structures, such as home pages of large corporations and the search engines we so rely on, are made persistent, reliably available facilities. In architecture, banks, galleries and museums increasingly have their supporting structure on the outside, so that internal partitions and passageways can be easily reorganised. Buildings are made more adaptable to their current users while URLs become more corporate and fixed.

We suggest that it is the common semiological basis of informatics and architecture that makes for these similarities. Human activity involving information in computers is not so special. It is little different from activity involving other media of communication and representation. We thus gain an understanding of how the choices for future significant action of each individual, in and through symbols of all types, are influenced by the past and current activity of others. This gives rise to the "hermeneutic circle" of representation, interpretation and action. We have tried to make clearer how these foundations lie deep – almost hidden – beneath the theory and practice of informatics, and to show how we can build on them in new ways when we better comprehend their origins, strengths and weaknesses.

Antoni Gaudi, architect, said that "originality is going back to origins". In this chapter we have looked back through informatics to mathematics, in order to make more "ready to hand" our materials and tools for building new approaches to information representation. We consider that contemporary mathematics and philosophy show us that activity, language and the social are at the centre and origin of informatics, rather than on the periphery of a core of objectivity and positivism. This view does not make us discard positivist approaches for those based on structuralist, or even post-structuralist ones. As was pointed out by Ricoeur, positivist approaches allow access to open semiological systems while structuralist approaches, such as the path model, involve a closed, relative system of symbols [23]. Structuralism needs positivist approaches to connect each new symbol into its representational systems, and so begin another turn of the hermeneutic circle: taking a new symbolic representation, interpreting it in terms of how we use it with others, and creating new patterns of action and signification. Thus, while path-based systems and related collaborative filtering systems may be able to represent and support more of the complexity and subjectivity of information than positivist information access approaches, the former are dependent on the latter.

We believe that the understanding gained from re-examining the origins of informatics opens up new space for originality in informatics theory and systems design. This shift towards what Wegner called the interactive paradigm will let us handle information that seemed intractable, combine tools that seemed unrelated and interweave theories that seemed incommensurable. This can be done if we face up to informatics' limitations and historical biases, and see what Paul Churchland calls the "deeper unity" of modern intellectual discourse all the way from neurophysiology to philosophy of science. Informatics can then step forward to take its place with contemporary mathematics, philosophy and semiology.

Acknowledgements

The author wishes to thank the editors for discussion and comment related to this paper, and also Dominique Brodbeck, Paul Dourish, Luc Girardin, Christian Heath, Aran Lunzer, Marissa Mayer, Kerry Rodden, Lisa Tweedie, and those who directed the laboratories where the bulk of this work was done: Hans-Peter Frei of UBS Ubilab, and Yuzuru Tanaka of the Meme Media Laboratory, Hokkaido University. And Jeff Mills (Liquid Room), and Mogwai (Young Team).

References

1. Chalmers, M., Rodden, K. and Brodbeck, D. The order of things: activity-centred information access. *Proceedings of the Seventh International World Wide Web Conference (WWW7)*, Brisbane, 1998, pp. 359–367.
2. Hillier, B. *Space is the Machine*. Cambridge: Cambridge University Press, 1996.
3. Foucault, M. *The Archaeology of Knowledge*. Translated by A.M. Sheridan Smith. Routledge, 1972.
4. Nardi, B. and Barreau, D. "Finding and reminding" revisited: appropriate metaphors for file organization at the desktop. *ACM SIGCHI Bulletin* **29**: 76–78, 1997.
5. Berners-Lee, T. World-wide computer. *Communications of the ACM* **40**: 57–58, 1997.
6. Resnick, P. and Miller, J. PICS: internet access controls without censorship. *Communications of the ACM* **39**: 87–93, 1996.
7. Nardi, B. (editor). *Context and Consciousness: Activity Theory and Human–Computer Interaction*. Cambridge, MA: MIT Press, 1996.
8. Wegner, P. Why interaction is more powerful than algorithms. *Communications of the ACM* **40**: 80–91, 1997.

9. Churchland, P. Folk psychology. In *On The Contrary: Critical Essays 1987–1997*, P.M. Churchland and P.S. Churchland, editors. Cambridge, MA: MIT Press, 1998, pp. 3–15.
10. Everitt, N. and Fisher, A. *Modern Epistemology: A New Introduction*. New York: McGraw-Hill, 1995.
11. Karatani, K. *Architecture as Metaphor: Language, Number, Money*. Cambridge, MA: MIT Press, 1995.
12. Wittgenstein, L. *Philosophical Investigations*, 3rd edn. Translated by G.E.M. Anscombe. Oxford: Oxford University Press, 1958.
13. de Saussure F. *Course in General Linguistics*. Translated by W. Baskin. New York: McGraw-Hill, 1959.
14. Gibson, J. *The Ecological Approach to Visual Perception*. Hillsdale, NJ: Erlbaum, 1979.
15. Chalmers, M. Comparing information access approaches. *Journal of the American Society for Information Science*, 1999 (in press).
16. Suchman, L. *Plans and Situated Actions: The Problem of Human Machine Communication*. Cambridge: Cambridge University Press, 1987.
17. Schmidt, K. Of maps and scripts: the status of formal constructs in cooperative work. *Proceedings of the ACM Group 97*, Phoenix, AZ, 1997, pp. 138–147.
18. Colomina, B. *Privacy and Publicity: Modern Architecture as Mass Media*. Cambridge, MA: MIT Press, 1996.
19. Major, M. (editor). *Proceedings of the First International Symposium on Space Syntax*. University College London, 1997.
20. Bellotti, V. and Sellen, A. Design for privacy in ubiquitous computing environments. In *Proceedings of the Third European Conference on Computer-supported Cooperative Work (ECSCW'93)*, G. de Michelis, C. Simone and K. Schmidt, editors. Dordrecht: Kluwer, 1993, pp. 77–92.
21. Chalmers, M. A linear iteration time layout algorithm for visualising high-dimensional data. *Proceedings of IEEE Visualization 96*, 1996, pp. 127–132.
22. Brodbeck, D., Chalmers, M., Lunzer, A. and Cotture, P. Domesticating Bead: adapting an information visualization system to a financial institution. *Proceedings of IEEE Information Visualization 97*, Phoenix, AZ, 1997, pp. 73–80.
23. Ricoeur, P. *Hermeneutics and Human Sciences: Essays on Language, Action, and Interpretation*. Translated by J. Thompson. Cambridge: Cambridge Univerity Press, 1981.

Chapter 5

Information that Counts: A Sociological View of Information Navigation

R.H.R. Harper

Abstract

This chapter presents two empirical examples of information navigation work in organisational contexts. They show that information gains its relevance to any individual organisational actor in proportion to how actors in other organisations use that information. In this respect, it will argue that information is socially organised. The chapter will then discuss what implications this has for providing new sources of information and new information retrieval techniques with the World Wide Web.

5.1 Introduction

In this chapter I will analyse the structure that supports and is embedded in the information navigation work of individuals who are called "desk officers". These individuals are economists who work for the International Monetary Fund in Washington, DC. Their task is to be the conduit between the private information space of the Fund (as it is known) and the public information space of the world outside. Specifically, I will offer a sociological analysis of two topics: first, the institutional motivations that drive the navigational activity of these desk officers; and second, the parameters of relevance that guide their work, parameters that are institutionally set and yet endlessly re-negotiated. I will argue and assume that these motivations and parameters are constituted and made observable by the methical ways that desk officers operate. Following the sociologists Garfinkel and Sacks, I will take it that desk officers structure their behaviour to achieve, accomplish and display the rationally motivated, institutionally relevant nature of the information work [1]. Description and examination of those methods can be used to make recommendations for the design and use of computer-based information navigation systems [2, 3].

The chapter will be structured in the following way. First, I will say something about the setting studied and report the fieldwork programme used in its study. I then present empirical examples related to the two areas of interest. I conclude by considering what implications these have for the role and design of web-based information systems.

5.2 The Fieldwork and the Setting

The Fund may be thought of as a financial "club" whose members consist of most of the countries of the world. Member countries contribute to a pool of resources that can then be used to provide low interest, multi-currency loans should a member find itself facing balance of payments problems. The Fund has some 3,000 staff, of which 900 are professional economists. These economists analyse economic policies and developments – especially in the macroeconomic arena. They have particular interest in the circumstances surrounding the emergence of financial imbalances (including those that lead to a balance of payments crisis), the policies to overcome such imbalances, and the corrective policy criteria for making loans. Key to this work is undertaking "missions" to the country in question. A mission normally involves four or five economists, supported by administrative and secretarial staff, gathering and discussing macroeconomic data with key officials in the authorities of the member country. This leads to the creation of a picture or view of that situation which in turn is used in policy discussions between the mission and the authorities. Member authorities are obliged to agree to Fund policy recommendations if Fund resources are to made available to them. Missions normally last about two weeks. A mission team's view of a member country, the outcome of policy dicussions and any recommendations for the disbursement of funds (or otherwise) are documented in "staff reports" prepared by the mission teams once they return to Washington. These reports are used by the organisation's executive board for its decision making.

Desk officers are a subset of the Fund's economists. Each desk officer works on a "desk". A desk is a label for a number of individuals whose responsibility is to gather, analyse and represent information about particular member countries. But for the sake of simplicity, I will treat these individuals and their roles as more or less one under the title of desk officer.

The work of these individuals was studied ethnographically. Ethnography is a much used term by sociologists and anthropologists – something of a catch-all word for a range of methods and approaches, just as long as they involve direct observation of one kind or another [4]. In this case, the ethnography involved six months' fieldwork.

The structure of the fieldwork involved focusing on the *career* of staff reports, from the first draft of what is called the "briefing paper" prepared before a mission commences, through the mission process itself, to the post-mission review, and then to translation, printing, and circulation processes of the final staff report.

This was accomplished in two ways. First, I followed a hypothetical staff report around the organisation and interviewed parties that would be involved in its career. In all, 138 personnel, including 90 economists were interviewed. Second, I observed a Fund "mission" and its allied document production practices. I observed meetings between the mission team before the mission commenced, between the member authorities and the team during the mission, and observed post-mission meetings. Once this process was completed, a full ethnographic text was prepared and published [5]. Some of the materials presented here have been analysed within that text, although I have attempted to bring a different slant on them here.

5.3 The Empirical Examples

5.3.1 Motivation

The first example relates to what motivates desk officers. Here I am not concerned with what motivates them personally (i.e. whether they want to become managing director of the Fund or have any other similar ambition). What concerns me is what motivates them by dint of their organisational role. Moreover, and following Garfinkel and Sacks, I am concerned with the methodical things that desk officers do which makes these motivations demonstrable.

As remarked, desk officers are meant to be at the juncture between the outside world and the private world of the Fund. Their role as "experts" means that they will necessarily receive a great deal of information related to their member country that they will have to sort through, manage and share with the Fund as whole at certain times during the year (and most especially when financial crises show themselves).

There are a number of ways of looking at this work and what motivates it, some of which are more appropriate than others for design. This can be illustrated by outlining what would appear to be a useful way but which in practice is misleading. This will enable me to more effectively convey what I mean by motivations and thus more accurately describe what role technology might have in supporting it.

One approach, the wrong one I believe, would be to simply describe and catalogue the information desk officers collect and then to assess its comprehensiveness and its timeliness. This could then be measured against some metric for information as a whole that one might devise, i.e. against some notion of what would be a full or comprehensive data set.

With this approach, one would probably start off with a list. This list would include the following sorts of facts: that desk officers routinely collect and receive newspaper editorials in the *Financial Times* and articles in the *Economist* aswell as analyses of similar organisations to the Fund, such as the Organisation for Economic Co-operation and Development (OECD) and the World Bank. In addition, they receive *government bills* for financial or economic legislation, ministry of finance *budgetary statements*, and *financial statements* from commercial banks, all from the member country in question. Most desk officers also receive quarterly statements of *foreign currency holdings* from central banks, and *estimates of government fiscal revenues* are sent to them.

One could easily add to this list. The point is that an analyst is then able to make a comparison between what desk officers have and what may be thought of as what they would want/need in ideal circumstances. That is to say, the analyst may then ask such questions as "Is the information they have enough?"; "Is it the right sort of information?", and so on. If not, a system could be designed that would deliver it. For example, a system could automatically search for and deliver all newspaper articles referring to the member country. It could automatically notify desk officers of any new academic articles, or it could even automatically update spreadsheets when budget statements and balance sheets are released. In other words, the designer would want to expand the kinds of information desk officers can have access to and would want also, perhaps, to help expedite or ease the difficulty desk officers may have in sifting through that information. One might suggest that such design considerations reflect or are typical of the orthodox metaphor in human–computer interaction. (As it happens this is pretty much the way the economic profession examines work of the Fund. One result of these analyses is the claim that the Fund needs to broaden the scope of the materials its desk officers receive. For a review see Bird [6].) This view takes for granted and assumes that the goal of any and all desk officers is to increase their informational resources.

Now, in contrast, the view I think much more appropriate puts those sorts of questions aside. That is to say, instead of asking whether desk officers have *enough* information, the approach I think appropriate involves asking what motivations lie behind the way desk officers organise themselves to deal with the information they have at hand. According to this view, it is not the extent, scope, or speed with which information is processed that is at issue, it is how someone like a desk officer *treats* it. And if one does this one will see that the nature of their task is more complex and subtle than it might at first appear. (For some more examples of this see Button and Harper, *op cit.*)

An example will illustrate this. The desk officer for a country I will call for the sake of confidentiality, Arcadia, received most of the materials I have just listed (and hence was like any other desk officer). This included newspapers. In particular, he received (via the embassy in Washington) copies of two "dailies" from Arcadia. But, as he explained to me, the utility of the information that these newspapers conveyed was quite limited:

Desk officer: *"You've got to understand, Richard, that in Arcadia there is no real opposition. This paper [he pointed to one] is the official party paper. That tells the official party line. This one [he pointed to another] is meant to be the opposition paper but it isn't really."*
Interviewer: *"So what do you do with them?"*
Desk officer: *"Oh, I still read them. You get some feeling for events in Arcadia. Also you can get a feel for the views of the authorities. That can be useful in policy discussions. You can't trust the figures in them though."*

Of course, in relation to the kinds of materials one finds in newspapers – official party ones or otherwise – it may be remarked that they do not convey the kind of "stuff" the Fund would be primarily interested in: namely macroeconomic data. As a result, it would hardly be surprising that a desk officer would not be too interested in what they find. The contents of newspapers may just be too inconsequential – though useful for getting a "feel" for a country. But here, I think, is precisely the thing that is of salience: the ability to distinguish between information that provides "a feel for something" and information that does something else. Let me explain.

To begin with, what one can learn from this is that the desk officer was able to "use" information because he already knew something about it. In particular, he knew a great deal about the institutional provenance of the information he received in those newspapers. There was then a reflexive relationship between the information searched for and what was already known. Reflexive in the sense that in knowing something about the information, a desk officer could see why that information might be useful. But the interesting thing here is not so much this general observation; instead what is interesting is how the desk officer in question used the particular set of information thereby identified.

For the desk officer had a methodical way of dealing with this information. Not in the sense that the desk officer received the papers arrive every day or so, it was rather that he *methodically allocated* the information he received into a broader context. He recognised that the information in the newspapers was far from factual and instead treated it as a resource for getting a "feel" for "better understanding" the authorities.

Thus, the relationship the desk officer had to information was not one similar to, let us say, the relationship a fisherman has to his nets – hoping simply to expand the methods by which he captures what he is after. Rather, the desk officer had a relationship with information wherein his task was to know where to place or allocate the information. This is what motivated the desk officer; he was concerned to put things in the right place, to allocate information into a broader context, a "field of relevance" if you like (or as the ethnomethodologists put it, into an "occasional corpus of knowledge" or a "topical contexture of meaning").

As it happens there was an incredibly rich field into which things could be placed. In this case, the desk officer's methodical activity involved a sort of taxonomising which depended on multiple criteria. For example, it involved an ability to place information within what one might call a moral order and all that that implies: not just facts and figures but also intentions, motivations, perspectives, hopes aspirations, and the denials of the Arcadian authorities. In this sense, the desk officer was seeking to understand the political situation in which the Arcadian authorities find themselves, not just their economic situation.

5.3.2 Institutional Relevance

Knowing where to put information – irrespective of the technological tools to support that task – is itself based on another set of concerns, related to institutional relevance. This is the second area of concern for me. Again, I will illustrate my interests with an empirical example but this time taken from the information work that desk officers undertake on a mission. The mission in question was to Arcadia.

During the mission, a meeting between the desk officer and one of the Arcadian officials led to a discussion of an economic analysis institute that had recently been set up. Thereafter, the desk officer and a colleague arranged a visit to the institute to find out what it had been doing. They discovered that it had undertaken a range of studies on recent macroeconomic developments in Arcadia. The desk office reported this in a meeting they had with the rest of the mission the following day. During this meeting, the desk officer made available two of the reports that the institute had given him which the team browsed through. The team discussed whether they could use the information in their own analytical work. Essentially, what they wanted to determine was where they could place it in the already mentioned field of relevance. In this case, however, the mission decided that it would be inappropriate to use the information despite the fact that it was "in one sense obviously relevant", as one put it. They decided not to do so because the institute's views were not taken on board by the Arcadian authorities themselves. What mattered was whether the information was "in sight" of the authorities, not whether it ought to be or was "naturally relevant". The outcome of the meeting was that it was decided that in the future the institute might get a higher profile in Arcadia and that therefore its work might well become more relevant. Accordingly, the desk officer decided to keep a note of what the institute did in the future and to track whether it started to have any influence with the authorities.

In practice, this meant that the desk officer did not so much ignore the institute's work as *put it aside* until such time that it became something the Arcadian authorities themselves started to use. Here, one might make a paradoxical contrast between the desk officers' use of newspapers and the decision of the mission to disregard the macroeconomics analyses of the institute. It is paradoxical because the important point is not what the information is in itself, but whether and where it can be placed in the field of relevance. In this case information that seemed a long way from macroeconomics (namely information in the newspapers) was viewed as "more relevant" than information that was quintessentially macroeconomic. This judgement was based on what the mission knew to be the concerns and perspectives of the authorities. Key to this was defining what was included in the authorities "sight" and what was out of their view. In this respect, the information-gathering work that the desk officer and the mission team undertook was built around the assumption that they had to understand the world-as-known-in-common by both the authorities and themselves rather than as might be understood by just one side. It was not enough that information is somewhere out there in the world and simply has to be gathered by some clever processes of human agency and technology. Rather, information was relevant when it was seen by all the parties concerned.

These two examples show then two aspects of information. On the one hand, the nature of information work involves the methodical allocation of information into a field of relevance. Connections between the items in this field can be complex and often orthogonal. The second example indicates that it is not simply the nature of the information itself that is crucial to its categorisation and allocation to a field of relevance, it is also whether other parties include that information or not in their field of relevance. Above all, both or rather all parties need to have a common field of relevance – they need to have more or less the same taxonomies of information with the same sets of information. Furthermore, exclusion or as in this case, the putting aside of information until the future, is one of those "natural" categories in the methodic taxonomising activities of these information workers.

5.4 Implications for Information Navigation on the Web

So, with these examples in mind, let us now start to think about a technology that might alter or impact on information work. The World Wide Web obviously comes to fore. One of the problems that is beginning to emerge with the web is that it is difficult for users to be sure of the exact provenance of the information they download. Bound up with this is the problem of how to appropriately interpret and use that information. As Nunberg has noted, the Web offers a kind of anarchic information world [7]. Some commentators – web propagandists if you like – claim that it is precisely this that makes the web so exciting: those who access web sites are themselves defining what counts as good information and what is not, and furthermore anyone can set up a web site. But what authors like Nunberg are drawing attention to is how all information is bound up with broader systems of use that are not limited to the precise point of use at any moment in time.

So for example, to understand what it "means" to use an article in an academic journal cannot be defined by what happens at the point of reading (i.e. the cognitive processes of understanding the contents of the article). Instead it is to understand that activity as one that occurs within broader institutional practice. That is to say, an academic journal article does not just convey information, its meaning is wrapped up in such things as what is known about the status of the journal; whether articles in that journal are reviewed; the status and expertise of those reviewers; and finally what the reader knows about the general impact of the article (i.e. whether it is viewed as an article that is expected to or has had a major impact in the relevant academic community). It is knowledge about these kinds of things that provides the context for "reading".

Now, with the web, similar "systems of interpretation" have not been developed yet, though of course with time they might evolve. At the moment the existence of informal ratings of web sites that are freely available on various bulletin boards is an indication of how web surfers are trying to create for themselves what one might call institutional processes that enable people to determine what is or is not of value. The fact that these ratings are informal is viewed as one of the things that makes the web so exciting: for here is an information world in which the participants are defining for themselves the institutional processes of use.

Nonetheless, it may be that the institutional processes of using the web will slowly become similar to already existing institutional patterns of information use. Thus it might be that the reading of a web site will become analogous to the reading of, say, an article in an academic journal. Readers will assume that the web site they have downloaded has been through processes of review, that the findings in question have been previously disseminated elsewhere, and so on. This seems unlikely, however. A more probable scenario is that the institutional practices that give meaning to and provide a context for the use of the Web will be diverse, supporting a whole range of different systems of use and, furthermore, that these will be dependent on other, already existing, systems of interpretation and use.

And this leads us back to the examples of the information work of the Fund's desk officers. In the first example I noted that a desk officer knew enough about the institutional context of particular newspapers to know how to place the information they conveyed. Similarly, if the desk officer were to surf the web to get to sites purporting to present information about Arcadia, he would want to know who produced those sites. In particular, he would want to know things like whether the sites in question were "official party" sites or "genuine opposition" sites. Knowing these differences would enable the desk officer to determine what would be the appropriate use of the information he found or, as I have put it, how to place that information in the field of relevance. What would be key to the desk officer's information work wouldn't be simply the gathering of information, as if he (or she) was "surfing the web" to see what they could find; it would be being able to place that information somewhere. Information would be useful insofar as it could be allocated to some point in the field of relvance.

This leads on to the second example. This illustrated a somewhat different set of issues. It drew attention to the question of getting new information. With regard to the Web, it would appear that a desk officer may be able to use the Web to locate a great deal of "new" information that might look directly relevant to his or her concerns. But the problem the example highlighted is that a desk officer needs to distinguish between the information he can use in the "here and now" and that which he has to "put aside until the future". This in turn relates to such things as whether that information is viewed as "counting" by the authorities of the member countries themselves. As it happens, and to offer a little bit more emprical material, it may well be that in some cases member authorities disregard some information that a desk officer would prefer they took heed of. Here there are issues to do with the various pressures that a desk officer and the mission he or she is part of can bring to bear on the authorities and what resistance the authorities can offer to such pressure. In the case of Arcadia, the relationship between the authorities and the mission was in part a didactic one. Therefore the mission was in a position whereby it could direct the authorities to take account of information the mission thought appropriate. Even so, in this case and as I say, the mission decided to let the Arcadian authorities make the decision about the new economics institute for themselves.

What this is leading me to say is that what is interesting about new information resources like the World Wide Web is not how individuals like the Fund's desk officers will familiarise themselves with the "information tools" of the web – its various browsers, informal standards and so on – so much as how they will place the information they find thereby within a field of relevance. Partly they will be able to do so because they will already know, by dint of their knowledge of previous information delivery mechanisms, what position some information resource already has. So for example, many newspapers are beginning to release their articles on web sites. (This may be happening with the newspapers in Arcadia, I do not know.) But since the desk officers already know the provenance of that information (i.e. they know something about who owns the newspapers and so on), they will be able to allocate that information to the right place. The medium of its delivery will make no difference to their effective use of that information.

In contrast, they will have considerably more difficulty if they use the web to get new information resources, especially if they do not have sight of what those with whom they want to work have as part of their own (i.e. the other groups') field of relevance. In other words, there is need for the web-based searching activity of desk officers to be somehow linked to the similar processes of those individuals with whom they work within the governments they deal with so that when "new information" is found, it can come into view in a way that is acceptable and useful to all parties. Put another way, increasing access to information for desk officers via the web will be a wasted opportunity if that information is not presented in such a way as to be demonstrably identifiable as part of those relevances that all accept. The information runs the risk of being unusable if desk officers cannot tell how it is perceived and used by those they deal with on missions and elsewhere. There needs to be a way of integrating the navigation work of the Fund's desk officers with the ongoing establishment of a view by the member authorities themselves.

5.5 Conclusion

The main thesis of the chapter has been that the conception of which factors are important to the design of any new "information navigation" technology needs to include the institutional context in which the technology will find its place. What is meant by this context is both what motivates individual information workers by dint of their organsational role, and second, how these motivations are bound up with knowledge about institutional relevance. Key to both these issues is the methodical allocation of information into a field of relevance.

Although the emergence of technologies like the World Wide Web is suggesting to some that the methods for the electronic delivery of information are now so robust and comprehensive that that there will be a radical change in the nature of information gathering and use, I have been implying, then, that the social and institutional processes that have hitherto supported information gathering and use will continue to be fundamental. Furthermore, to the extent that this is so, then the conception of the ways in which "information-based" work practice is supported needs to be broadened. Instead of devising technologies that can support *individuated* information gathering and use, technologies will need to be developed that are supportive of collaboratively defined information in the following way:

- Technologically mediated information needs to indicate the socially organised provenance of that information. This will include indicating such things as which institution produced it; who owns that institution; what is the known perspective of that institution (political view, etc.), and so on.

- It means also devising technology which ensures that when information is used by any one individual, this can be indicated to other parties in the relevant institutional process. Information use needs to be tracked, logged and, ideally made concurrent with its use by others. Or if that concurrency of use will not occur (being rejected for whatever reasons) or if that information is "put on hold to another time", this too needs sharing somehow and made demonstrable.

To reiterate: the thesis has been that information work in organisations is socially organised. Individuals use information as part of their membership of work groups and organisations and indeed as part of society as a whole. Information tools need to be designed with this in mind. And doing so, I think, will shift the basis away from what has been the traditional metaphor underscoring human computer interaction – the individual and the machine – towards a view that asssumes that it is people in roles interacting with the machine; where it is organisations with particular concerns that converse with the machine; where it is, in a phrase, a conception where it is *society* interacting with the machine.

References

1. Garfinkel, H. and Sacks, H. On formal structures of practical actions. In *Theoretical Sociology*, J.C. McKinney and E.A. Tiryakian, editors. New York: Appleton Century Crofts, pp. 338–366, 1970.
2. Button, G. and Dourish, P. Technomethodology: paradoxes and possibilities. *Proceedings CHI'96*, Vancouver. New York: ACM, 1996, pp. 19–26.
3. Button, G. and Harper, R.H.R. The relevance of "work-practice" for design. *CSCW: An International Journal* 4: 263–280, 1996.
4. Harper, R.H.R. The organisation in ethnography. *CSCW: An International Journal* (in press).
5. Harper, R.H.R. *Inside the IMF: An Ethnography of Documents, Technology and Organisational Action.* London: Academic Press, 1998.
6. Bird, G. The International Monetary Fund and developing countries; a review of the evidence and policy options. *International Organisation* 50: 477–511, 1996.
7. Nunberg, G. (editor). Farewell to the information age. In *The Future of The Book*. University of California Press, 1996, pp. 103–138.

Chapter **6**

Screen Scenery: Transposing Aesthetic Principles from Real to Electronic Environments

Monika Buscher and John Hughes

Abstract

The link between material arrangements, time, and people's practices is complex. It has long been a concern for architects, landscape architects, and artists. Now, it has become a concern for the designers of virtual spaces in electronic media. However, there are substantial differences. The design of material arrangements in real spaces draws on the laws, patterns, and aesthetic principles of the real world with all its physical and cultural characteristics. "Material" arrangements in virtual space do not have to face many of these constraints. Yet, in order to be intelligible, virtual worlds have to exhibit at least some familiar features. An ethnographic study of people's interactions with and in the real and virtual spaces of a media art exhibition in Germany has shed some light on possible principles for the design of electronic spaces. We combine this study with an analysis of empirical studies, and theoretical considerations of architectural design and its relation to use with a view to informing the design of electronic, "inhabitable information spaces".

6.1 Introduction

Three-dimensional online electronic environments allow people from globally distributed locations to "meet" (e.g. Alphaworlds and VWWW, http://www.alphaworlds.com and http://vwww.com/hub/3dpage.htm). The character of such meetings can be anything from casual meetings between strangers in public spaces to the focused encounters of participants in a business meeting. The environments themselves differ with regard to the visual and physical qualities of the "material" structures they provide, the embodiments or avatars that users can employ, and the means of communication that are available: video, audio or textual chat. The potential use of such environments is still a matter for discovery, but there are indications that a number of areas would find interest in them.

There are commercial interests, for example, that seek to transform web sites "into a growing online community, enabling highly interactive real-time communication" [1]. Creating such a "home" for customers reflects the fact that "corporate site owners have recognised that by creating a community around their products, they are able to greatly increase ... the loyalty of the customer for their company" [2]. Urban design and planning is another field where the application of three-dimensional electronic environments promises some benefits. It is argued that allowing people to see and even move around a model of designs for public spaces – inserted into the context of the existing environment in a virtual reality visualisation – enhances their understanding of the design and thus facilitates a more democratic planning process (see, for example, [3]). The flexibility and the realism of such visualisations of design ideas by-pass constraints of traditional visualisation techniques and planning procedures, such as the conventions of architectural drawings that are intelligible only to the trained eye and the limits to the exploration of alternatives inherent in the time-consuming procedure of producing traditional visualisations.

It is also possible to visualise "information" in three dimensions. The contents of databases and archives, and people's movements in this "infoscape" [4] – whether at work or for entertainment – could develop into "inhabitable" information spaces. Moreover, the concept of inhabited information spaces can also be extended to encompass *several* single environments. This paper is based on research undertaken with a focus on the design of such "electronic landscapes" (e-scapes).[1] An ethnographic study of people's engagement with a number of interactive multimedia art installations suggested that people orient towards visual and narrative strategies of organising electronic environments. Such strategies can be found in the fine arts, but also in film, and architecture. There is a large body of literature that deals with a diverse range of these strategies. With a view to structuring large-scale spaces as envisaged for e-scapes, architectural theory and practice, in particular, promise some interesting insights. After a brief glance at the motivation, nature, and extent of the ethnographic study, Section 6.4 explores a selection of architectural approaches to achieving spatial legibility. Sections 6.5 and 6.6 outline some aspects of the ethnographic study. In the conclusion we draw the threads of these two different perspectives together.

6.2 Electronic Landscapes

Over recent years there has been a proliferation of online environments which has resulted in a great number of systems whose insularity and diversity causes both technical and orientational difficulties [5]. Different standards mean that software has to be downloaded and machine settings have to be configured before a user can enter a world for the first time. This is particularly awkward in view of the fact that there are few other means of finding out about a world

---

[1]ESCAPE: Electronic Landscapes, a project funded under the EC i3 programme.

without entering it. In effect, users might go through time-consuming preparations only to find themselves in a place that does not match their expectations, or whose structure and appearance they find confusing. Standardisation is one solution to allowing users to move between them that is being pursued by the virtual reality community (most notably in the case of VRML). Another possibility is to support translation between environments. We envisage an "electronic landscape" (e-scape) as a collaborative virtual environment (CVE) that provides a container – a landscape or universe – where translation across a diversity of worlds can take place. We can consider an e-scape as a technical common ground and, at the same time, a "physical" space where people can move around and gauge some information about a world before entering it. This information may include world-specific information such as the quality of display available from the user's respective platform, the structure and appearance of worlds, the current population of, and activities in, different worlds. It could also include information about the relationships between different worlds (expressed, for example, through their position relative to each other in the e-scape) and the orientation of people inspecting these worlds. Such information provides users with a sense of overview and *transition* and improves on the current abruptness of teleporting between environments.

However, there are some difficulties that impinge on the design of CVEs of this kind. "Community", for example, is not, as it is often tacitly assumed, an automatic corollary of giving people a place where they can meet – neither in the physical world nor in an electronic environment. The "material" arrangements of a virtual environment – whether they exist for their own sake or represent actual informational content – have to be, above all, *legible*. It is here that principles of real-world urban design can be useful. The complex links between material arrangements, time, and people's practices that have to be considered in making physical environments "legible" and conducive to specific social uses have long been a focus of attention for architects, landscape architects, and artists [6–12]. If we want to create three-dimensional information *spaces* that can be "inhabited"; that provide a socially and aesthetically rich, intuitive, and pleasurable experience, the insights gained by an examination of the relationship between people and material arrangements in the physical world may be a source of inspiration.

But an exploration of a body of work concerned with "spatial grammars" cannot yield a catalogue of principles that can be literally translated into the electronic medium. Although current theoretical and experimental approaches to the design of electronic environments almost univocally (cf. [13]) suggest that legibility can be achieved through ever more sophisticated "realism" with regard to both sensory access to and visual appearance of the architecture and landscapes in the environments [14–17], there are serious limitations to this. There are, first, technical limits to realism. Secondly, a consideration of the "physical" character of electronic environments suggests that realism is not the only option. There is, for example, no topography, no natural light, no gravity, and no friction in the electronic medium unless it is artificially introduced. How and in what combination such factors could and should be introduced is a matter of investigation. The prospective use of a CVE should have an influence on the inclusion of such factors. Moreover, studies of people's engagement with existing virtual environments are a way of finding out what kinds of organising principles are oriented to and in what ways they are utilised.

6.3 The Ethnographic Study

We explore some of these orientation strategies drawing on fieldwork undertaken in the media museum of ZKM, where a number of interactive art works made shared electronic environments available to the public. The centre for art and media technology (ZKM) in Karlsruhe, Germany, was founded as a new type of institution that brings together art and technology in an unprecedented and unrivalled way. It combines two research institutes – the Institute for Music and Acoustics, and the Institute for Visual Media – a Media Library, and three museums: the Museum for Contemporary Art, the City Gallery, and the Media Museum. 12 years after its conception, ZKM was opened to the public in October 1997. This event was celebrated with a number of events and the Fifth Multimediale (a biannual exhibition organised by ZKM since 1989). Permanent and temporary exhibitions at the Media Museum contain around 30–40 works. The ethnographic study reported on in this chapter was carried out over a total of nine weeks, covering the whole of ZKM's opening, and three weeks of 'normal' opening time in December 1997 and March 1998.

One means of informing an intuitive and legible design of CVEs and to allow people to experience a sense of presence within them are studies of the strategies people employ in moving around real-world space. We draw on observations from these studies here, but note that such observations in physical space settings alone do not suffice to inform the design of CVE's. Mediating technologies, while in many respects extending the possibilities for human interaction, inevitably alter and impoverish the richness of interactionally relevant information we take for granted in our everyday face-to-face interactions. Therefore, it is necessary to observe people's interactions in and with such environments. However, despite the emergence of online electronic environments these still remain a specialist concern and are seldom used by general citizens – there are few opportunities to observe real-world situations in which people begin to inhabit electronic spaces (but see, for example, [18]). To overcome this problem, we undertook a series of studies at ZKM. Field notes derived from participant observation, and documentation of people's engagement with the art works in audio and video recordings were transcribed and analysed.

In our analysis we explore a set of issues that emerge from our observations. Essentially, video, audio, and observational data suggest that a sense of presence is not an individually owned, private quality, but embedded in the sociality of our existence in the world [19]. People adapt familiar practices and interpretations entwined with a sense of presence in the physical world to the affordances of electronic environments [20]. Everyday practices of orientation, movement, and interaction in space are drawn upon, but *transposed* rather than transplanted in order to fit in with the affordances of the environment. In addition to the limits to realism in virtual environments outlined above, such transpositions suggest that a more creative, symbolic approach to structuring electronic environments could be useful. However, this does not give the designers of electronic environments complete artistic freedom. On the contrary, once we turn our back to realism as an organising principle, the question of what other structuring principles could be employed becomes crucial. The exploration of spatial grammars in relation to the practice of engaging with existing electronic environments aim to develop some ideas for a more symbolic approach to the design of legible electronic environments.

6.4. Spatial Grammars

According to a number of design theorists, the design of material arrangements in physical spaces is and should be rule governed. There are studies that try to define the elements of spatial design and the rules of their combination, ranging from algorithmic approaches that describe a variety of systems of rules for the combination of elements [21], to Alexander's universal pattern language [7], and Kevin Lynch's observations of structures of urban agglomerations as they feature in people's mental image of the city [6]. Another approach is to place an emphasis on the generative laws of spatial configurations, such as the relationship between strangers and inhabitants and its expression in the permeability or restrictiveness of buildings and cities [12]. A third perspective focuses on the detail of features that make material arrangements legible [6, 8, 10]. Due to constraints of time and space, the review of spatial grammars in this section is selective, brief and opportunistic in the sense that the focus lies on issues related to the legibility of large-scale spatial configurations. It is meant as an exploration, providing a glimpse of the kinds of means that underpin the legibility of physical environments.

6.4.1 Some Elements of Spatial Design

Christopher Alexander's *Pattern Language* is a hierarchical, networked list of 253 components of material arrangements in the environment [7]. The individual patterns are arranged in clusters. The highest order pattern is *independent region*, the smallest one is *things from your life*. Each pattern is structured to incorporate links to higher order patterns, a formulation of the problem it addresses, an analysis of the problem, the essential features of the pattern and the solution it provides, and links to lower order patterns. The first 94 patterns are of such high order that no individual or group could build what they describe in one sweep. They are to be understood as principles that must be *shared* in order to create coherent large-scale structures in a piecemeal manner. One example is *identifiable neighbourhood*. Based on the premise that a neighbourhood must provide some opportunity for people to organise to participate in local politics, Alexander draws on anthropological and other empirical studies to specify a maximum number of inhabitants (500) and a maximum area (not more than 300 yards across) (p. 82). Further, he argues that there has to be some protection from heavy traffic. Through his patterns Alexander attempts to describe how the physical constraints and possibilities that our environment provides, social and cultural conventions, and the qualities of different materials can be combined in a way that "can make people feel alive and human" (p. xvii). However, his theory of what exactly it is that constitutes such architecture is based on a world view that draws on eastern philosophies, the nineteenth-century English arts and crafts movement, and some romantic ideas of community. As a result it is "to say the very least dogmatic, vitalistic, and nostalgic" [22]. While the idea of a pattern language is attractive, Alexander's particular version of principles thus comes with an ideological baggage that makes it unacceptable for anyone who does not share his views.

Kevin Lynch [6] identifies large-scale structures and their characteristics with reference to empirical studies. His research is based on people's "mental images" of three different American cities. Interviews, sketches, and the researchers' own explorations of the cities are combined in order to gain insight into the structure of large-scale urban agglomerations. The distinctions people make allow Lynch to specify five elements and to suggest ways in which they could be enhanced in order to serve the overall legibility of a city. Here, a brief definition of each of the elements is followed by some examples of possible means to enhance it:

- *Paths* – roads, bridges, tunnels, etc. that link different parts of the city. Since people tend to perceive a path as "going toward something", the designer could respond by providing directional indicators: "a progressive thickening of signs, stores, or people may mark the approach to a shopping node; there can be a gradient of colour or texture of planting as well; a shortening of block length or a funnelling of space may signal the nearness of the city centre (p. 97).

- *Edges* – linear boundaries that impede, divert, or require a change in the mode of, movement. For a pedestrian, for example, a major road can be an edge, whereas the motorist would find the beginnings of the pedestrian zone an edge. There are also "softer" edges that are like seams that visually divide and hold together different areas. An edge may be improved by giving it continuity and lateral visibility (p. 100).

- *Districts* – areas with a common, identifiable character. Homogeneity can be achieved through "a continuity of color, texture, or material, of floor surface, scale or facade detail, lighting, planting, or silhouette ... It appears that a 'thematic unit' of three or four such characters is particularly useful in delimiting an area" (pp. 101–102).

- *Nodes* – points at which paths cross or converge, or a change of transport takes place, but also public squares, street corners, and other points at which people gather. Apart from giving a node a clear identity through the means already outlined, it can be more clearly defined "if it has a sharp, closed boundary, and does not trail off uncertainly on every side; more remarkable if provided with one or two objects which are foci of attention" (p. 102).

- *Landmarks* – objects that stand out from their surroundings and can typically be seen from many different positions. The "contrasts with its context or background" is where the designers' efforts to enhance a landmark's features should focus (p. 101). Moreover, it is "stronger if visible over an extended range of time or distance, more useful if the direction of view can be distinguished. If identifiable from near and far, while moving rapidly or slowly, by night or by day, it then becomes a stable anchor for the perception of the complex and shifting urban world" (*ibid*).

Lynch's work provides a repertoire of structures, but his work also indicates that the legibility of the elements that make up a large-scale material arrangement and the intelligibility of the whole lie in the *detail* of the design. While Lynch takes an interest in people's perceptions of a city with a view to orientation, the detailed design of material structures is also related to the fit between the design and its use.

6.4.2 Sociability, Mystery and "Visual Gravity"

William Whyte and his colleagues studied the use of public squares in New York to find out what it is that makes one place popular and another deserted [8]. Through time-lapse photography and observation they specified some factors that could be incorporated in the zoning regulations for the creation of public space. Perhaps not surprisingly, the majority of these relate to the embodied nature of our experience of physical space: the role of light, warmth, and weather conditions, the seating arrangements, the provision of food, or the joy of experiencing water are some examples of issues that proved to be significant for the success of a space. However, there are also some insights that could be more easily seen to be relevant for the design of virtual spaces. One example is the fact that people were seen to seek out the company of other people. Quite simply, "what attracts people most ... is other people" (p. 19). The favourite places to sit or stop for a conversation were those that were close to a flux of passers-by (p. 21). Evidently, there is pleasure in sociability – in seeing others, being seen, and taking part in the sociality of even fleeting encounters. In fact, enabling people to watch the "show" of other people's movements and activities is one of the main factors for the success of a space: "The activity on the corner is a great show and one of the best ways to make use of it is, simply, not to wall it off. A front row position is prime space; if it is sittable, it draws the most people" (p. 57). Whyte's studies also reveal that allowing people to use the space flexibly – through, for example, the provision of moveable chairs, or sitting space that could be used in different ways rather than enforcing face to face positions or other configurations – had a favourable effect on the level of use. Moreover, the studies show that planning regulations should be more rather than less specific (p. 30). Rather than specifying *that* sitting space should be provided and leaving the detail to the respective architects and builders, Whyte et al. argue that what counts as valuable sitting space should be specified in detail.

The study of the relationship between use and design detail aims to shed light on generic features of perceptions of spatial features. The importance of sociability is one such feature. Rachel Kaplan and Stephen Kaplan point to another characteristic that is important with a view to the legibility of spaces [10]. *Mystery* is an informational factor that draws a person towards a point in the landscape. Through the promise of further information if one ventures a little further, *mystery* attracts attention and directs movement. The means of achieving mystery "include the bend in the path and a brightly lit area that is partially obscured by foreground vegetation" (p. 56).

Rudolf Arnheim takes a similar approach to the study of the compositional features of paintings, sculptures, and architecture [11]. He describes a visual force field whose balance can be manipulated to communicate meaning (p. 123). One of the features of this force field is "visual gravity". Being constantly exposed to physical gravity in our everyday experience, we perceive it not only through our muscles, but also visually. Essentially, our perception of visual gravity means that some part of the composition becomes a centre of attraction. The effect of visual gravity is that "an element in the top part of a composition has a greater weight than elements in the lower parts. If the element in the top part is meant to counterbalance the element in the lower part, it has to be smaller" (p. 30, our translation). There are three sub-laws that further define the qualities of this force (pp. 34–35). First, Arnheim states, "weight increases attraction". The bigger the visual "weight" of either element, the greater the eye's tendency to perceive the attraction between them. This can lead to a problem where physically correct proportions, if literally translated on the canvas, or into a sculpture, give the impression of being top-heavy. Secondly, "distance increases visual weight". Visual gravity functions like a rubber band that ties different elements together. If the distance increases, the stress on the band grows, and the element is perceived as becoming heavier. Thirdly, "distance decreases attraction". If the "rubber band of our imagination breaks" and an element is far enough away from its centre of attraction, "the dynamic is no longer rooted in the centre of attraction, but has moved to the element itself" (p. 35), thus allowing the element to float away.

6.4.3 The Relationship Between Material Form and Forms of Sociability

In *The Social Logic of Space* Bill Hillier and Julienne Hanson address the question of how certain material arrangements foster certain forms of social interaction, less with an emphasis on the perceptions of design detail but the larger underlying structures of spatial design [9]. They outline a complex theory and provide analytical tools for the measurement and definition of the factors they find to affect this relationship. One of the main characteristics is the permeability or degree of restrictiveness of spatial arrangements to movements of inhabitants and strangers. Hillier and Hanson identify axial and convex elements of spatial design, which have, they argue, their equivalent in people's movements along axial lines and the spatial extension of interactions between people which take the form of convex spaces ([23], p. 676). Very crudely, the number of access points to a spatial element, and the nature of its relationship to other spaces (e.g. whether one can get to it directly or whether one has to pass through another space), express and have an impact on social characteristics such as "control" or "integration" ([9], p. 108). There is no direct correlation between "integration" (a high degree of different access points) and positive social forms such as neighbourliness, but the right kind of integration – for example a public space constituted through the fact that a number of houses open towards it – can foster "good" social relations.

6.4.4 Discussion

This brief glimpse into the variety of means employed by designers of physical spaces to map social practice and material form, and to achieve legible configurations serves as a repository that we will return to after descriptions of two examples from ethnographic fieldwork undertaken at the multimedia art museum of ZKM. However, a few remarks about the spatial grammars delineated here are in order before we proceed.

There are, first, considerable epistemological differences that underpin these studies. While some take an empirical, almost ethnographic approach (Whyte, Lynch), others build on a substantial – albeit at times implicit – theoretical edifice (Hillier and Hanson, Alexander, and, in a different way, Arnheim and Kaplan). This is not the place to debate the intricacies of these positions. However, it must be made clear that there is more overlap, from an epistemological point of view, between Lynch's and Whyte's findings and the ethnography undertaken at the ZKM. There are many kinds of ethnography. Ours is an ethnomethodologically informed study. Ethnomethodology takes an interest in the activities involved in creating, understanding, and maintaining the order of social life. Rather than being guided by a "theory" or interested in developing a "theory", ethnomethodology is interested in description. Careful observation provides ample material that can help to understand "ordinary" everyday action (see, for example, [24, 25]). Building on the experience of traditional research within the field of computer-supported cooperative work (CSCW), where similar ethnographic studies of the sociality of work have been used to inform the design of collaborative systems, we undertake ethnographic studies of the social organisation of space in real and virtual environments to inform the design of electronic landscapes [26–28]. From this point of view Lynch's and Whyte's studies are more easily adapted to the overall aim of informing the design of CVEs.

Secondly, the examples chosen are neither representative nor exhaustive of research undertaken with a view to visual and narrative strategies employed within architecture, let alone the fine arts, film, or other forms of expression that combine aesthetic concerns with a desire to communicate meaning or facilitate use. There is a huge repository of reference material which has only been tapped into at the surface here. However, what this brief exploration makes clear is that there is potential for such approaches to sensitise us to issues of importance in the design of electronic spaces. It also begins to delineate the dimensions that we need to consider:

- *Landscape elements*: the material environment is *inter alia* structured through an arrangement of distinctive artificial structures or clusters of such structures that dissect spaces into distinguishable "areas" with different uses and practices associated with them. Kevin Lynch's *Paths, Nodes, Edges, Districts, Landmarks,* are one way of categorising such elements.
- *Quality*: some of the qualities of such structures (e.g. the exact height of "sittable space"), as well as their visual appearance impact on how they are perceived and used. *Mystery* and *Visual Gravity* are examples.
- *Composition*: the composition of clusters of structures also relates to their perception and use. Alexander, Lynch, and Hillier and Hanson (in different ways, of course) formulate "rules" that can improve this relation.
- *Action and environment*: human action is entwined with the material environment. *Sociability*, for example, seems to be a pervasive feature of the use of public space. There are ways of fostering, as well as controlling such tendencies (Whyte and Hillier and Hanson).

6.5 Presence and Orientation in Electronic Environments

The main questions for an ethnographic study that is intended to inform the design of CVEs are "How is presence in electronic environments experienced?", "How do people find their way around in such environments?", and "How can interaction between people and between people and the environment be supported?". The following two sections present examples from the field work. The first example is concerned with the legibility of "material" arrangements in the virtual environment, while the second example looks at people's interactions in the space.

6.5.1 Example 1: Reading the City

In *The Legible City* (Jeffrey Shaw) a single visitor can cycle through representations of three cities (Figs 6.1–6.3): Amsterdam, Manhattan, and Karlsruhe. Each city is based on the street plan of the respective real city, while the buildings are made up of letters and words. An LCD display of a street map with a blinking dot indicating the user's position is mounted on the real bicycle. The following excerpt from the field notes describes how a succession of people moved around the Legible City:

> A is cycling. His friends B, C in the back laugh when he can't stay on the street. A: "where am I?". Checks on plan. Two women behind him look over his shoulder onto display. B, C leave. A evades letters, gets off, leaves. Two women take over, one cycling, one checking where they are on the display. A woman (D) and boy (E) enter. She explains what can be done in this installation to the boy, referring to what a friend told her yesterday. A couple and a boy. The boy wants to try, but his father

(F) gets to the bike first, gets on and cycles really fast, switches to Manhattan, cycling. D comments: "there's going to be a storm" (referring to the gloomy sky). E: "it's not raining" D: "mhm seems to just stay the same". F cycles up closely to letters, then goes through ... A boy gets on, goes through letters carefully, then backwards, then through them again ... a girl, audience laugh when she goes through letters. When she comes to a row of red letters, she can't get through. Turns round, goes through blue letters opposite, then tries the red ones again. Her parents look over her shoulder. A small child and father. Father doing the pedalling: "No not there it doesn't go anywhere, don't go through letters, stay on the road ... this is Karlsruhe now, let's go to the castle, past the tax authorities, that's where the castle is."

There are a number of issues relevant to the experience of presence and practices of orientation that can be drawn out from this extract. First, people were concerned with finding answers to the question "where am I?", which in the course of movement through the city turned into "where have I been and/or where am I going?". Visual and physical features of the environment were employed as orientational aids. They range from urban infrastructural elements such as roads, junctions, buildings, landmarks, to natural characteristics such as

Figure 6.1 The Legible City.

Figure 6.2 Reading between the lines.

Figure 6.3 Discovering a different way of travelling.

the sea, or the weather, to physical laws such as the solidity of materials. In our everyday interaction with the material world, such features are mainly understood and interpreted tacitly. Some of the features in the Legible City give rise to surprise or puzzlement and thus draw to our attention, not only what kinds of material features form part of our tacit interpretation of the physical world around us, but also point us in the direction of how such features are employed differently within electronic environments.

- *The urban metaphor*: the combination of the bicycle as an interface and the roads and buildings on the screen that adjust to the cyclist's moves in real-time allow the user to travel through the cities. People readily recognise this as the function of the installation. Whether people recognised the cities as based on the ground plans of real cities or not, the urban metaphor allowed them to see *at a glance* what the rudimentary workings of this piece of art are. Most carried an urban interpretation further by initially trying to stay on the roads. This is sanctioned by the audience. The father instructing his daughter not to go through the letters but to stay on the road makes public just how strong the thrust of such an interpretation is. Similarly, the laughter that greets the difficulties people have in staying on the roads indicates that there is something "wrong" in deviating from these paths "meant" for traffic.

- *Collision control*: most people inevitably overshoot junctions and pass through a row of letters sooner or later, as the speed facilitated by the bicycle is quite high and the corners narrow. Thus the lack of collision control within the Legible City is discovered as the result of "accidents" (see also [33]). It proved a source of enjoyment: after having crossed through letters once, people visibly and audibly enjoyed cutting through them. This was pleasurable for its own sake, but it also had practical consequences. It meant that the roads were no longer the only channels for traffic, and that different types of journeys through the cities were possible (Fig. 6.3). While slow movement allowed people to read the text that stretched alongside the roads, they could also decide to "just go" or orient with the help of the street plan and steer towards their destination regardless of the "buildings". The degree to which the immaterial character of the letters is taken for granted once it has been established is illustrated through the fact that when a row of letters is discovered that is impenetrable, people react with disbelief. Several trials are made to establish that *this* row (which is the end of the model) is not like the others.

- *Landmarks*: People's use of landmarks is most pronounced in situations where visitors who are familiar with Karlsruhe choose this city in the Legible City. The castle, as the main architectural focal point of the city is sought out and located with the help of other landmarks in the above example. Other, more sculptural landmarks in Karlsruhe, which are set (in the real and the simulated city) in the middle of major lines of sight helped people locate themselves within the whole of the environment, and were also used as short term navigational aids: on their way to a point further down a road that was adorned by such sculptures, people cycled through them to stay in the middle of the road.

- *Weather*: A woman states "there's going to be a storm", as a cyclist is proceeding along a broad road towards a vista of dark clouds on the horizon.

However, the fact that the "weather" does not worsen even though they should be getting closer and closer to the centre of the storm, exposes it as governed by different physical laws than weather in our familiar physical environment. In this case the everyday practice of reference to large-scale orientation clues in the sky fails, because the way the storm follows the cyclist makes it unusable as such a resource.

6.5.2 Example 2: The Hunter's Perspective: Labyrinthos

Labyrinthos (F. den Oudsten) is a collaborative environment. It is a game that allows eight players to move through a model of the real museum and "shoot" each other. Each player controls a coloured sphere through a simple control. The terminals have different colours that correspond with the colours of the avatars. If an avatar is repeatedly hit, its face changes expression. Eventually it becomes transparent and immobile. The successful "hunter" on the other hand gets a crown. Here, three boys are hunting in a "pack". Y is sitting at the yellow terminal, O at the orange one to his right, and the third boy (T) is moving from one to the other (the transcript is translated from German; some expressions and overlaps (...) do not exactly correspond with the original) (see Figs 6.4 and 6.5).

1		O:	hey I'm alive again!
2		T:	eh cool (d'y' know)
3	→		[where you are right now?]
4	→	Y:	[eh? Where are you?] (0.) eh ((glances across)) where are you?
5		O:	I don't know either I'm back at (the square there)
6		?:	(face)
7	→	Y:	WHERE?

Figure 6.4 Y, T and O at their terminals.

Figure 6.5 Y bending to see O's screen.

8	O:	there comes purple
9	T:	ah!
10	O:	eh!
11 →	Y:	((bends across and back while saying)) I I'm coming where are
12		you? Oh there
13	?:	(inaudible)
14	T:	blue!
15	Y:	I've got gre- I see red I SEE RED I I'm coming to help you I've
16		got green under attack ((starts shooting))
17	O:	me too I'm shooting
18	Y:	eh? (but you said)
19	O:	shit
20	T:	he! (inaudible)
21	Y:	die you (damn)
22	?:	attention!
23 →	T:	eh! Red is behind you
24 →	O:	red is behind you (0.)
25 →	Y:	where? (.) ni- attention (.)

At the beginning of this excerpt of talk, O's avatar suddenly responds to his manipulations of the control after a period of immobility. Immediately, he makes public that he is once more an active player in the game. Now his position in the space becomes an issue. In order to "attack" their opponents effectively, the three have decided to "pack up" and pursue them. They need to assume strategic positions and coordinate their actions, and it is therefore crucial to know *where* they are with respect to each other, possible targets, and other hunters. In their approach they take advantage of the fact that they are distributed across a large distance in the electronic space, yet co-located in the real space. Y asks "where are you", but the answer is too vague for him to find O. His third request "WHERE?" is cut short by the fact that O discovers "purple" and is put on the spot. Shots are exchanged and help seems required. Y, in one move, bends across, and glances at O's screen. Having seen what O sees, Y is able to work out where he is. However, on his way he encounters first "red", then "green". He attacks "green". What he fails to realise is that while he is busy shooting "green", "red" has moved behind him and is attacking him. By looking at O's screen the others notice and warn him. A little later on in this game, O "kills" purple. He switches to plan view, where the avatars are represented as coloured arrow-heads, and notices that "yellow" is being attacked by "green".

6.6 Everyday Practices of Orientation and Perception and Spatial Grammars

Two issues that are relevant with a view to an exploration of spatial grammars are highlighted by these observations of people's interactions in and with electronic environments. First, people employ resources that are provided *within* the environment. Secondly, the interpretation and use of these resources is based on everyday practices of orientation, interaction, and perception. Both apply to the interaction with other people and the interaction with the environment itself.

6.6.1 Affordances and the Practice of Interpretation

Objects and people within an environment, and the environment itself can be seen to provide ample information about themselves. What we usually orient to in this informational field are not the *qualities* of entities, but their *affordances* [20]. This term refers to the physical and interpretative constraints and possibilities that impinge on our engagement with people, animals, objects, and spaces. Which affordances we perceive at a certain moment in time depends on the situation. The fact that the letters in the *Legible City* are permeable and afford passage, for example, is taken for granted when people steer by orienting to the LCD display of the map or just deviate from the roads because they can, while it is disregarded when the activities focus on reading the words made up by the letters. With regard to interactions between people, face to face situations are characterised by a rich tapestry of interactional information that allow us to gauge information about other people's activities, that could be seen as the language of sociability. Through our position in relation to others, our posture, movements, the direction, intensity, and duration of our gaze, and other finely tuned embodied actions we occupy a place in the encounter, where we make available to others what actions they can reasonably expect us to take within the frame of an encounter. Face-to-face situations thus afford the dynamic and flexible display of, for example, the degree of our involvement in the interaction, how we understand others" actions and our own in its context, whether we are listening or want to say something. Similarly, there is a large-scale spatial dimension of interaction between people in spaces. Moving in public places, we routinely gauge information about who else is there and what their activities are, and weave our own actions into the flow – not least to avoid "collisions", but also to see who we could talk to, how long we would have to wait for our turn if we were to join a queue, or where there is something to see [29, 30]. Usually we have a whole host of clues that indicate *at a glance* "what sort of people" there are and "what they are doing", reaching from clothing to movements, gestures, and facial expressions. We use this information to interpret and categorise what we see [29]. Sacks delineates the rules that govern this activity. For a person there are a number of categories that might apply. She can be a "female", "mother", "teenager", the "head of a queue", the "captain of a rowing team", and so on. How we choose categories depends on the context and the cluster of categories that apply in a particular situation. These clusters of categories are, in Sacks' terms, "membership categorisation devices" or MCDs. In an MCD – for example "family" – the different categories "mother", "father", "grandfather", etc. are mutually exclusive. Moreover, there is a "consistency rule" that implies that once a category and its MCD is chosen we tend to hear or see the surrounding talk, activities, or features as belonging to that MCD (unless explicitly stated otherwise). In the mediated variation of encounters, the tapestry of information relevant to our interaction with others and the environment is considerably impoverished and altered. Yet people manage to conduct orderly encounters. An examination of how this is achieved can provide some insight into which features of interaction between people should be represented in a more symbolic approach to visualising electronic environments.

6.6.2 The Hunter's Perspective: Transpositions of Everyday Practices

In his move in line 11 of the transcript Y rightly assumes that the space of the game is the same for everybody involved and thus projects a principle we routinely apply in our everyday actions into the electronic space. As a "hunter" in the game Y does and, in fact, needs to take for granted that views from different positions in the game are interchangeable. The space *must* be the same for everyone, otherwise one would not be able to aim at targets. "Reciprocity of perspective" [31] is necessarily reflected in the visual display of one's position in the space on the respective screens of the players. Y exploits this fact. Seeing what O sees allows Y to locate him on his own screen. If the game took place in physical space, Y would not be able to put himself in O's place so easily. Here he makes use of the affordances of the situation and transposes his knowledge of the interchangeability of perspectives without hesitation.

This "reciprocity of perspective" is one aspect of a more general principle of *intersubjectivity* that underpins interactions between people and between people and their material environment. What we say, perceive, or do is part and parcel of a world known in common [31]. In our actions, we assume that others know the world and the situation at hand in ways that are similar to how we know them. This includes people's distribution, appearance, and activities in space. In the context of *Labyrinthos*, position, colour, shooting or not shooting are the only clues available to map people's activities into the MCD of the game. However, this information is sufficient to allow the players to monitor the state of affairs. In lines 21–23 competence is enacted in a way that mirrors everyday practices of seeing. From O's screen, O and T see that Y is being attacked from behind his back. Later on in the game, similar information is used in a different way. On the plan view, O discovers that Y is being attacked. Here, O does not see the bullets hitting Y's avatar, but a semi-circular configuration of coloured arrowheads pointing at Y. Such an abstract, animated, real time plan view is a device usually unavailable to us. It is an affordance that exceeds everyday environmental conditions. O transposes and combines competencies of categorising people ("hunter", "target", "us" and "them") drawing on a minimal and abstracted set of clues, and his knowledge of real-world plan views to fit in with the environmental affordances of the game.

The example of Y and O's use of different screen views and abstract, animated, real-time plan views illustrates how everyday practices of seeing and categorising people's activities can survive in an environment where only a fraction of the interactional information usually available is present.

6.6.3 The Legible City: Grammars of Perception

Practices of categorising people and their activities have their equivalent in practices of categorising the material arrangements that we encounter. The urban metaphor employed in the design of the *Legible City* provides a membership categorisation device that is initially extended to cover movement and orientation within the "cities". People stick to roads and look out for familiar landmarks. However, through the (accidental) experience of some of the physical and structural laws in the cities, and the letters' invitation to literally read "between the lines", this MCD is altered to fit the engagement with the altered affordances of the cities. In practice interpretation and perception are *one, situated* activity [20, 25, 32]. This activity draws upon the resources provided within the electronic environment and its artistic context to form a "grammar" for its perception [25]. Transpositions of everyday practices of movement and orientation – such as the strategy of cycling through landmarks to stay in the middle of the roads – reveal the creativity that is inherent in our efforts to make sense of/perceive the affordances of a new environment.

The design of structures or objects is inextricably tied to such practical perceptual grammars. In ways that are similar to the intelligibility of texts, the legibility of the material world is "both a feature and an outcome of a set of practices inscribed and evident in, and relied upon" [37] by the material environment. The categories that we employ with regard to the material world are intersubjective – albeit in a mediated fashion. Living in the physical world we assume that architects, landscape designers, and urban planners are members of the world known in common and likely to have imbued their designs with features that are known to us all. They, on the other hand, may apply their creativity, but part of their professional skill is to know, for example, what features make a door recognisably a door, and when those limits are exceeded, leading people to mistake a glass door for air and hurt themselves, for example ([20], p. 142); or when people are likely not to see a door at all but a continuous wall. The designers of objects and spaces provide legible affordances and thus act at a distance (by anticipating and inscribing certain forms of engagement with these affordances). Together, the situated interpretative freedom we have with respect to the objects and spaces that we encounter, and the well-trodden trajectories of everyday practice that both designers and users rely on make for both the intelligibility and the flexibility and creativity of our relationship with the world.

6.7 Conclusion

The observation of the actual use of electronic environments clearly illustrates that realism with a view to visual appearance, but also physical laws, and sensory access is not as critical as the dominant thrust of the current debate suggests. The examples illustrate that the affordances of the electronic environment – with a view to both "material" structures, and people's activities within them – are interpreted with reference to the physical "world known in common" but submitted to review and learnt about in light of the experience of moving around in the environment. The crucial issue with regard to an exploration of architectural principles is that people need very little persuasion to adopt an urban metaphor or to extrapolate from what could potentially be seen from a real-world real-time plan view for their approach to the space. The letter-buildings in the *Legible City* are based on the original street plans of the respective real cities. But this is all there is that really is urban or "realistic". Equally, the perspective of the plan view in *Labyrinthos* may be similar to a real-world aerial view, but what is seen *in* it, is very different. Yet people readily applied and adjusted these metaphors. Given that so many of the rules of a perceptual grammar that one would expect to be part of these activities are relaxed, what is it about the *Legible City* or the plan view in *Labyrinthos* that makes it possible for people to transpose everyday practices of orientation to these environments? Spatial grammars provide some insight into this question and some inspiration for features that could be used to enhance the legibility of electronic environments. The remaining text indicates some possible avenues with a view to the design of large-scale electronic landscapes. E-scapes are likely to attract people who want to search for a world that suits their current interests. It therefore needs to display some rudimentary information about a world that can be gauged at a glance, and/or some more detailed information on closer inspection. What would this information be and how could it be visualised?

6.7.1 Sociability

If "what attracts people most ... is other people" ([8], p. 19) this should motivate the visualisation of other people's movements in an e-scape. In order to make the e-scape itself a "place", people's tendency to seek out the company of others suggests that it is not enough to display the number of people that have already entered a world. In order to make available how visitors of the e-scape itself distribute their interest in different worlds, it would be desirable to also visualise their movements around it. By making embodiments directional visitors could see where other people are heading, where there is a crowd of people taking an interest in a specific world, or where there are some special interest worlds that only attract a small group of people. Allowing people to choose or create an individual avatar could increase the amount of information available in a way that is similar to how we express aspects of our identity through clothing.

6.7.2 Reciprocity of Perspective

In order to be able to coordinate actions, people must be able to assume that others see the same space and the same objects as they do. But this does not mean that the principle must not be relaxed under any circumstances. It can be relaxed – as long as this is made available as a feature of the environment – or provided for in ways that deviate from familiar strategies. Based on the transpositions of everyday practices that build on the interchangability of perspective, it could be possible to provide:

- an abstracted, animated, real-time plan view
- a screenshot of "this is what I see from where I am"
- a vehicle into which others can be invited to share a particular perspective on a joint space [33].

6.7.3 Intersubjectivity

Some of the spatial grammars outlined here explicitly draw on common-sense knowledge of, for example, large-scale urban structures and seek to enhance their design by extrapolating from and refining already existing and commonly known features. These elements could be used to structure an electronic landscape. There have been attempts to introduce Lynch's structural features into the design of visualizations of information [34]. However, Benford et al. impose, for example, edges as an *additional* feature into the environment (a plane that separates two areas). This clutters the environment rather than contribute to its legibility. In order for Lynch's (or Alexander's) structures to be useful in the visualization of information, the patterns have to underpin the distribution and appearance of information in the space.

6.7.4 Planning Principles for a "Metaverse"

There are two different ways in which an electronic landscape could be realized. The first is static – the number of worlds, their appearance, and the visualization of information about them would be determined before construction, and people would be allowed in after completion. The other possibility is to provide a space that could be incrementally populated with worlds. The "metaverse", invented by Neal Stephenson in his novel *Snow Crash* [35] is a fictitious cityscape that grows in such an organic fashion. In order to ensure the legibility of an environment based on the "metaverse" metaphor planning principles would have to be inherent in the tools given to people to add their own worlds. Alexander's patterns, and Lynch's structures could be a starting point for such principles. Equally, Hillier and Hanson's "Social Logic of Space" can and have been used to build planning principles into the tools that generate an electronic environment [36].

References

1. Blaxxun, http://www.blacksun.com/
2. Rockwell, B. From chat to civilization: the evolution of online communities. http://www.blaxxun.com/company/vision/cmnty.html, 1998.
3. Levy, R.M. Visualization of urban alternatives. *Environment and Planning B: Planning and Design* 22: 343–358, 1995.
4. Benford, S., Snowdon, D. et al. Informing the design of collaborative environments. *Proceedings of GROUP'97*, Phoenix, AZ. New York: ACM, 1997, pp. 71–79.
5. Trevor, J., Palfreyman, K. and Rodden, T. Open support for shared spaces based on e-scapes. *eSCAPE Deliverable 1.1*. Lancaster University. This report is available from http://escape.lancs.ac.uk/
6. Lynch, K. *The Image of the City*. Cambridge, MA: MIT Press, 1960.
7. Alexander, C. *A Pattern Language. Towns, Buildings, Construction*. New York: Oxford University Press, 1977.
8. Whyte, W.H. *The Social Life of Small Urban Spaces*. Washington, DC: Conservation Foundation, 1980.
9. Hillier, B. and Hanson, J. *The Social Logic of Space*. Canbridge: Cambridge University Press, 1984.
10. Kaplan, R. and Kaplan, S. *The Experience of Nature*. Cambridge: Cambridge University Press, 1989.
11. Arnheim, R. *Die Macht der Mitte. Eine Kompositionslehre für die bildenden Künste*. Cologne: Dumont, 1996.
12. Hillier, B. *Space is the Machine*. Cambridge: Cambridge University Press, 1996.
13. Bridges, A. and Dimitrios, D. On architectural design in virtual environments. *Design Studies* 18: 143–154, 1997.
14. Benedikt, M. *Cyberspace*. Cambridge, MA: MIT Press, 1992.
15. Heim, M. *The Metaphysics of Virtual Reality*. Oxford: Oxford University Press, 1993.
16. Zahorik, P. and Jenison, R.L. Presence as being-in-the-world. *Presence* 7: 78–89, 1998.
17. Slater, M., Steed, A., McCarthy, J. and Maringelli, F. The influence of body movement on presence in virtual environments. *Human Factors* (in press).
18. Becker, B. and Mark, G. Social conventions in collaborative virtual environments. In: *Proceedings CVE '98*, D. Snowdon and E. Churchill, editors, University of Manchester, Manchester, 1998, pp. 47–55.
19. Büscher, M., O'Brien, J., Rodden, T. and and Trevor, J. Red is behind you: the experience of presence in shared virtual environments. In *Presence: Teleoperators and Virtual Environments* (in press).
20. Gibson, J.J. *The Ecological Approach to Visual Perception*. Hillsdale, NJ: Erlbaum, 1986.

21. Stiny, G. and Gips, J. *Algorithmic Aesthetics: Computer Models for Criticism and Design in the Arts.* Berkeley, CA: University of California Press, 1978.
22. Stiny, G. Review: Alexander, C. The timeless way of building. *Environment and Planning B: Planning and Design* **8**: 119–122, 1981.
23. Hanson, J. "Deconstructing" architects' houses. *Environment and Planning B: Planning and Design* **21**: 675–704, 1994.
24. Garfinkel, H. *Studies in Ethnomethodology.* Cambridge: Polity Press, 1967.
25. Wittgenstein, L. *Philosophical Investigations.* Oxford: Blackwell, 1997.
26. Bannon, L. and Schmidt, K. Taking CSCW seriously. Supporting articulation work. *Computer Supported Cooperative Work* **1**: 7–40, 1992.
27. Hughes, J.A., Randall D. and Shapiro, D. Faltering from ethnography to design. *Proceedings of the Conference on Computer Supported Work.* New York: ACM, 1992, pp. 115–122.
28. Shapiro, D. The limits of ethnography: combining social sciences for CSCW. *Proceedings of CSCW '94: Transcending Boundaries; the Fifth International Conference on Computer Supported Cooperative Work, Chapel Hill, NC, 22–26 October.* New York: ACM, 1994, pp. 417–428.
29. Sacks, H. *Lectures on Conversation* Vols 1 and 2, 1992, pp. 81–94.
30. Sudnow, D. (editor). Temporal parameters of interpersonal observation. *Studies in Social Interaction.* New York: Free Press, 1972.
31. Schutz, A. *On Phenomenology and Social Relations.* Chjicago, IL: University of Chicago Press, 1962.
32. Coulter, J. and Parsons, E.D. The praxiology of perception: visual orientations and practical action. *Inquiry* **33**: 251–272, 1991.
33. Murray, C. The Cityscape: theory and empirical work. eSCAPE Working Paper, 1998.
34. Benford, S., Ingram, R. and Bowers. J. Building virtual cities: applying urban planning principles to the design of virtual environments. *Proceedings of the ACM Conference on Virtual Reality Software and Technology (VRST'96),* Hong Kong. New York: ACM, 1996.
35. Stephenson, N. *Snow Crash.* New York: Bantam Books, 1992.
36. Bowers, J. The social logic of cyberspace. In: COMIC Deliverable 4.3. Lancaster University, 1995. This report is available via anonymous FTP from ftp.comp.lancs.ac.uk.
37. Jayyusi, L. Values and moral judgement: communicative praxis as moral order. In G. Button (ed) *Ethnomethodology and the human sciences.* Cambridge: Cambridge University Press, 1991, pp. 227–51.

Chapter 7

Navigating the Virtual Landscape: Co-ordinating the Shared Use of Space

Phillip Jeffrey and Gloria Mark

Abstract

Collaborative virtual environments, such as multi-user domains (MUDs), chatrooms, or three-dimensional graphical environments, provide a common space for people to interact in, independent of geographical location. In this chapter we examine how the different metaphors used to represent two- and three-dimensional environments might influence interpersonal behaviours. We focus on behaviours related to navigation and positioning: (1) proxemics – the maintenance of personal space, (2) the signaling of private space and (3) the effects of crowding. We discover that the design of the three-dimensional space offers sociopetal spaces that encourage interaction, make clusters of actors easily visible and provide cues so that people maintain a sense of personal space. In both environments, adverse reactions to crowding occur. We suggest that differences in interpersonal behaviours may be influenced by embodiment (avatar) design features of the space and the number of other actors present. In a three-dimensional environment, these factors appear to influence navigation and positioning in the environment.

7.1 Introduction

Collaborative virtual environments (CVEs) have been attracting interest for their potential to support communication, collaboration and co-ordination of groups. We are already seeing cases where such environments are proving effective as shared information spaces for collaborative tasks, e.g. multi-user domains (MUDs) [1] and graphical virtual worlds [2]. A special case of CVEs employ a three-dimensional graphical representation of space in which users are represented as a figure (i.e. avatar). Examples of diverse approaches to these types of CVE designs includes the DIVE system [3], MASSIVE [4], and Active Worlds [5].

The advantage of using such systems is that group members located in different geographical places can therefore occupy the same virtual space, and thus have a common frame of reference for collaboration. Many CVEs, whether text-based or graphical, are synchronous, which offer actors the advantage of receiving immediate feedback; one can present one's position, field questions, and receive immediate answers. Such immediate feedback is useful for clarification of an issue and, it has been argued, may lead to group cohesiveness [6]. Users' interaction and navigation within a virtual space may be influenced by a variety of factors, among them, how the space is designed, the presence of other users, and the positioning of artefacts. This result was found in an empirical study involving the rearranging of artefacts in a virtual CVE room; the choice of targets was to some extent determined by other participants' behaviours [7].

The purpose of this chapter is to examine the question of how a three-dimensional representation of space in a CVE might, if at all, influence interaction among the participants. To answer this question, we observed interaction in two CVE's distinguished by how their space is represented; we contrasted interaction in a text-based chat environment that employs a three-dimensional representation of space, with a purely text-based chat environment. In this exploratory study we were particularly interested in how the design of space and the presence of others affect navigation and positioning. Our selection of interpersonal behaviours to observe was motivated by choosing those that concern spatial positioning and navigation among actors. This led us to choose behaviours that show how personal space is maintained, how privacy is indicated, and reactions to crowding. In the following, we provide a framework in which to understand the relevance of such types of social interaction in a CVE.

7.2 Shared Information Spaces

A CVE is a kind of shared information space. Bannon and Bødker [8] describe that shared information spaces are characterised not only by the information present, but also by the interpretation and the meaning derived from it by the users who inhabit the environment. A digital library, chatroom, or even graphical virtual world must provide a meaningful context for collaborative and co-operative work. We view an information space as an environment such as the World Wide Web (WWW) that presents information to be interpreted to the user and that can be extracted or previewed based on how one navigates through the space. Enhancements to CVEs already enable the capability to link to web pages [9], shared virtual workspaces [10], and even physical places [11].

7.2.1 Spatial Metaphors and Connotations

How the information is found and interpreted in such a space is important. Spatial metaphors have been shown to be valuable devices to aid navigation in virtual environments. The experience of the WWW has been described using the metaphor of a mansion, filled with rooms, continually expanding [12]. People have been found to use spatial metaphors while navigating the WWW for information, perceiving it as a landscape that they move through physically, or as navigation towards information [13]. Maglio and Matlock [14] found in a later study that experienced and inexperienced users differed in their descriptions of their web experience. Both experts and novices used language representing physical movements and actions towards information (e.g. "go", "went") rather than simply passively receiving it. However, when language usage involved trajectory movement, experts perceived themselves as an active agent, navigating through the information space (i.e. "I *went into* this thing called Yahoo"). In contrast, beginners were more likely to see the web as the agent, with the information moving toward them (i.e. "It *brought me to* the Anthropology page"). Maglio and Matlock [14] suggest that these results indicate that WWW users view the virtual space within the familiar context of a physical space.

Other metaphors used in CVEs include a city metaphor applied in a MUD environment [15] and the results suggest that spatial metaphors can be helpful tools for navigation. Some chatroom environments use metaphors from the physical world such as a house or extended hallway, with different rooms having different functions or purposes. A "homepage" represents a personal location on the WWW, further reinforced with its location marker as an "address". Language associated with the WWW such as "surfing", "explorer", and "navigator", and with CVEs such as "teleporting" all convey the notion of movement and traversal through a virtual landscape. In [16], Dieberger describes navigation in the information spaces of graphical-user interfaces and WWW hypertext in terms of "magic features" which provide shortcuts for increased efficiency. If the source of a metaphor is something in the physical world, then features such as teleportation that disrupt this metaphor may be seen to originate from a magical world.

The contextual cues of a place may suggest appropriate behaviour within that space. Dieberger [17] refers to these cues as social connotations. Certain behaviours that may be appropriate within the privacy of one's bedroom are forbidden in a public workplace. In addition, the nature of navigation may also be influenced by the social connotation conveyed by a place [17]. The same space in fact may connote different interpretations of appropriate behaviour at different occasions, such as a community hall that may be used for a music concert, a wedding or for a local sports event. Spaces that don't provide what Harrison and Dourish [18] refer to as a *sense of place* may adversely affect conversations and the participants' behaviour. Thus, it appears that the design of a CVE, and the cues it contains, can influence the nature of the interaction and movement through that space.

This idea is intuitively obvious when we think of physical environments, such as workplaces, which have different spatial layouts. Architects and interior designers have long been concerned with the spatial positioning of offices, interiors and public areas in designing environments for work. Not only can the interior layout influence the degree of interaction between employees, but it can also define the hierarchy within the workplace, and distinguish areas for work from those for socialisation. Even the placement of shared artefacts is crucial since they can provide peripheral awareness information about other people's activities [19]. Osmond [20] classifies space as being sociopetal or sociofugal.

Sociopetal spaces (i.e. a cafeteria) encourage interpersonal communication. Sociofugal spaces, such as a lobby waiting area, are designed to restrict or discourage social interaction. A working environment where people perceive the space as sociopetal should have expectations that social communication will take place.

Although we may exist in *space*, we actually refer to artefacts and architecture within the context of a *place*. Tuan [21] believes that we navigate through space until we pause, at which point an awareness of our positioning with respect to artefacts transforms our location into a place. This perception is similar to that proposed by Altman and Zube [22] who state that as individuals become more familiar with a space, they learn to associate meaning with it, and thus through experience, a space is perceived as a place.

7.2.2 Social Navigation

It may be that the perception of a space as a place is facilitated as interacting partners develop a shared understanding of the environment [17]; this creates a dynamic process as interaction within a space fosters meaning. Social navigation has been described as navigation through collected information that is enabled due to the activity of others. Dieberger classifies social navigation as being direct or indirect [17]. Direct social navigation involves active and direct interaction between users such as pointing out information required. Indirect social navigation is more passive and may involve recognising and identifying navigational cues left by users found in the information space.

In our definition, we believe that navigation within an information space such as a virtual environment could be considered "social navigation" when the presence of others within the shared space influence one's direction of movement or choice of position. Social navigation involves an awareness of other users who are currently present or who have been there in the past. This may involve the process of recognition of others; cues such as a username or unique personal identifiers signal familiarity. The user who is navigating within the environment is completing a goal-oriented task – finding others to interact with or seeking information from others. This view is similar to that proposed by Dourish and Chalmers (discussed in [18]) who describe that clusters of individuals serve as focal points to where people navigate. The type of environment may determine the type of questions asked, the behaviour that is appropriate, and how users will respond [17].

Many virtual environments are designed as spaces primarily intended for socialisation and interaction; in this context, navigation serves to find others. In the physical world, public spaces such as local pubs or neighbourhood parks are used as areas for meeting new people and reacquainting relationships with friends. Similar places may also be provided within virtual space. Thus, the process of socialisation with others in a shared space may involve both communicative as well as navigation behaviours.

7.2.3 Social Norms for Shared Spaces

Thus far we have discussed the notion of how one's perception and behaviour in a shared information space is affected by the presence of others. As the level of interaction increases, shared norms, values and expectations emerge in a group. Expectations about behaviour during interpersonal interaction exist as social guidelines. This is to ensure that group members behave in a regulated manner as participants within the group specifically, and more broadly, as members within an organisation or society [23]. Shared expectations of behaviour also have sanctions, which lead to uniformity in the behaviour of group members. These standards of behaviour can be commonly referred to as norms [24].

Group norms are informal regulations that are usually unwritten and implied [25]. They have two main purposes: to create a frame of reference for understanding within the group, and to identify appropriate and inappropriate behaviour [26]. Norms are not static but rather dynamic. Determinants such as the environment, one's culture, and the composition of the group will influence the emergence, acceptance and effectiveness of norms by the members [27]. The existence of group norms is an important part in helping members feel a sense of integration within the group.

Carry-over behaviour, according to Feldman [25], is one method that enables social conventions to develop within a workgroup. Individuals, through past experience as participants in other groups or similar working environments, are believed to transfer those previously learned conventions into their current group. Violations of social norms may produce conflict between the violator and the other group members, increase the level of interpersonal communication directed at the violator about appropriate behaviour, or result in dismissal from the group [24].

Thus, social norms regulate behaviour within society as well as one's social group and play an important role in guiding one's own behaviour as well as shaping expectations about the behaviour of others. The members of a workgroup or social group have dual roles as societal members in the physical environment. Therefore, it is expected that conventions are not only transferred from group to group, but also are transferred from society to a group. It seems reasonable then that interpersonal behaviours and social conventions might transfer from experiences with the physical use of shared space to a virtual shared space. In fact we may view CVEs as "social spaces" as an awareness of other users may produce conditions that foster social conventions in order to regulate interactions. Virtual environments are still a novelty compared with face-to-face interaction. With any technological invention, users develop new usage conventions that are appropriate for the new technology by transferring over and modifying familiar metaphors [28]. However, familiar norms such as when to answer a telephone call must be modified when using mobile phones, due to the often public nature of such conversations. Dix et al., [29] suggest that success with a new media environment may be dependent on the success of transferring these norms.

Behavioural research on social interaction in CVEs has suggested that social conventions are formed and used as common communication systems in text-based MOOs [17, 30, 31], newsgroups [32], as well as in graphical environments [33–35]. In the following, we introduce a set of interpersonal behaviours which people use to co-ordinate interaction in physical shared spaces. We then examine how they function in virtual environments.

Personal Space

Proxemics is the study of personal space, a field founded by Edward T. Hall [36]. It focuses on the societal use of space to attain comfortable conversational distances and obtain preferred levels of interpersonal involvement. The study of proxemics focuses on theories related to the distances expressed during social interaction. In this sense individuals are regarded as active participants within their environment, rather than passive observers. Personal space may be defined as an area with invisible boundaries surrounding an individual's body which functions as a comfort zone during interpersonal communication [37, 38]. Violations may result in adverse and emotional reactions [39].

According to the principle of proximity, individuals who are physically closer develop a stronger attraction to each other than when they are further apart [26]. According to Hall [36], one's preferred distance for comfortable communication has societal (contact vs non-contact), social (stranger vs friend) and conversational intimacy (business vs casual) determinants. Face-to-face communication is often seen as the ideal setting that computer-mediated forms of communication strive to emulate [29]. Therefore, the question of whether a sense of personal space might also exist in CVEs, as it does in face-to-face interaction, was examined, keeping in mind that a sense of personal space in a CVE may be a function of the nature of the embodiment and degree of immersive experience.

Private Space

In social interaction, privacy is the selective control by individuals or groups, of personal information, the degree of interpersonal communication, and the level of social interaction [40, 41]. An imbalance results when one's perceived level of social interaction differs from one's optimal level. The need for group and individual privacy and its related social conventions are common throughout physical world societies, with studies showing that the preference for privacy, disclosure or social interaction is cultural and situation-dependent [41]. This desire for private spaces for interaction is reflected in the design of CVEs, which provide password-protected private rooms or whisper commands, which create a private, shared communication space. Following this idea, we investigated how behaviours indicated a desire for a private space in a CVE.

Crowding

Crowding is a psychological perception characterised by feelings of personal space violations if one's current level of social interaction is higher than preferred [40]. Stokols [42] refers to social crowding as feelings of being crowded due to the presence or awareness of others, which can lead to stress. However, stress due to crowding was found to be lower when individuals received positive group feedback [43] and where groups were in an atmosphere of getting to know each other rather than evaluating one another [44].

The feeling of crowding in virtual environments has been intentionally simulated to help people overcome phobias [45] which suggests that such a simulation can provoke genuine feelings of being crowded by others. A technical solution to support crowds in CVEs has been implemented in MASSIVE-2 [46], although the psychological effects have not been examined. We examined here how users react to crowding in CVEs.

The commonalities linking personal space, private space and crowding are related to people's shared use of space. Each of these social norms involves physical spaces that may be perceived as personal although interaction occurs within public environments. We now examine these behaviours within virtual space.

7.3 Methodological Approach and Research Setting

To examine the relationship of users in shared virtual spaces, we have explored two contrasting online, virtual environments: Active Worlds (AW)[1] and WebChat (WBS)[2] in order to examine social conventions that may exist in interaction and navigation. All environments are multi-user and are accessible from the Internet. The environments differ in terms of their means for navigation and their representation in space, which enabled us to detect differences that might be due to the design of the virtual space.

WBS is designed in a flat two-dimensional space using the spatial metaphor of a house, each with different "rooms" serving as a different chat environment (Fig. 7.1). Together the rooms form what is called "a community". Communication is text-based.

AW is designed using the spatial metaphor of a three-dimensional physical world or landscape. Navigation occurs within and between "worlds" where users communicate with text and use graphical representations (i.e. avatars) (Fig. 7.3).

The basic functionality available for navigation, communication and representation in these virtual environments is:

- WBS: communication is text based in an input box. All public messages appear in a large text area and individuals can alter their size and colour. Individuals are represented with user names and can attach pictures, images or WWW links to messages. Private messages can be sent using an internal e-

---

[1]Copyright © 1995–1998 Circle of Fire Inc. http://www.activeworlds.com
[2]Copyright © 1995–1998 Infosek Corporation, Inc. http://wbs.net

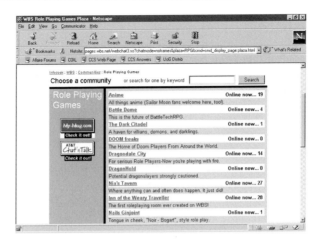

Figure 7.1 WBS: role-playing games chatrooms.

mail message system or directly to the desired user during conversation. Private rooms can be created for private conversations. Socialising occurs within a room and individuals can easily navigate within and between rooms using the mouse.

- AW: full-bodied avatars can walk and exhibit movements dependent on the particular world visiting and one's membership status (i.e. citizen or tourist). Possible movements include waving, dancing, and fighting activated by mouse clicks. Avatars can navigate in three dimensions by using the arrow keys or the mouse. Communication is text based from the keyboard. All public messages appear in a scrollable window and may also appear above the avatar's head with the avatar's name for 30 seconds or until the next typed message appears, if this functionality is desired. Citizens can send private messages by clicking on the telegram icon to access the internal message system.

Approximately 50 hours of time was spent observing the two different online environments. The observations were performed in a period spanning September 1997 through November 1998, although not continuous. Online recording occurred in AW in order to ensure accurate data.

We were interested to what extent notions of space in the physical environment might transfer into a three-dimensional graphical virtual environment, and whether references to these notions also appear in a text-based chat environment. We coded the following behaviours: (1) Is a certain personal distance kept? (2) What happens when this distance is violated? (3) When people navigate through the environment, do they disturb others' personal space? (4) Does personal space exist in a primarily textual environment? (5) How do people express privacy? (6) How do people respond to environments that they perceive are crowded?

7.4 Observations

In comparing our observations, we find differences in how the social behaviours we examined are expressed within each virtual space. In each environment, such behaviours appear to be influenced by the available functionality for representation and communication. We describe our observations as follows.

7.4.1 Personal Space: Social Positioning

In the text environment of WBS, instances were rarely observed that indicate that users behaved as though they had a sense of personal space. The following is one example that occurred in a role-playing chatroom called "Inn of the Weary Traveller".

> Steven: *edges closer and closer to DAYNA [another user] until within talking distance*hello...*Grin*

In contrast, behaviours involving personal space were regularly observed in AW. During interpersonal communication, avatars maintained a distinct personal distance. In areas with open spaces, such as the landing point in each world, which also functions as a common meeting place, the avatars of a number of users were present. Invasions of personal space occurred when the Observer (**O**) moved the avatar from a comfortable personal distance and positioned it face-to-face close to another (Fig. 7.2). Reactions to personal space invasions were generally expressed through verbal or non-verbal behaviour and were observed to signal discomfort, as in the following example.

O met *Laracat* in the AlphaWorld supply dock getting building materials. O moved face-to-face and immediately *Laracat* moved back maintaining a slightly larger distance than had been previously observed and remarked:

Laracat:	*actually ... It's funny ... but it does make me uncomfortable when another avatar gets "too close" in my avatar's face.*
O:	*[moves forward] why?*
Laracat:	*[moves back] just does*
Laracat:	*This is a nice distance to keep ... :)*
Laracat:	*[I backed up] the same way I'd back up if a "real" person got up that close ... someone I don't know very well*

On several occasions in AW, O experienced a personal space invasion. When O was teaching a user named "*trooper*" to fly, "*trooper*" passed through O's avatar and then apologised.

"trooper":	*sorry about that ... you sure do speed up with the ctrl key! :)*

Having an avatar navigate into one's personal space was unappreciated, as the comments below suggest:

Wepwawet:	*something about personal space ... we carry the conventions of [real life] into AW*

(a)

(b)

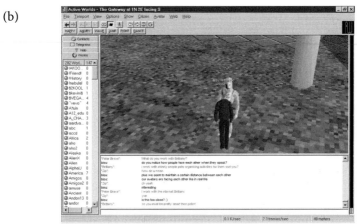

Figure 7.2 (a) Comfortable personal distance. (b) Personal space invasion.

22much:	*... i need my personal space here*
Geophrey:	*Netiquette will be amended to include "personal space" in [virtual reality] I'm sure :)*
Dizzy:	*I try to walk around [avatars]. Courtesy you know.*
Danaidae:	*[avatars] need [their] personal space too*
Blade:	*... it feels wrong/strange like we stare at each other if we are [too] close don't you think?*

Thus, we observed that within the three-dimensional graphical environment AW, individuals preferred to maintain a comfortable personal distance during conversation. When the O removed the personal distance and moved face-to-face in close proximity to another avatar, behavioural reactions indicated that the other participants realised their personal space had been violated. Individuals preferred to navigate around rather than through other avatars.

As AW uses a metaphor of three-dimensional space, it appears to produce a place conducive for behavioural conventions such as personal space to emerge. The AW observations parallel those found in physical shared spaces. Physical proximity plays a significant role in how workgroups communicate, collaborate and co-ordinate tasks [47]. In face-to-face interpersonal communication, people change their positioning to maintain their personal space [36], and use non-verbal communication such as gestures and turn-taking for assistance in regulating conversation [48]. Reactions to personal space invasions have been perceived in mediated environments as threatening [49, 50].

There may also be a practical reason for adverse reactions to personal space invasions in AW. Participants used the first person perspective for communication and viewing the immediate surroundings. During face-to-face personal space invasions, one's viewpoint is obscured if using the first person perspective. This may explain negative reactions towards personal space invasions, namely, that standing in front of an avatar will block the viewpoint, similar to real life. Personal space may then affect people differently depending on their avatar perspective (i.e. first or third) or where their vision is primarily focused (i.e. viewscreen or text window).

Geophrey: *... when you stand in someone's face, you're taking up most of their viewing area ;-)*

We attribute the low referral to spatial behaviour in WBS to the text-based nature of the system. This behaviour may be due to the lack of a distinct spatial location provided for participants of textual environments such as chatrooms or MUDs [51]. A perception of personal space may also be due to the result of avatar embodiment. What it may mean is that personal space, as a non-verbal cue for communication, may not have an equivalent in a purely textual environment such as WBS. According to Fletcher [52], non-verbal behaviours such as personal space and gestures complement one's communicative message and are ineffectively transferred in computer-mediated communication such as e-mail, the Internet or chat groups. This should not be surprising. In MUD studies of non-verbal communication [53, 54] personal space seems to be redefined in terms of private places rather than the proxemic definition defined by Hall [36] for physical world spaces. Therefore, it may be that personal space does in fact exist in text-based WBS, when we re-define it in terms of private areas of the metaphorical rooms.

7.4.2. Private Space: In the World but Not of It

In WBS, messages are typed into a chat-input window and appear immediately on the message screen. Functionality exists to send private messages and images to other users instantaneously, to leave messages in users' personal WBS e-mail box, or to create a private room. Often more people were present than those participating; this could be due to using ICQ, a private real-time chatting system simultaneously with WBS, or they may have been passive observers rather than active participants. Sometimes people would express in their tagline attached to their message, or through a public message, whether private messages are desired, as the following user's tagline indicated.

Lizabeth21:   *(no PMs [private messages] please!!!): I dislike private messages. Because there's nothing that anyone can say to me that can't be said in public*

Lizabeth21:   *Plus in my experience, I only get nasty messages in PM ... not many have the guts to be crude, rude, crass in public*

Similar to WBS, functionality in AW exists for citizens to send private messages to each other, but avatar positioning sometimes also indicated that a conversation was private. Avatars were observed positioned high above others watching the avatars below or two avatars were positioned face-to-face in close proximity. Other examples that may also be indications of privacy markers include avatars that were observed separated from the main gathering points either above ground or on the same plane. Attempts at communication usually resulted in non-responses or the avatars leaving to another location.

Our observations indicate that a desire for private spaces exists in virtual environments. In both environments, people used functionality to engage in private conversations and to create privacy markers. Westin [41] classifies privacy into four different definitions: solitude, intimacy, reserve and anonymity. Solitude and intimacy may be most relevant to the environments observed. Westin defines solitude as choosing to separate oneself from observation by other people. In both environments this behaviour, referred to in CVE vernacular as "lurking", occurred as individuals observed the social interaction of others without participating themselves. In WBS, this was noticeable, as the population of the world was usually higher than the number of active participants:

vista-958:   *Thirty-five people in here and only 15 are talking*

Intimacy is commonly associated with close knit friends or small workgroups. Personal messages in WBS and telegrams in AW provided a means for more intimate conversations. In AW, dyads and groups were observed face-to-face sometimes during conversation, creating, perhaps, a privacy marker by their avatar positioning. Thus people create private spaces in three-dimensional environments.

One reason for this privacy behaviour that we observed may be that people, especially new users, are trying to understand appropriate behaviours from watching others. By removing oneself from the main activity centre, one can become an observer, without the social pressure to interact with others.

7.4.2 Crowding: Navigation and Design Constraints

In each environment there was a limit on the number of users that could comfortably communicate, and in AW, navigate. Reaching these upper limits created technical and social problems. If the environment is perceived as crowded, individuals have a number of options available. They could remove themselves from the environment, limit the numbers of conversational partners in order to minimise the number of messages missed, or use direct, private methods of communication such as telegrams or personal messages.

Indicators in WBS list the most populated public rooms that can be immediately accessed. Communities are listed by current population, e.g. Thirtysomething, 61 or Inn of the Weary Traveller, 26. Communication is often disrupted due to crowding. Crowding is not only related to the number of people but also the size of messages or pictures that are sent. When there are too many people present in a room the server may be affected causing messages to first stall and then to rapidly speed past. This requires extra effort for the user who must scroll back to read missed messages.

In other instances, a large number of users made it difficult to be "heard". In a WBS chatroom with 137 users present, O repeatedly tried to initiate conversation with others present for 45 minutes without one reply. In both WBS and AW, individuals expressed displeasure when ignored, such as:

SneakinSam: *i dont know what it is, but it happens all the time ... WHY IN THE HECK DOES EVERYONE IGNORE ME?! oh well, i'll never know, adios*

In AW, users must not only attend to the flow of the text as in WBS, but must also be aware of the positioning of other avatars. Functionality exists that enable people to communicate with either the 12 or 50 nearest avatars. A large number of avatars concentrated in one's immediate vicinity led to a number of observable difficulties. The ability to effectively navigate without violating the personal space of others decreases significantly as the density of avatars increases. Again, it is extra effort for the users, some of whom report strategies on where to focus their attention:

Lixx Array: *it's a two way thing for me ... either chatting, or exploring/building ... where my focus on the screen is*

Geophrey: *When it's slow like it is now the chat window can be narrower. But when it gets busy I stretch so there's only a sliver for viewing.*

When text was displayed above the head of each avatar (an option), then crowding of avatars made the text difficult to read since it overlapped. Monitoring one's own conversations, understanding the thread of other conversations, or maintaining conversations with more than one person became quite difficult with crowding. Concerning the central interaction area, one user expressed that she never stays there since it is too crowded.

In WBS, pictures that consumed a large area of the screen would cause difficulties for some users whose computers were unable to handle the increased bandwidth. In AW, some individuals would maximise the text screen and minimise the viewer screen in order to view more messages at one time in crowded environments.

The effect of crowding, a type of information overload, may have cognitive, technical, and visual implications. Messages streaming too fast are difficult to read; it is easy to miss personal messages. In both virtual worlds, a high volume of users and messages combined with a low baud modem or slow Internet browser resulted in a system slowdown or users being disconnected, which disrupted social interaction. Therefore, crowding made both navigation and communication within the environment more difficult:

Darrs: *Bob don't know how those people talk in those crowded rooms*

7.4.3 Group Space: Positioning and Movement Within a Group

In WBS, group membership was indicated by a common thread of conversation. Similarly in AW, group membership was also indicated through common threads of conversation as well as sometimes the positioning of avatars (Fig. 7.3). However, it was also observed that moving together or using the same motions conveyed group membership in the environment, e.g. simultaneously flying up and down in rhythm (i.e. bobbing vertically), walking together, dancing, or teleporting together. Such simultaneous movements served to create a weak "social boundary", separating this group from others.

Common language also created a form of social boundary. In AW, when O was speaking to *Stine*, a native Norwegian speaker, another disrupted the conversation by speaking to *Stine* in their native language. The shape of the group changed from a dyad to a triad, although O could not participate in the Norwegian conversation.

The face-to-face positioning of avatars indicated visually a group formation. We observed also in AW that when the avatars were positioned face-to-face, the relative distance between them would increase as group size increased. The physical shape was determined by group size: two avatars formed a line; three, a triangle; four, a diamond; and larger groups adopted a circle-shaped form. As

Figure 7.3 Group conversation in AW.

additional members navigated into a group, group members would reposition, expanding the social space and relative personal space of the group.

Was this lack of embodiment in WBS compensated for in any way by the text medium? The answer seems to be that to some extent, users tried to portray embodiment through text. For example, participants created a form of virtual embodiment using self-portraits, filling out their user profile, or attaching links to homepages. And as mentioned, in role-playing chatrooms, users took on identities consistent with the themes.

7.4.4 Movement Through Space

In each virtual environment, users must first determine which particular world to enter. In WBS, worlds are grouped into categories, which contain related themes. Either one navigates through categories and themes to reach a room, e.g. destinations and travel, or the user navigates directly to a community using a menu command. Users can also navigate room to room. Individuals sometimes would leave the room temporarily to navigate to another's homepage or to temporarily view pictures sent to them.

Although seldom occurring in WBS, using text to describe navigation occurred occasionally in role playing rooms. For example, in the WBS "Star Trek: Nexus Bar", text descriptions which were encased in asterisks indicated navigation, thoughts, and gestures.

Lady Milldorf2: ... decides to leave the bar *walks along the wall towards the door*

In AW, in contrast to using the first person view during communication and when scanning the surrounding area, third person view was used more for navigation and situations where it was important to see one's avatar, e.g.

"trooper":   *talking is better in 1st and moving is better 3rd :)_*
Danaidae: *only use third when checking out avatars or trying to position myself precisely*

Seeing oneself during navigation helps minimise personal space and group space violations of other avatars and provides a better view of what personal and group space boundaries are. In other words, it enables one to navigate around invisible boundaries rather than pass through other avatars. From user comments, it appears that they see this as impolite behaviour, and in fact a violation of personal space.

7.5. Summary and Conclusions: Co-ordinating the Shared Use of Space

Virtual information spaces are being designed along physical world spatial metaphors, e.g. digital libraries [55], CVEs [56], and information spaces such as StackSpace and InfraSpace [57]. They enable the user to visualise information and can provide an immersive environment for collaboration of work activities as physical objects are represented as virtual artefacts, as people become virtual embodiments and representations, and as workplaces become virtual landscapes. A key to the effective use of these spaces should therefore be that the information space is designed to be both meaningful and logically organised for the task.

Familiar spatial concepts may not be sufficient to guide navigation within these new forms of shared information spaces, even though they are found to be used [12–15]. Consistent with the spatial metaphor of a physical world, users walk through AW. However, the ability to teleport within and between worlds may represent navigation that deconstructs this spatial metaphor [16]. In WBS, it can also be argued that the teleport functionality in chatrooms, and the ability to navigate through the WWW and back is also inconsistent with a spatial layout of rooms.

The design of the two- and three-dimensional CVE spaces, in our opinion, helped guide the nature of the interaction. Navigation in WBS occurred mostly between chatrooms and the WWW rather than within a particular chat community. Information artefacts, such as user profile icons, contain selective personal information which enable others to immediately access and navigate to someone's personal home page or send e-mail. The degree to which this information was meaningful and navigable may have assisted users in developing a mental image of others that enriched conversation and facilitated social interaction. We observed users coming back after visiting another user's homepage or requesting others to accompany them to another room. Navigation often relied on the assistance of others, suggesting it was a collaborative effort [57].

The design of the virtual space in AW also influenced interaction by offering sociopetal spaces conducive for interaction [20]. In most worlds (i.e. divisions of the virtual space), the highest concentration of individuals was found in the landing point in each world, called "Ground Zero". Similar to communicative behaviour associated with a courtyard, individuals would congregate there. We also observed examples of worlds having a meeting place other than the "Ground Zero" location such as in a beach bar, which may also offer specific connotations for interaction.

In the two-dimensional space WBS, rather than providing an embodiment through visual forms, usernames and linked images seemed to provide a degree of embodiment. But we also see differences with respect to the portrayal of space. In contrast to the flat two-dimensional plane, a three-dimensional graphical environment conveys a landscape for interaction. As the user navigates through the space, the environment and artefacts change as one passes by. Also, one can change perspectives, from first person for communication purposes to third person for navigation. Distance perception exists: a small avatar or a small house is far away. Thus, in AW, the three-dimensional metaphor offers certain cues that influence users to behave and interact in a virtual space in ways that are different to how people interact with a two-dimensional virtual space metaphor.

These similarities to a real environment may translate into navigation playing a stronger role in a three-dimensional space for social interaction. In a two-dimensional space, conversation rather than movement is the primary focus. In the three-dimensional space, movement and conversation both play a role and differ in relevance depending on the user's task (e.g. house building vs. group interaction). The visualisation of the three-dimensional space provides a visible awareness of others; a sense of their presence is not only based on conversation but also on their movement and positioning. Each user then maintains a visible location within the three-dimensional world. From a cognitive viewpoint, one can easily grasp the information of who is interacting with whom (within a close distance) simply by observing the clusters of avatars. In a two-dimensional chatroom, gaining this information requires more effort; one must observe the conversational threads. However, this ease of awareness breaks down when the avatars do not change position as they change conversation partners [33].

In a three-dimensional space, proximity of avatars may be a factor in determining who one converses with and whether others may enter a conversation (i.e. by signalling privacy). In a two-dimensional space, conversational partners are not linked by proximity. In addition, the three-dimensional landscape influenced specific behavioural actions (i.e. groups collectively viewing houses or dancing) and often entered into the conversation (e.g. speaking together about the view while suspended in space). In contrast to the more varied landscape, the rooms in WBS are fairly uniform.

In each environment, social conventions involving the use of the space exist. This was observed in AW from the personal distance separating interacting avatars, the reluctance of avatars to invade another's personal space, especially during navigation, the positioning of an avatar in the environment to indicate privacy, and in the adverse reactions towards avatars when their personal spaces were violated. In WBS, the conventions were less obvious, but some references in conversation were made by users concerning navigation and the personal space of other users. In addition, reactions to crowded environments occurred, illustrating the cognitive difficulties of interacting with large numbers of users.

The existence of social conventions may support Harrison and Dourish's [18] notion of place. They consider *placeness* as an evolved social understanding of the type of behaviour and actions that are appropriate within a space. The architecture and inhabitants may have collectively functioned to enable conventions to emerge. Without the particular concept of place that developed, we would not have expected some of the conventions in AW to exist, such as personal space. Similarly, the theme chatrooms of WBS also appear to promote an idea of place, since conversations appropriate to the theme occurred in them. While both AW and WBS also have conventions that are more of a linguistic nature, such as the use of acronyms, AW appears to have additional conventions associated with an idea of a three-dimensional space.

Although we cannot draw strong conclusions from this exploratory study, we can suggest that such interpersonal behaviours may be influenced by factors such as the presence of an embodiment (avatar), expectations about the space, design features, and the presence of others. But as new forms of shared information spaces continue to emerge, corresponding new notions of placeness and appropriate behavioural conventions will also need to develop. Our basic results that shared virtual environments are regarded as social spaces, with corresponding socially acceptable behaviours, is also consistent with results found from studies of other virtual environments [1, 54, 58]. CVEs are an exciting new form of virtual communication, and we hope that our results can stimulate further research into their behavioural aspects.

Acknowledgements

We thank Andreas Dieberger, Kaisa Kauppinen, Alan Munro and Mike Robinson for their useful feedback during earlier drafts of this chapter. We thank Deborah Stacey for her generous support at the University of Guelph.

References

1. Bruckman, A. and Resnick, M. The MediaMOO project: constructionism and professional community. *Convergence* 1: 94–109, 1995. (http://asb.www.media.mit.edu/people/asb/convergence.html)
2. Neal, L. Virtual classrooms and communities. In: *Proceedings of ACM Group '97*, S.C. Hayne and W. Prinz, editors. New York: ACM, 1997, pp. 81–90.
3. Fahlén, L.E., Brown, C.G., Stahl, O. and Carlsson, C. A space based model for user interaction in shared synthetic environments. *ACM Conference on Human Factors in Computing (InterCHI '93)*. New York: ACM, 1993.
4. Greenhalgh, C. and Benford, S. MASSIVE: a virtual reality system for tele-conferencing. *ACM Transactions on Computer Human Interfaces (TOCHI)* 2: 239–261, 1995.
5. Active Worlds: http://www.activeworlds.com
6. Gómez, E.J., Quiles, J.A., Sanz, M.F. and del Pozo, F. A user-centered cooperative information system for medical imaging diagnosis. *Journal of the American Society for Information Science* 49: 810–816, 1998.
7. Hindmarsh, J., Fraser, M., Heath, C., Benford, S. and Greenhalgh, C. Fragmented interaction: establishing mutual orientation in virtual environments. *Proceedings of CSCW '98*. New York: ACM, 1998, pp. 217–226.
8. Bannon, L. and Bødker, S. Constructing common information spaces. In: *Proceedings of the 5th European on Computer Supported Cooperative Work (ECSCW '97)*, J. Hughes, T. Rodden, W. Prinz and K. Schmidt, editors. Dordrecht: Kluwer, 1997, pp. 81–96.
9. Fuchs, L., Poltrock, S. and Wojcik, R. Business value of 3D virtual environments. *SIGGROUP Bulletin* 19: 25–29, 1998.
10. Huxor, A. An active worlds interface to BSCW, to enhance chance encounters. *Proceedings of Collaborative Virtual Environments '98: CVE '98*, University of Manchester, 1998, pp. 87–93.
11. Benford, S., Brown, C., Reynard, G. and Greenhalgh, C. Shared spaces: transportation, artificiality and spatiality. In: *Proceedings of CSCW '96, ACM Conference on Human Factors in Computing*, M. Ackerman, editror. ACM: New York, 1996, pp. 77–86.
12. Goldate, S. The "Cyberflaneur" – spaces and places on the internet. Art Monthly Australia, 1997. (http://www.geocities.com/Paris/LeftBank/5696/flaneur.htm)
13. Maglio, P. and Matlock, T. Constructing social spaces in virtual environments: metaphors we surf the web by. In: K. Höök, A. Munro and D. Benyon, editors. *Workshop on Personalised and Social Navigation in Information Space*. SICS Technical Report 98:01, Stockholm, 1998, pp. 138–149.
14. Maglio, P. and Matlock, T. The conceptual structure of information space. In *Social Navigation of Information Space*, A.J. Munro, K. Höök and D. Beynon, editors. London: Springer, 1999, Chapter 9.
15. Dieberger, A. and Frank, A.U. A city metaphor for supporting navigation in complex information spaces. *Journal of Visual Languages and Computing* 597–622, 1998.
16. Dieberger, A. On magic features in (spatial) metaphors. *SigLink Newsletter* 4(3): 8–10, 1995.

17. Dieberger, A. Social navigation in populated information spaces: social connotations of space. In *Social Navigation of Information Space*, A.J. Munro, K. Höök and D. Beynon, editors. London: Springer, 1999, Chapter 3.

18. Harrison, S. and Dourish, P. Re-Place-ing space: the roles of place and space in collaborative systems. In: *Proceedings of CSCW 1996*, Boston, MA, M. Ackerman, editor. New York: ACM, 1996, pp. 67–76.

19. Heath, C. and Luff, P. Collaboration and control: crisis management and multimedia technology in London Underground line control rooms. *Computer Supported Cooperative Work (CSCW)* 1: 69–94, 1992.

20. Osmond, H. Function as a basis of psychiatric ward design. *Mental Hospitals (Architectural Supplements)* 83: 235–245, 1957.

21. Tuan, Y. *Space and Place: The Perspective of Experience*. Minneapolis, MN: University of Minnesota Press, 1977.

22. Altman, I. and Zube, E.H. *Public Places and Spaces*. New York: Plenum, 1989.

23. McCormick, E.J. and Ilgen, D. *Industrial and Organizational Psychology*, 8th edn. Englewood Cliffs, NJ: Prentice-Hall, 1987.

24. Field, R.H.G. and House, R.J. *Human Behaviour in Organizations: Canadian Perspective*. Scarborough: Prentice-Hall Canada, 1995.

25. Feldman, D.C. The development and enforcement of group norms. *Academy of Management Review* 9: 47–53, 1984.

26. Vecchio, R.P. *Organizational Behavior*, 2nd edn. Orlando, FL: Dryden, 1991.

27. Goodman, P.S., Ravlin, E. and Schminke, M. Understanding groups in organizations. In: *Research in Organizational Behavior* 9: 121–173, 1987.

28. Carroll, J.M. and Thomas, J.C. Metaphors and the cognitive representation of computing systems. *IEEE Transactions on System, Man, And Cybernetics* SMC-12(2), 1982.

29. Dix, A.J., Finlay, J.E., Abowd, G.D. and Beale, R. *Human–Computer Interaction*, 2nd edn. Hemel Hempstead: Prentice-Hall, 1998.

30. Bruckman, A. Identity workshop: emergent social and psychological phenomena in text-based virtual reality. 1992. (ftp:media.mit.edu:/pub/asb/papers/identity-workshop.ps)

31. Raybourn, E. An intercultural computer-based multi-user simulation supporting participant exploration of identity and power in a text-based networked virtual reality: DomeCityTM MOO. Dissertation. Department of Communication, University of New Mexico. Albuquerque, NM, 1998.

32. Baym, N. The performance of humor in computer-mediated communication. *Journal of Computer-Mediated Communication* 1(2) (online), 1995. (http://shum.cc.huji.ac.il/jcmc/vol1/issue2/baym.html)

33. Becker, B. and Mark, G. Constructing social systems through computer-mediated communication. *Virtual Reality* (in press).

34. Jeffrey, P. and Mark, G. Constructing social spaces in virtual environments: a study of navigation and interaction. In: Workshop *on Personalised and Social Navigation in Information Space SICS Technical Report 98:01*, K. Höök, A. Munro and D. Benyon, editors. Stockholm, 1998, pp. 24–38.

35. Kauppinen, K., Kivimäki, A., Era, T. and Robinson, M. Producing identity in collaborative virtual environments. *VRST'98: Symposium on Virtual Reality Software and Technology*. New York: ACM, 1998, pp. 35–42.

36. Hall, E.T. *The Hidden Dimension*. New York: Anchor, 1966.

37. Aiello, J.R. Human spatial behavior. In: *Handbook of Environmental Psychology*, Volume 1, D. Stokols and I. Altman, editors. New York: Wiley, 1987, pp. 389–504.

38. Knapp, M.L. *Nonverbal Communication in Human Interaction*. New York: Holt, 1978.

39. Altman, I. and Vinsel, A.M. Personal space: an analysis of E.T. Hall's proxemics framework. In: *Human Behaviour and the Environment*, Volume 2. *Advances in Theory and Research*, I. Altman and J.F. Wohlwill, editors. New York: Plenum, 1977, pp. 181–259.

40. Altman, I. *The Environment and Social Behavior*. Monterey, CA: Wadsworth, 1975.

41. Westin, A. *Privacy and Freedom*. NewYork: Atheneum, 1970.

42. Stokols, D. On the distinction between density and crowding: some implications for future research. *Psychological Review* 79: 275–278, 1972.

43. Freedman, J.L., Heshka, S. and Levy, A. Crowding as an intensifier of the effect of success and failure. In *Crowding and Behavior*, J.L. Freedman, editor. San Francisco, CA: Freeman, 1975, pp. 151–152.

44. Stokols, D. and Resnick, S.M. The generalization of residential crowding experiences to non-residential settings. *Annual Conference of the Environmental Design Research Association*, Lawrence, KS, 1975.

45. Strickland, D., Hodges, L., North, M. and Weghorst, S. Overcoming phobias by virtual exposure. *Communications of the ACM* **40**(8): 34–39, 1997.
46. Benford, S., Greenhalgh, C. and Lloyd, D. Crowded colloborative virtual environments. *Proceedings of CHI '97.* New York: ACM, 1997, pp. 59–66.
47. Citera, M. Distributed teamwork: the impact of communication media on influence and decision quality. *Journal of the American Society for Information Science* **49**: 792–800, 1998.
48. Okada, K., Maeda, F., Ichikawaa, Y. and Matsushita, Y. Multiparty videoconferencing at virtual social distance: MAJIC design. *Proceedings of CHI '94.* New York: ACM, 1994, pp. 385–393.
49. Persson, P. Towards a psychological theory of close-ups: Experiencing intimacy and threat. *KINEMA: A Journal of History, Theory and Aesthetics of Film and Audiovisual Media.* Waterloo: University of Waterloo Press, 1998.
50. Meyrowitz, J. Television and interpersonal behaviour: codes of perception and response. In: *Inter/Media: Interpersonal Communication in a Media World,* G. Gumpert and R. Cathcart, editors. New York, 1986.
51. Dieberger, A. Personal communication, 1999.
52. Fletcher, A. The rhetoric of synthetic vs analytic truth on the internet as it relates to nonverbal communication, 1997. (http://www.geocities.com/Paris/Metro/1022/nonverbl.htm)
53. Masterson, J. Nonverbal communication in text based virtual realities. MA thesis. University of Montana, 1996. (http://www.montana.com/john/thesis/)
54. Turkle, S. *Life on the Screen: Identity in the Age of the Internet.* New York: Touchstone, 1997.
55. Nilan M. Ease of user navigation through digital information spaces. Proceedings of the 37 Allerton Institute 1995, Monticello, Illinois, (on-line), 1995. http://edfu.lis.uiuc.edu/allerton/95/s4/nilan.html
56. Munro, A. Inhabiting information space: work artefacts and new realities. In: *Exploring Navigation; Towards a Framework for Design and Evaluation of Navigation in Electronic Spaces,* N. Dahlbaeck, editor. Stockholm: SICS Technical Report, T98:01, 1998, pp. 91–114.
57. Waterworth, J.A. Personal spaces: 3D spatial worlds for information exploration, organisation and communication. In: *The Internet in 3D: Information, Images, and Interaction,* R. Earnshaw and J. Vince, editors. San Diego, CA: Academic Press, 1997. (http://www.informatik.umu.se/%7ejwworth/spaces.pdf)
58. Curtis, P. MUDding: social phenomena in text-based virtual realities. In: *High Noon on the Electronic Frontier: Conceptual Issues in Cyberspace,* P. Ludlow, editor. Cambridge, MA: MIT Press, 1996, pp. 347–373.

Chapter **8**

Spaces, Places, Landscapes and Views: Experiential Design of Shared Information Spaces

John A. Waterworth

Abstract

This chapter focuses on the World Wide Web (web) as a provider of shared information landscapes. It reviews our work to design three-dimensional spaces for information navigation and social interaction, and suggests an approach to such design based on an experiential theory of meaning. The increasing use of virtual three-dimensional space in information environments is noted, and personal spaces are contrasted with public places. Earlier work on information islands, vehicles and customisable views of such information spaces is also presented. The experiential approach, as applied to information landscape design, is contrasted with the traditional view of human–computer interaction (HCI) design as a means of conveying system functionality from the mind of the designer to that of the user. This experiential approach seems promising, if we assume that we do not know in advance what the functions of interactions in shared information spaces might be. As with life in general, such interactions mean what they are experienced to be.

8.1 Introduction

There are many ways in which the world's most popular hypermedia system (by far), the World Wide Web (web), does not reflect the hypermedia usability research that preceded it (see [1] for a catalogue of what were considered the key research issues at that time). Perhaps the most unexpected thing about the web as a whole is that no-one is designing it. Three other ways in which it has not conformed to what was expected of hypermedia, are: first, the use of three-dimensional graphics to give a sense of space; second, the fact that there is one web which all users cohabit; and, third, the fact that we can communicate with each other from within the web.

This chapter develops three themes that follow from these unexpected characteristics of the web: personal spaces versus public places, the notion of vehicles with views, and the potential for presences and concealment. These are

132

illustrated with some recent examples of our work and which adopts an experiential approach, which contrasts with the traditional view of HCI design as facilitating the communication of functionality between designer and user. This work has been motivated by the realisation that a profound change is taking place with the evolution of information and communications systems into self-contained virtual environments. Inhabiting a virtual world is very different from using electronic tools in the real world: hence, the importance of virtual presence.

8.2 Personal Spaces versus Public Places

An increasingly popular approach to the representation of information on the web is to use three-dimensional rendering techniques to convey a sense of space and apparently solid structure. This means that information explorers can bring their innate skills for spatial navigation into play, in addition to those few sensori-motor abilities utilised by the familiar direct manipulation (WIMP) interface. However, because no-one is designing the web, and because of the simple linking mechanisms underlying its evolution, there is no way to make sense of its structure as a whole. There can be no three-dimensional representation capturing its whole structure which, as is implied by our approach, means that people simply cannot make sense of its structure as a whole.

However, space is powerful as a means of representing the structure of designed environments, such as personal file systems and the intranets of organisations. Personal environments can be happily represented as personal spaces – three-dimensional structures apparently containing stored and current items of interest to the individual user (e.g. StackSpace [2]). Figure 8.1 gives an indication of the StackSpace environment.

The need for multi-threadedness, chronology and currency-tracking are taken care of in StackSpace in the following ways. Multiple stacks develop as the user

Figure 8.1 A space in StackSpace.

explores, and the top slices are the most recent or current. The items that are furthest in (away from the viewer) are of least interest or relevance to the current task. Items decline gracefully in interest, by moving away from the viewer, but then fall over a cliff (although the retentive user could, in principle, extend the horizon to infinity). Contextual bridges (shown in Fig. 8.2) and cords show relationships between items in different stacks. For example, the bridges shown in Fig. 8.2 indicate that stacks A and B, B and C, and C and D all have at least one "slice" in common – the same web page has turned up on both stacks.

Computer searches, for example the results of an agent carrying out a collecting task on a topic, also produce stacks. Users then manipulate, edit and label spaces, stacks and slices to customise the material found. Edited stacks comprise views of a topic, hot-lists and traveller's tales reflecting experiences navigating in pursuit of a particular purpose. They are exchangeable with others and thus provide a means of communicating about the process and results of information navigation.

We can distinguish the idea of space from that of place. In a sense, everywhere on the web is currently a public place (some have restricted access, but I will disregard that for present purposes), even the humble single-screen individual-user home page. They are public because anyone can go there and, often, several people will be there at the same time. But they are not aware of each other. In this sense cyberspace is unlike reality. We cannot generally see where people are in cyberspace. We have public places and personal spaces but no public spaces, because users do not share a sense of each other's presence in those three-dimensional structures. (Here I am ignoring a few emerging social spaces, such as The Palace, specifically designed for some kind of social interaction via the web.) We need to use three-dimensional space to convey aspects of human presence, to represent the people, not just aspects of the available information; but three-dimensional space alone will not be sufficient.

My use of the words "space" and "place" here is differs from that of Harrison and Dourish [3]. By my usage, "Public spaces" are shared three-dimensional places where we can and do interact with others in real-time, which reflects normal use of this term. I recognise that socialising in places implies conventions of appropriate behaviour (behaving in or out of place), a topic thoroughly

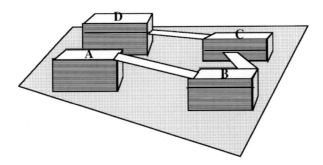

Figure 8.2 Contextual bridges in StackSpace.

addressed by Meyrowitz [4], and that social places need not be spatial. However, and contrary to Harrison and Dourish, "a place" in normal usage does not of itself imply what they mean by "placeness". A place can simply be a location. To avoid this confusion, we need to be explicit about whether we mean by "place" a particular location, or a social occasion to which a set of conventions applies. By "place" I mean the former, and "public places" are spatial or other locations open to everyone. A "public space" is then a spatial place, which may or may not be used for social occasions to which a set of conventions applies. To use the term "place" to mean the latter seems to me confusing, since we are really talking about appropriate behaviour for a social occasion, wherever it takes place. The expression "behaving out of place" refers metaphorically to a time when different social occasions took place in different places (in the sense of location in space). We cannot, I think, design "placeness" in Harrison and Dourish's sense, we can only design locations that may or may not be spatial. Real or virtual space may reduce the tendency to behave "out of place" (as suggested by Meyrowitz's thesis); the absence of such space in newsgroups may account for the frequency of socially inappropriate behaviour there – despite their being "a place".

The web provides a marvellous medium for information exchange, for contacting others, for sharing opinions, for finding out about events, and for keeping in touch with recent developments. But as we explore the web, we stay "at home". People can send us messages, can search for things posted with our names attached. Once they have our address they can write to us. Maybe they can send e-mail from a page of ours they came across. But they don't know where we are at any given moment, they know only the address we use for sending information (and not always that) and the information we make available. If we have a camera set up in the office and linked to the web, they can see when we are in our office. But they probably already knew where we worked. If they see us at the terminal we might well be navigating around the web, but where? Bodily presence is no longer as important as where our attention is located.

We don't always want to have to go and look for things ourselves, and search engines of one kind or another are increasingly used to locate information on the web, especially by more experienced users. The notion of software "agents" (also known less misleadingly as personal digital assistants), which can carry out tasks for us in the background while we get on with other things, is much talked about and complimentary to the idea of using space. Agents provide services for their "masters", but the real agents in cyberspace – the people – remain unrepresented. Because of this, we cannot search for people, only for the things they have left in cyberspace. It should also be possible to enter the attributes of people we might want to locate and have the system report back where they are in cyberspace, where they have been recently, and so on, adding value to everyday reality. Some of the things we might wish to track about people include when and where social groups form, the navigational paths of individuals (or their agents), and their interests (which their agents would know about). Agents meeting with other agents could provide some of this information. Whether this is seen as threatening or not depends on the level of confidence one has in one's agents.

In summary, public three-dimensional places are proliferating but they suffer from the same limitations as two-dimensional places – they are inflexible yet changing. There is no space between public three-dimensional places, because there is no context between sites. As the web is today, this is an insoluble problem. Personal spaces are a promising way to make sense of material gathered from the web. They also provide mechanisms for editing and sharing materials. Public places are shared, but public spaces need *3Dspace* and *visible* people.

8.3 Information Cities, Islands, Vehicles and Views

About 10 years ago, the idea of a virtual three-dimensional information city – a way of presenting sets of information to tap people's skills in urban navigation, was raised in Singapore, itself a highly "wired" city aiming to deal largely in information in the future. This idea was circulated in internal research reports in 1988–1989, and published in a brief form [5].

An "information city" was also suggested early on by Dieberger [6], who has since published many papers on a textually-described three-dimensional city (see Dieberger's chapter in this book, Ch. 3). Navigating a space that exists only as a text description, as in early networked "adventure" games, is an interesting task to study, especially when it is carried out as a social activity. But interpreting text is a very different skill from navigating a three-dimensional structure, whether real or virtual. It is hard to imagine that one could use textual descriptions to create spaces that could be used for purposes other than their own exploration, since that in itself will be very demanding of cognitive resources. The point of the virtual three-dimensional structure is to remove the need for people to use linguistic interpretation to make sense of interfaces and to tap their largely unconscious sensori-motor skills, thus freeing cognitive capacity for tasks other than navigation.

A later development of our basic info city idea was the "information islands" model for the Singapore National Computer Board's National Information Infrastructure Project; this work was carried out in 1993–1994 and aspects of the model were published soon after (see [7–9]). Again, this was a "natural" idea to arise on an "intelligent island" in close proximity to the giant archipelago of islands that is Indonesia. This section of the chapter summarises the information islands model, to set the central idea of views in its original context.

Under the "information islands" model, the world (through which the structure of a set of information is represented) is seen as a group of archipelagos, each composed of information islands. Each archipelago represents a set of broadly related entities, providing a clear, top-level classification of what is available in this world and where it is to be found – an overall orientation that is easily accessible to both the novice and the experienced user. Each major class of service or application exists as an archipelago. Examples might be entertainments, government services, information services, communications, medical, and financial services. Archipelagos are collections of information islands. The size of an archipelago depends on the number (and size) of the islands of which it is composed.

Each island generally contains only one subclass of service. Users will become familiar with this world mostly by learning the location of islands with the kinds of services they use or are interested in. Each island contains one or more buildings. Some islands may be representations of the services offered by particular providers – provider islands. An example might be a particular information provider's island located near other information services islands.

Each building contains a set of information sources or services related to a particular topic or application focus. Examples might be weather building, sports building, stocks and shares building. Buildings on a particular island will have distinctive appearance (shape, colour, graphics, text). All buildings have common features including a store directory and an information counter (see Fig. 8.3). The store directory allows users to browse and select from what is available in a Building. The information counter is a public agent that searches for information in response to requests from users. Buildings contain standard features to assist in navigation and item location (cf. [10]).

Archipelagos, islands, and buildings become bigger the more items they contain. Each archipelago is formed by placing a boundary around the islands from which it is composed. Each archipelago has a distinctive colour that provides a context and reminder to the user of the focus he has chosen. As the user zooms in for more detail, the view of archipelagos is replaced by a view of the islands from which the selected or central archipelago is composed. Intermediary views provide realism and orientation as the user zooms down.

When a single archipelago is shown, the islands from which it is composed are represented separately. A view of a single Island is a map of the collections of services provided and which are represented as buildings (see Fig. 8.4). Buildings that are related are clustered together into no more than ten villages. Each

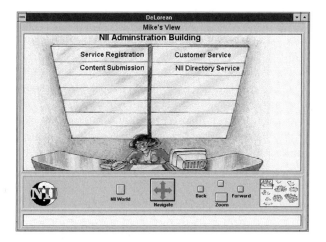

Figure 8.3 The store directory.

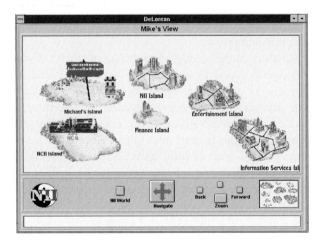

Figure 8.4 Islands and buildings.

building contains no more than 20 floors, and each floor generally contains a set of related services.

The user views the available services by zooming down and selecting a particular building, which is of a different colour from that of its neighbours. He or she enters the foyer (the background retains the colour of the building to which it belongs) and can then either browse the store directory or consult the public search agent at the information desk. The store directory presents a list of the service types available on each floor of the building. at each floor there is a "lift lobby" where users consult a floor directory (like the store directory, but listing individual services), to invoke the service they require.

An important part of interacting with this world of information is the exploration, selection and collection of items of interest to the individual user. These items may be services, information or particular configurations of applications. One common way of catering for this need for a personal selection from a public world (a set of public places) is to demarcate part of the world as personal, and allow the user to collect items and configure that private area. This is one of the key ideas behind the rooms concept [11]. However, such an approach is limiting. Users must navigate to their own area frequently, bringing back items they want to collect, then venture out again into the world-at-large. In such a case, the disadvantages of a spatial metaphor can outweigh the advantages: because the users' personal space is part of the global information space, they frequently have to move around to switch between their own perspective and the higher levels of organisation. Use (which always involves a user) is confounded with level of structural organisation (which includes a user level). Use should be possible at any level, at any time. A private area at a particular location in the informational world may not be the best way of supporting individual customisation.

8.3.1 Vehicles with Views

To overcome these problems, the concept of private vehicles was developed; these can be thought of as transparent, mobile, personal workspaces. They combine the idea of a private collection of information and configuration of services (customised workspace) with that of multi-level navigational device and customised information viewer. Users are always in their own vehicle, and therefore always have access to both public and private worlds. Items can be transferred between these two without navigating space. A key aspect of the model is that the user has a filtered way of looking at the same spatially-arranged world that occupies public space. The private "world" is actually a manipulable way of viewing rather than a specific place (cf. [12]). It assumes that there is no one true view of the world, but always many possible ways of looking at things.

In the original information islands model, users in their own vehicle had two views of the world outside – a (somewhat ironically named) public "god's eye" view that includes everything that is available, and a personal view showing only those items that the user has selected as of interest or use (see Fig. 8.5). The user has only one set of navigation and viewing controls; the user chooses on which view or views they act. Although there are two views, there is only one world. The private view and the god's eye view are different perspectives on the same world; the former is filtered and limited, the latter is a complete display at the level of detail on which it is focused. The user can choose to have a split screen showing both views simultaneously, or alternate between the two. Views have some similarities with the idea of "magic lenses" [13]. However, a key aspect of views is that the three-dimensional structural integrity of the world is always maintained (the philosophy of "one world, multiple views").

The user can also "yoke" the two views together so that the public view and the private view are then both from the same viewpoint (viewpoint: literally, the

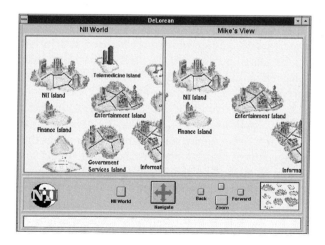

Figure 8.5 Two views of the same part of the world.

position in virtual space from which views are taken), changing together as the user navigates or inspects information at different levels. This can be useful when he or she wants to know what else is available at a place, other than the things the user has already chosen to include in his or her view. This is also useful during customisation, when the user can fly around the world-at-large and select things that he or she will then see included in his or her own view. At other times, the two views are "unyoked" and the user will select one or other of the two views to be updated as the vehicle moves, but not both. The view that is selected (private or public) will be the one that is affected by the navigation controls, the other will remain focused on where it was when last selected. The user can use the public view as a navigational overview while exploring in detail with the private view. Alternatively, he or she can have their own private view as an overview and move around the world via the public view collecting items to add to their private world. Selecting "yoke" will cause the less-recently-selected view to be updated to match that of the more-recently-selected view. This means that navigation can be done on either view, and the other view aligned to that viewpoint when required.

The provision of both a personal and a comprehensive public view means that the user has access to a customised world, as well as the world-at-large. This customised world is a subset of the world-at-large, selected by the user but retaining the layout and grouping inherited from the larger world. Items are simply dragged from the public view to the private view window. Apart from this simplified view, the user may want instant access to a few frequently used services and applications. Two mechanisms are provided for this: the vehicle's memory and the glove compartment. The vehicle's memory is a list of places the user wants the vehicle to remember, so that they can be rapidly revisited without the need for navigation. This is essentially the same as what became known as "bookmarks". When at a particular location, at whatever level in the hierarchy, users may select the "memorise" option, resulting in that location being added to the memory list. At a later time, they simply click on that item in the list to instantly move from wherever they are to that location. A simple "forget" option allows locations to be removed from the list.

There may also be particular applications or services to which users wish to have instantaneous access, and/or which may be used at a variety of locations. Such items can be stored in the vehicle itself and so are always with users wherever they may be in the world-at-large. The glove compartment is located to the side of the navigation controls. When closed, the "open" option is displayed. Selecting this causes a moveable window to appear, displaying the contents of the glove compartment. Applications are stored in the glove compartment by the user dragging their icon from a navigation window onto either the open window or the glove compartment feature on the dashboard. Items are removed by dragging out of the open window and dropping anywhere else.

Views become more interesting when applied in the social sphere. I may want to see only items visited by members of my research team recently. Or I might want to compare one view I have (or my agent has) compiled of interesting sites, with the view a colleague (or his agent) has collected. My view is a way of looking at cyberspace where only things of interest to me exist, and the same applies to him and his view. We can combine these two into another view that shows only those items that are of interest to both of us, or we can create a difference view which shows only those things chosen by only one of us. So the collection of public places that currently comprises cyberspace is filtered to give a socially-shareable and customisable view of cyberspace. This is arguably quite close to the way different groups and individuals hold different views of cities in the real world. The obvious next step is to include representations of cyberspace inhabitants in selective views. I might want a view that conveys the number of people present in the regions I explore, but I am unlikely to want to see all available information on all the people there. I might want only to see people if they are known to me. I might want to see them differently if they are business colleagues rather than competitors. In general, I will want different attributes of people represented in cyberspace according to their relationship to me. Increasingly, interacting on the web will become like participating in an online multi-user game. Of course, the privacy issues raised are quite daunting.

There are several unanswered questions arising from this work. Is a single hierarchical structure realistic? What are the advantages of information islands versus other world models? Would forests, trees, and leaves have been any different (e.g. the "dataforest"; [14]). Would more than two views give additional benefits? If we assume a hierarchy of ten archipelagos, with 20 islands per archipelago, 20 buildings per island and 20 floors per building, we have the necessary scope for a large number of individual information items to be located in the world. With 20 items per floor, we have 1.6 million items. Relaxing the restrictions on items per floor by having subsets of items accessed by two submenus after the initial floor directory selection, and expanding the world to a maximum of 20 archipelagos, would allow us to accommodate over a billion individual items. Can users navigate in such a world? Almost certainly not if by "navigation" we mean that users can easily find what they want by self-directed wayfinding. The problem of classification is not solved by using three-dimensional space, but the nature of the environment is radically changed. Social and personal habitation of large virtual information spaces may require assessment by criteria other than those we usually apply to information retrieval systems. There is also a good case to be made for the use of an "event horizon" to reduce the complexity of the environment. Waterworth [2] suggests a flat-earth metaphor for personal spaces where older items gradually move towards the "edge of the earth", and then disappear.

But in any case this view of expansion is unrealistic. The information islands model was designed to meet a particular need. It was assumed that the world would start life relatively empty and would then gradually expand, as providers offered information and other services. In this sense, it is rather like a plan for a city. But it is not clear to what extent development will match the original planning. As Alexander [15] has pointed out, "a city is not a tree" – not a simple hierarchy that grows according to predictable rules. As providers offer services, and users gravitate to the things they are interested in (i.e. willing to spend time and/or money on), the original plans are likely to be heavily modified by market pressures. Like a pleasant city, the world-at-large should evolve to meet the needs of users and providers alike. But not all cities are pleasant, and the balance between central planning and market-led evolution is not easy to strike. The vital question is whether users can find their way around such an evolved model, by whatever means (both agent-mediated and self-directed), to a degree that suits them. Success might be better assessed by the nature of "traveller's tales" [2] reflecting experience of the system, than from more familiar objective measures such as time to "solve a problem" or number of "bad solutions".

8.4 The Experiential Approach to HCI Design

The problem of interface design has traditionally been characterised as one of communication between the designer and the users. Norman's [16] well-known account of HCI design centres on three kinds of model: the design model (in the head of the designer), the user's model (in the head of the user) and the system image (as presented in the designed interface). The system image serves as the medium of communication between the designer and the user (see Figure 8.6). In the ideal case, the user's model comes to match the design model closely. The common approach to facilitating this process has been to incorporate one or more metaphors in the system image. It then becomes of great importance that the designer chooses appropriate metaphors that convey relevant aspects of the functionality of the system in terms that are understandable to the user [17]. A good metaphor is supposed to permit the user to apply knowledge of the source domain of the metaphor to the unfamiliar target domain of the interface [18].

According to the traditional, objectivist approach to interface design, an interface metaphor is some kind of specialised device for conveying a complex of concepts, based on speaking of, or presenting, one thing as if it were another. However, there is considerable confusion between metaphors and models. A metaphor is not a model, and metaphors are not unambiguous.

In several books published over the last two decades, George Lakoff and Mark Johnson have presented an alternative view of meaning, one that casts a completely different light on the role and importance of metaphor from that assumed in traditional HCI design [19–22]. At the same time, they manage to avoid the problems of both objectivism and pure subjectivism.

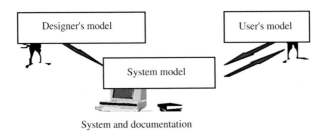

Figure 8.6 The traditional view of HCI design.

According to Lakoff and Johnson, metaphor is much more than a linguistic and rhetorical device. They argue that we always think metaphorically, that our everyday experiences are shaped by three kinds of metaphor: structural, orientational and ontological. Structural metaphors are found when one concept is structured in terms of another, for example that argument is war: "Your claims are indefensible", "He attacked every weak point in my argument", "His criticism was right on target", "He shot down all of my arguments", etc. We not only speak of argument as if it were war (and very pervasively, so that many statements about argument reflect this underlying structuring – although we don't actually think of them as metaphorical), we think about argument as if it were war, and often act according to the same, unconscious, assumption. Orientational metaphors structure experience in terms of spatial orientation. For example, down is negative, up is positive: "I am depressed", "I am really down", "I feel low", "Things are looking up", etc. Ontological metaphors structure our experiences of abstract phenomena in terms of concrete objects and forces (see also [23]). Efforts to visualise information as shapes, colours, and textures can be seen as reflecting the operation of underlying ontological metaphors, as, arguably, can the application of general interface metaphors such as direct manipulation. The term "synaesthetic media" [24] refers to the presentation of the same information in different modalities, which could be guided by the adoption of appropriate ontological metaphors.

If every concept is metaphorically structured, are we not stuck in some kind of infinite regress? If every concept is structured in terms of another, we are indeed (which is, of course, a fundamental problem with objectivist accounts of meaning). Lakoff and Johnson avoid this infinite regression by suggesting that, at bottom, meaning is rooted in basic, bodily, experiences of life as animals with a certain physical configuration residing on a planet with certain characteristics (notably, gravity). When we use expressions like, "I fell asleep" or "Wake up!" we use metaphor in a way that reflects the physical nature of life on earth. Our body configuration, combined with gravity, makes it necessary for us to sleep in a horizontal position. Johnson [19] provides more detail on the grounding of the (fundamentally metaphorical) conceptual system in corporeal, earthly existence. He proposes the existence of image schemata, which are basic structures of experience. These structures are then projected metaphorically on to more complex experiences. Lakoff ([21], pp. 271–278) suggests that image schemata (i) are based on bodily experience, (ii) have structural elements, (iii) have a basic logic and (iv) are manifested in actions and expressions. He gives many examples of image schemata, including the container, the centre-periphery, and the verticality schema.

8.4.1 Designing Experiential Information Landscapes

Our current approach to information landscape design [25], based on this experientialist account of meaning rather than the usual objectivist cognitivism of the traditional "mental model" approach, rests on the fundamental premise that *to design HCI is to design the conditions for possible users' experiences.* In the traditional approach, the metaphor is part of the interface. This need not be the case with experientialism since, by this account, metaphor is everywhere. Taking an experientialist view of interface design suggests that a meaningful interface is one that is experienced in a way that supports the metaphoric projection of image schemata. This is done by users in the same way that they make sense of all the other experiences of daily life, by unconscious projection of bodily image schemata. If the experientialist designer is primarily a creator of user experiences, the traditional interface designer is primarily a communicator of mental models, using metaphor as a useful device.

While I am not arguing that all traditional interface metaphors should be replaced, I do suggest that for several application areas – and these are areas that are at the forefront of current HCI research and development – an experiential approach to HCI design may be more appropriate. A notable example is that of information visualisation and exploration. If we revisit the information islands interface wearing our experiential sunglasses, we see that what matters is not so much the metaphor itself, as the experiential features we chose to take from the real world and incorporate in the virtual.

Conklin [26] argues that "there is no natural topology for an information space", and this claim could be extended to include other aspects of the interactive experience, such as how an information space sounds or how quickly one travels through it. However, an experientialist designer would argue the opposite; that there are, in fact, not one but many natural topologies for such a space, topologies ultimately grounded in human bodily experiences, and projected as image schemata. As mentioned in Section 8.2, Waterworth [2] outlines a design for a web browsing environment – a personal space (StackSpace) – that was informally based around considerations of human bodily experiences in real, physical spaces.

In adopting the experiential approach, a valuable source of design insights is that of language. How do users talk about their experiences? Utterances can be gathered at two stages of the design process: user requirements analysis early on and, later, as corroboration that a particular design is producing the kind of experiences the designer intended. It could be argued that we cannot effectively describe experiences with words but, as Samuel Beckett remarked, they are all we have. The approach to understanding these words is somewhat akin to psychoanalysis; we are looking for the unconscious structures (image schemata) that lie behind the chosen way of describing an experience.

The traditional approach to HCI design uses metaphor to communicate the functionality of the system to the user. The designer draws on users' experiences in another domain to assist their understanding of the system. As Erickson [17] has pointed out, this implies that designers know what the system really is. Despite its problems this approach has been successful in encouraging the widespread use of computers, at least for certain classes of application. The experientialist approach to design also draws on users' prior experiences, but there are several fundamental differences. First, from the traditional perspective, metaphors are useful (usually) but not essential. A traditional user interface metaphor can always be paraphrased into a literal interface. From the experientialist perspective, however, metaphoric projection is essential to the way people make sense of the world, including a user interface. Secondly, that metaphoric projection is essential to sense-making does not mean that we live in a world of metaphors. If we design from an experiential perspective, this does not mean that the interface need be a virtual world of metaphoric objects. Such a world is more likely to be the outcome of the traditional approach. Experientialism can, however, provide the basic elements of a natural and flexible HCI design pattern language (cf. [27]).

Even though I consider the experientialist approach to user interface design to be a new approach, the experientialist theory of meaning has already attracted attention in fields related to user interface design. For instance, Clay and Wilhelms [28] present a linguistic interface for placement of three-dimensional objects which focuses heavily on spatial relationships as discussed by Lakoff and Johnson. Maglio and Matlock (Ch. 9 of this book) demonstrate the usefulness of an experiential analysis in understanding how people conceptualise their explorations of the web.

There has also been criticism directed towards the experientialist view of metaphor. Although he recognises some merits of experientialism, Coyne [29] claims that Lakoff and Johnson put too strong an emphasis on the primacy of bodily experience and that there are non-embodied and non-spatial uses of concepts like containment and balance. However, Coyne's criticism seems to illustrate, rather than contradict, Lakoff and Johnson's main point; that is, that we project our spatial experiences (embodied as image schemata) to abstract, non-spatial domains of experience.

8.4.2 SchemaSpace: An Experientialist Environment

Andreas Lund is currently engaged in a more thoroughgoing attempt at the experiential design of an environment called SchemaSpace [25]. The approach can also be seen as a development of the idea that HCI design is mostly a matter of sensual or perceptual ergonomics rather than the "cognitive ergonomics" that follow from the traditional, cognitivist approach [23, 24]. SchemaSpace is a three-dimensional virtual environment in which a potential user may organise and browse a collection of references to information sources, located on the web or elsewhere. As such SchemaSpace is a personal information space [2].

How should an information space like this be designed? An answer to this question, from a traditional point of view, would in part be formulated in terms of functionality and ways to convey that functionality to the user through the system image. If we instead try to answer the question from an experientialist point of view we first have to reformulate the question as: *what kind of experiences does the user want to get from the interface?* By posing the question this way we put emphasis both on the designer's role as a creator of meaningful experiences and on the role of the user interface as a source of meaningful experiences.

The intention with SchemaSpace has been to design the interface in such a way that it allows the user to have four different kinds of experiences that each informs the user about different qualitative aspects of the information space:

- *Distinctiveness* – which of the information references belong together, for example, fall under the same subject or category?
- *Quantity* – how does the number of references in a sub-collection compare to other sub-collections found in the information space?
- *Relevance* – given that a collection of information references belong together, of what relevance is each individual reference in relation to the subject or category?
- *Connectedness* – how do different sub-collections of references relate to each other?

Obviously, these qualitative aspects are by no means all-encompassing and we can see a whole range of other aspects that a user might want to experience from a personal information space. One important dimension not yet addressed by SchemaSpace is that of time; items should show their age, as they tend to do in the real world [2]. However, our purpose here is not to design the ultimate application, but rather to illustrate what we understand to be features of practical experientialist design.

As already mentioned, a meaningful experience is an experience which allows for structuring by means of metaphoric projection of image schemata. Thus, one important step in the design process is to identify image schemata that are associated with the qualitative aspects of the information space we want the user to experience. This identification is by no means arbitrary, on the contrary it ought to be informed by empirical enquiries.

Distinctiveness Through Containment

In the particular instance of SchemaSpace described here we have about 300 different web references to information on very disparate subjects, ranging from modern literature, via architecture, to computer graphics. Even such a relatively small collection calls for some kind of categorisation, a way to organise and order the information in sub-collections consisting of references belonging to the same category. Put differently, we have to provide for the possibilities of experiencing *distinctiveness*, that is, an experience that informs the user that some information references are in some respect different from other references. In order to provide such an experience we have to identify an image schema that is involved in our general understanding of ordering objects and activities in our everyday life.

Our encounter with containment and boundedness is one of the most pervasive features of our bodily experience. We are intimately aware of our bodies as three-dimensional containers into which we put certain things and out of which other things emerge. Not only are we containers ourselves, but our everyday activities in general – and ordering activities specifically – often involve

Figure 8.7 Distinctiveness through containment and quantity through verticality.

containment in some respect: we live in containers (houses, shelters, etc.), we organise objects by putting them in different containers. Our frequent bodily experiences of physical boundedness constitute an experiential basis for a *container schema.*

A plausible way of providing for the experience of distinctiveness is to present the information references that belong together in a way that allows for a projection of a container schema. There are countless ways of visualising containment and folders and rooms are probably the most familiar user interface containers. However, in our design of SchemaSpace we have as much as possible avoided elements which are – like folders and rooms – heavily metaphorically laden, in order to stress the experientialist features of SchemaSpace (although it is our strong belief that experientialist design need not by necessity exclude "ordinary" user interface metaphors). Instead, the elements of SchemaSpace consist largely of simple geometric shapes that are not closely associated with a specific source domain. We have chosen to visualise containment by means of semi-transparent cones (see Fig. 8.7). A cone contains information references visualised by stacks of slices (similar to StackSpace [2]), each with a descriptive textual label. By using semi-transparency it is possible to see that a cone actually contains information references, at the same time as it is apparent that they are bounded by the cone and are thus distinct from other references.

Quantity Through Verticality

Each cone contains a sub-collection of the totality of information references in SchemaSpace. Some of the sub-collections will contain more or fewer references in comparison to other sub-collections. Even though the cones are semi-transparent, viewed from a distance in the three-dimensional environment it will be difficult to judge the quantity of each cone. In order to provide for a meaningful experience of the quantity of each cone's contents we have to identify an image schema which is involved in our general understanding of quantity. Our basic experiences of quantity are closely associated with verticality (examples from [19]):

> Whenever we add *more* of a substance – say, water to a glass – the level goes *up*. When we add more objects to a pile, the level *rises*. Remove objects from the pile or water from the glass, and the level goes *down*.

Spatial experiences of the this kind constitute an experiential basis for a *verticality schema*, a schema which by means of metaphoric projection plays an important role in our understanding of non-spatial quantity. Our tendency to conceptualise quantity in terms of verticality reveals itself in everyday language used to talk about quantity:

> The crime rate kept *rising*. The number of books published *goes up and up* each year. The stock has *fallen* again. You'll get a *higher* interest rate with them ... .

In our design we have tried to exploit this verticality aspect of quantity. As shown in Fig. 8.7 the cones in SchemaSpace vary in height. The larger cones have a larger number of references inside compared to the shorter cones. Our intention has been to combine the container and the verticality schema in order not only to express quantity, but also to strengthen the experience of cones as containers of information references.

Degree of Relevance Through Centrality

As already mentioned, one of our goals has been to provide for the experience of distinctiveness. Even if a sub-collection constitutes a unity by virtue of belonging to the same category or subject, different references within a sub-collection may be of different importance or relevance in relation to that particular subject.

As pointed out by Johnson [19]:

> our world radiates out from *our bodies* as perceptual centers from which we see, hear, touch, taste and smell our world.

We also have very basic spatial and physical experiences of centrality as a measure of importance and relevance. Not only is that which is near the centre (the body) within our perceptual reach, but we also experience our bodies as having a centre and periphery where the central parts (trunk, heart, etc.) are of greatest importance to our well-being and identity [21].

In order for a potential user of SchemaSpace to experience some references as more important and relevant in relation to other references within a cone, we exploit a *centre-periphery schema*, which has its experiential grounding in perceptual experiences of centrality mentioned above.

As seen in Fig. 8.8, stacks of information references are organised along an arc. In those cases where there are a lot of references within a cone, the arc will eventually be closed and form a circle centred around the vertical axis of the

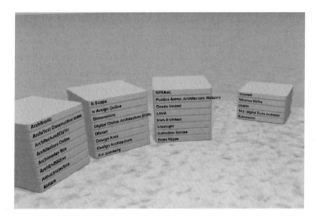

Figure 8.8 Degree of relevance through centrality.

cone. Information references can, however, be placed at varying distances from the centre; that is, some references will perceptually be closer to the centre and some will be more peripheral (see stack to the right in Fig. 8.8). Our goal with this arrangement is to invoke a metaphoric projection – on the part of the user – of the centre-periphery schema in order to experience those references which are perceptually central as conceptually central.

Connectedness Through Linkage

Finally, we want the user to experience that some sub-collections of references are related to each other, even though they are distinct from each other. In SchemaSpace we have a collection of references on the subject of the virtual reality modelling language (VRML). But we also have two categories with references to information on VRML-browsers and VRML-worlds. These two categories may be considered as distinct from VRML information in general, but not in the same sense as information on the writer Paul Auster is distinct from information on architectural magazines. There is a connection between general information on VRML and VRML-worlds and browsers, that does not exist in any obvious way between Paul Auster and architecture. In order to provide for an experience of this kind of connectedness we exploit a *link schema*. The link schema is often involved in our understanding of relations and connections of different kinds, not only physical connections, but also more abstract, non-physical connections like interpersonal relationships.

In SchemaSpace cones are connected with a path-like link if the sub-collections contained in the cone are considered to be connected, as is the case with VRML in general and VRML-browsers and worlds (see Fig. 8.9). As with the centre-periphery example above, our goal with this arrangement is to provide the user with perceptual cues that allow for structuring by means of projection of a certain schema, in this case the link schema. Of course, linking to show

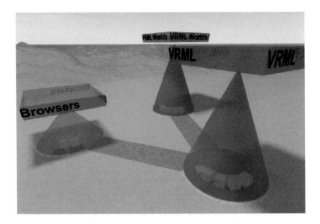

Figure 8.9 Connectedness through linkage.

connectedness depends on some knowledge of what is connected and why. This can be approached in several ways which will not be described here.

From the experientialist view, what is needed in HCI design is for the interface to be a source of experiences, designed in such a way that the experiences generated may be structured by the projection of basic bodily image schemata. As in poetry, metaphor is used to create an effect in the experiencer. What the resultant interface (or poem) means, *what it is* for a given user, depends on his or her unconscious reactions to the structures provided. If the interface feels right for its purpose, it is successful. No designer can know what the system really is, in general. It is what it means to individual users (often as members of social groups) and, like life, it means what it is experienced to be.

8.5 Presences and Concealment

Even if we don't want to be personally identifiable on the web (in the same way we sometimes don't want to be individually known to strangers in the real world) we currently don't even have presence as anonymous people in cyberspace. All we can tell, sometimes, is how many people have visited a site before us. Or rather, how many visits have been made to the site. For cyberspace to become real, we need a sense of people's presence (and absence), with suitable protection for privacy – if that is possible.

A limited sense of what shared presence in the web would bring is provided by experiences in "multi-user dungeons" (MUDs) and the internet relay chat services. The 1995 book by Turkle [30] gives an insight into those worlds, although the MUD and chat users are probably not typical of web users. Specifically, they are self-selected for their interest in role playing and/or a need to alleviate real life loneliness.

However, these environments, in so far as they use text descriptions of space rather than three-dimensional graphics, are also quite unlike a shared, as-if-real, virtual world. As already mentioned, a textually-described world of the kind investigated by Dieberger (for example, Ch. 3 in this book) is an unlikely candidate as a medium for shared exploration of information space. This is because conceptualising and navigating such a space is so demanding that little mental capacity remains actually to deal with the information located during navigation. The trend in the navigation of information spaces, and indeed in HCI in general, is to shift the burden of dealing with the environment from conceptual, linguistic processing and conscious decision making to direct perceptual processing within automatised sensori-motor behaviours. In other words, to allow people to deal with information while minimising the extent to which they have to deal consciously with information about how to deal with information.

Turkle [30] points to the ease of adopting multiple personae in cyberspace, to present the face we choose to present rather than the real-life person we have become over the years. This can be seen as partial or selective presence. We can think of degrees of presence, from totally concealed (invisible), through anonymous (featureless) but visible, to articulated personae one of which might be a representation of our real-world personality.

Should we be able to choose how we appear to others? Should we be able to appear present when we are not, and not present when we are? False presence arises when we appear to be somewhere, but are actually elsewhere. Each of us wants to know as much as possible about others, but to control what is known about ourselves. We achieve this in the real world by limiting the time we spend, or the visibility we have, in public places, retreating to personal spaces when we feel the need. Multiple personae multiply the scope for deception, and the creation of personae is, of course, much easier in the virtual world than the real one.

As more and more people migrate to cyberspace, both the amount of information and the number of sources of information multiply. But human attention is still singular. We are each aware of only one thing at once. I can watch television or read a book, but I cannot do both at any particular moment (of course, I can switch between the two, and if the rate of information transfer from the television is typically low I won't miss much).

In the same way, even though many browsers now offer some support for following multiple threads, we can only really attend to one thing at once, even with multiple display windows. This is how we can be said to follow links – to navigate – at all, and to be lost in cyberspace when we lose our (singular) way. This also helps give us the presence that is currently unrepresented in the web. In cyberspace we are where our attention is focused, but we have no presence until we are visible in public.

8.6 Conclusions

The web differs in several major respects from the hypermedia systems that were the focus of so much premature research in the 1980s. Like a capital city in a developing country, it is large and growing very rapidly, both in the amount of information available and the number of inhabitants. All those people use the same system, rather than having their own copy; in other words, they truly co-habit cyberspace. And no-one is designing it as a whole. Rather, we operate "locally" by introducing innovations that may or may not catch on in the electronic world-at-large. The design of such landscapes and features is more appropriately based on notions of meaning as experience, rather than traditional ideas of meaning as functionality conveyed through HCI models, since we do not know what the function of shared information landscapes might be. But such a situation is natural for us, since we don't know what the function of the real world is.

Individuals need personal space to make sense of the information they collect from the world at large. People (like all animals) are naturally equipped to deal with three-dimensional spatial environments without it imposing a heavy burden on their scant attentional resources. In fact, three-dimensional interactive environments are such a powerful way of presenting collections of information just because they allow people to explore virtual worlds of information in the same ways as they explore the real world, whether this is an individual activity or in groups. Views of spaces can serve as a powerful mechanism for social interaction because they can be compared, contrasted, and exchanged. Groups interacting in real time require a sense of presence in shared three-dimensional space.

In the future, people will be represented with degrees of presence, and we will search for each other in cyberspace, not only for the things we have created. This will allow us to behave more naturally as the social animals we usually are. Personae will multiply but attention (and thus true presence) is still singular. To be truly present implies to be communicable with, and visible. When we have such presence the web will gain social context naturally.

Experiential design captures basic, unconscious, animal reactions to physical environments and introduces them to shared virtual landscapes. We can design appropriate tools and environments, just as we can design churches, cinemas and houses, but we do not design societies or social behaviour. We are social (and spatial) by nature, not design.

Acknowledgements

The information islands model was originally designed in collaboration with Gurminder Singh of KRDL (formerly ISS), Singapore. The section on information islands, vehicles and views is a modified version of parts of [9]. StackSpace and some associated ideas were developed during a short-term research fellowship at BT Labs in England. At this department, Andreas Lund has done all the work on SchemaSpace, and has made large contributions to the sections of this chapter that deal with the experiential approach to HCI design. Eva Lindh provided insightful comments that helped me make the text less unclear. David Modjeska, on secondment from the University of Toronto, has provided useful feedback and stimulating discussions on this text and on related research issues.

References

1. Waterworth, J.A. and Chignell, M.H. A manifesto for hypermedia usability research. *Hypermedia* **1:** 202–205, 1989.
2. Waterworth, J.A. Personal spaces: 3D spatial worlds for information exploration, organisation and communication. In *The Internet in 3D: Information, Images, and Interaction*, R. Earnshaw and J. Vince, editors. New York: Academic Press, 1997.
3. Harrison, S. and Dourish, P. Re-Place-ing space: the roles of place and space in collaborative systems. *Proceedings of ACM Conference on CSCW'96*, Boston, MA, 1996.
4. Meyrowitz, J. *No Sense of Place*. New York: Oxford University Press, 1985.
5. Waterworth, J.A. *Multimedia Interaction: Human Factors Aspects*. Chichester: Ellis Horwood, 1992.
6. Dieberger, A. The information city – a step towards merging of hypertext and virtual reality. Poster at *Hypertext '93*, 1993.
7. Waterworth, J.A. and Singh, G. Information islands: private views of public places. *Proceedings of MHVR'94 East–West International Conference on Multimedia, Hypermedia and Virtual Reality*. Moscow, 1994.
8. Waterworth, J.A. Viewing others and others' views: presence and concealment in shared hyperspace. Presented at *Workshop on Social Contexts of Hypermedia*, Department of Informatics, Umeå University, Sweden, 1995.
9. Waterworth, J.A. A pattern of islands: exploring public information space in a private vehicle. In *Multimedia, Hypermedia and Virtual Reality, Lecture Notes in Computer Science*, P. Brusilovsky, P. Kommers and N. Streitz, editors. Springer, 1996.
10. Musil, S. and Pigel, G. Virgets: elements for building 3-D user interfaces. *Proceedings of the Symposium Virtual Reality*, Vienna, , 1993. Also available as TR 93/13, Vienna User Interface Group, Lenaugasse 2/8, A-1080 Vienna.
11. Henderson, D.A. and Card, S.K. Rooms: the use of multiple virtual workspaces to reduce space contention in a window-based graphical user interface. *ACM Transactions on Graphics* **5:** 211–243, 1986.
12. Nagel, T. *The View from Nowhere*. New York: Oxford University Press, 1986.
13. Fishkin, K. and Stone, M.C. Enhanced dynamic queries via movable filters. *Proceedings of CHI'95*. New York: ACM, 1995.
14. Rifas, L. The Dataforest: tree forms as information display graphics. Report of the Workshop on Spatial Metaphors at *ECHT'94 – the European Conference on Hypermedia Technology*, Edinburgh, 1994.
15. Alexander, C. A city is not a tree. In *Humanscape – Environments for People*, S. Kaplan and R. Kaplan, editors. Ann Arbor, MI: Ulrich's Books, 1982, pp. 377–402.
16. Norman, D. Cognitive engineering. In *User Centered System Design*, D. Norman and S. Draper, editors. Hillsdale, NJ: Erlbaum, 1986.
17. Erickson, T.D. Working with interface metaphors. In *The Art of HCI Design*, B. Laurel, editor. Menlo Park, CA: Addison-Wesley, 1990.
18. Gentner, D., Falkenhainer, B. and Skorstad, J. Viewing metaphor as analogy. In *Analogical Reasoning: Perspectives of Artificial Intelligence, Cognitive Science and Philosophy*, D.H. Helma, editor. Dordrecht: Kluwer, 1988.
19. Johnson, M. *The Body in the Mind*. Chicago, IL: Chicago University Press, 1987.
20. Johnson, M. *Moral Imagination*. Chicago, IL: Chicago University Press, 1993.
21. Lakoff, G. *Women, Fire and Dangerous Things*. Chicago, IL: Chicago University Press, 1987.
22. Lakoff, G. and Johnson, M. *Metaphors We Live By*. Chicago, IL: Chicago University Press, 1980.
23. Waterworth, J.A. Virtual reality for animals. *Proceedings of Ciber@RT'96, First International Conference on Virtual Reality*. Valencia, 1996.
24. Waterworth, J.A. Creativity and sensation: the case for synaesthetic media. *Leonardo* **30**(3), 1997.
25. Lund, A. and Waterworth, J.A. Experiential design: reflecting embodiment at the interface. *Computation for Metaphors, Analogy and Agents: An International Workshop*, University of Aizu, 1998.
26. Conklin, J. Hypertext: an introduction and survey. *IEEE Computer* **20**(9): 17–41, 1987.
27. Alexander, C., Ishikawa, S. and Silverstein, M. *A Pattern Language: Towns, Buildings, Construction*. New York: Oxford University Press, 1977.
28. Clay, S.R. and Wilhelms, J. Put: language-based interactive manipulation of objects. *IEEE Computer Graphics and Applications* **16**(3), 1996.

29. Coyne, R. *Designing Information Technology in the Postmodern Age: From Method to Metaphor.* Cambridge, MA: MIT Press, 1995.
30. Turkle, S. *Life on the Screen.* New York: Simon & Schuster, 1995.

Chapter **9**

The Conceptual Structure of Information Space

Paul P. Maglio and Teenie Matlock

Abstract

In this chapter we examine how people think about the information space of the World Wide Web. We provide empirical evidence collected in interviews with beginning and experienced web users to show that much of people's conceptual experience of the web is metaphorical and understood through the process of conceptual integration. We argue that designers of tools for navigation and collaboration in information space should consider how people experience web space, including the natural tendency to metaphorically construe information space in terms of physical space.

9.1 Introduction

Navigation is a basic part of human experience. Walking across a parking lot, driving to work, and searching for an item in a store or library all involve navigation: moving from one point to another in physical space. Empirical research has shown that while navigating, people rely on different types of knowledge: landmark knowledge, route knowledge, and survey knowledge [1, 2]. It has also been argued that people incorporate knowledge based on certain organizational principles [3, 4] and awareness of certain elements, such as paths, landmarks, districts, nodes, and edges. Navigation in electronic worlds has been compared to navigation in the physical world (e.g. [5, 6]), but navigation in information spaces is not as well understood as navigation in physical space. Although recent research shows interesting results, much emphasis lies in how people perceive the environment, rather than how people conceive of the environment [7].

In this chapter, we argue that people rely on experience in physical space to structure experience in virtual information spaces such as the World Wide Web (WWW). Specifically, we are interested in people's *natural* conception of information spaces. We report the findings of a web-use study conducted in 1996 in which we analysed how people talk about the web to get at their natural conception of information space. We found that both experienced and inexperienced web users naturally talk about the web in consistent ways. For instance, people see themselves as metaphorically moving toward information, rather than information as moving toward them. We also found some differences between experienced and inexperienced web users in the way they talked about web activities. In the end, we argue that (a) the particular language people use is based on conceptual metaphor and is motivated by basic image schemata, which emerge from natural embodied experience (e.g. [8, 9]); and (b) web users' experience is structured by conceptual integration [10–13].

Before presenting our data and argument in detail, we first discuss some prior research concerning the way people think about the web, along with some background on metaphor and thought.

9.1.1 How People Remember the Web

Based on data collected from people asked to recall specific web searches, Maglio and Barrett [14] argued that web navigation is conceived in terms of a cognitive map similar to a cognitive map of physical space, that is, in terms of landmarks and routes (e.g. [15]). In this study, experienced users searched the web for answers to specific questions. To identify key cognitive aspects of their activities, users were first asked about their plans, and their behaviour was tracked while they searched. Then a day later they were asked to recall the steps they had taken in each of their searches the previous day, and finally to retrace their steps. Participants were not warned on the first day that recall would be required on the second day. This method enabled Maglio and Barrett both to chart behaviour to uncover search tactics (using the behavioural traces) and to extract some of the structure of their internal representations (using the recall data).

The data showed that participants recalled only a few of the sites they visited. Specifically, they remembered key nodes that led to the target information. These nodes were called *anchor points* by analogy to the notion of anchor points in the cognitive map literature [16]. An anchor was defined as a node along a search path from which there is an unbroken sequence of links on successive pages that lead to the goal node (i.e. no URLs need to be typed in or explicitly recalled). Once traversed, anchor points are recognized as lying along the path to the goal – even if the same path is not followed to the goal in every case. For the participants in the study, searching on the second day often meant finding anchors encountered on the first day, rather than finding paths found on the first day.

A second observation that emerged from the behavioural data is that individuals relied on personal routines when trying to find information. For instance, some participants routinely used a particular search engine, such as AltaVista, whereas others routinely used a particular hierarchical catalogue, such as Yahoo! It is not merely that these searchers preferred to use one approach over another, but that they conceptualised their search tasks in terms of their favourite routines. It often did not matter what was actually done on the first day, the searchers remembered searching as if their personal routines had been followed. On the analogy to cognitive maps of physical space, personal routines correspond to the familiar routes that an individual uses to get from one landmark (or anchor point) to another.

If people mentally structure web use in this way, tools for web navigation ought to present the web in this way. Because individuals tend to use the same search patterns over and over, and because they recall their searches in terms of their standard patterns – almost regardless of what they actually did – Maglio and Barrett [17] built a personal web agent to identify repeated search patterns and to suggest similar patterns for new searches. Because people focus on key nodes or anchor points when recalling their searches, and because these structure memory for the searches, Maglio and Barrett [17] also built a web agent to identify the key nodes in finding a piece of information, and to maintain personal trails in terms of these.

9.1.2 How People Talk About the Web

The key to designing information navigation tools lies in discovering how people naturally conceive of information spaces. Technically, the web is part of a network of geographically distributed machines connected via wires. The information accessible by users of this physical network is organised in a conceptual network of hyperlinks among documents. Despite this actual structure, people's conceptual structure of the web is rather different.

Matlock and Maglio [18] found that web users often refer to the web as a multidimensional (most commonly two-dimensional) landscape. Obtaining information in this landscape is expressed as traversing interconnected paths toward locations that contain information objects, such as user homepages and commercial catalogue sites. Users say things such as, "I went to his homepage", and "I came back to where I saw that picture". Some of these information objects are talked about as two-dimensional and others, as three-dimensional; for instance, people say "in Yahoo!" which suggests a three-dimensional container, and "at AltaVista" which suggests a point on a two-dimensional plane.

In a follow-up study, Matlock and Maglio [19] asked experienced and inexperienced web users to judge the sensibility of sentences containing metaphorical language (specifically regarding motion) about obtaining information on the web. Using a scale of one to seven, participants rated the sensibility of sentences containing verbs of motion. For instance, "John went to a new web site today"; "Do you want to climb up to the UCSC home page?"; and "I waited for the information to come to me". Sentences in which the web user was viewed as an agent, actively moving along a horizontal path, were rated as significantly more sensible than those in which the web user moved up or down, and as significantly more sensible than those in which the web user was passive. These results suggest that both experienced and inexperienced participants have clear and consistent ideas about how motion does and does not occur on the web.

Though there are many ways in which people might talk about the web (see [20]), the fact that they naturally talk about it using particular metaphors is no accident. As Lakoff and Johnson [21], Gibbs [22] and others have argued, such language is motivated by metaphorical thought.

9.1.3 Metaphor and Thought

Prior to the seminal work of Lakoff and Johnson [21], metaphor was generally seen as nothing more than a literary device. Lakoff and Johnson radically changed this misconception, offering compelling arguments to show that metaphor is an integral part of thought and action:

> Metaphor is typically viewed as a characteristic of language alone, a matter of words rather than thought or action ... that metaphor is pervasive in everyday life, not just in language, but in thought and action. Our ordinary conceptual system, in terms of which we both think and act, is fundamentally metaphorical in nature ([21], p. 3).

Subsequent work in cognitive linguistics and psychology has continued to offer theoretical and empirical evidence to show that metaphor is ubiquitous and serves many functions relative to conceptual experience [9, 23, 24]. One of the functions of metaphor is that it helps people think about relatively abstract conceptual domains in terms of relatively concrete domains [22]. For instance, spatial concepts are often helpful when reasoning about time [25]. On the standard view of metaphor, a relatively concrete source domain maps on to a relatively abstract target domain. Consider the often-cited metaphor THEORIES ARE BUILDINGS. In this metaphor, elements of the conceptual structure of BUILDINGS (source domain) map onto THEORIES (target domain). Linguistic evidence to support the existence of this metaphor includes statements such as: "You need empirical evidence to buttress your arguments", "The foundation of the theory is shaky", "His entire theory was toppled by the claim that Basque is a language isolate", or "Construct a different argument to support your theory". It makes sense that this mapping progresses from BUILDINGS to THEORIES because buildings are common in everyday experience. In western culture, buildings serve an important function: namely, people live and work in buildings. In addition, buildings offer protection from adverse effects of nature, and so on. Theories, by contrast, are important in the academic or philosophical world, but not commonplace to most people.

Another example of a metaphor is the MIND IS A CONTAINER. In this case, the concrete conceptual domain of CONTAINER maps on to the more abstract conceptual domain of the MIND. Hence, we understand the mind as a storehouse. Ideas can enter the storehouse, can be processed there, stored in a specific location, or even misplaced. Linguistic evidence for this metaphor includes expressions such as "The thought suddenly came into my head", "It's in the back of my mind", or "She lost her senses". This metaphor underlies many psychological theories (see [22] for a discussion).

As pointed out by Coulson [26], the standard approach to metaphor arose in part to account for simple examples of analogical thinking, such as TIME IS SPACE. As such, the approach is parsimonious but cannot account for complex mappings requiring some degree of sensitivity [27]. Moreover, the standard approach falls short with respect to productivity: why do only certain elements of the source domain map onto the target domain? Consider THEORIES ARE BUILDINGS. As noted, foundation and support map onto the target domain, but doors and windows do not. Recent approaches have attempted to solve this problem by suggesting that there are a variety of types of metaphors, including primitive and compound (e.g. [28]).

Although more recent approaches to metaphor diverge from the standard model with respect to issues of mapping complexity, there is agreement that metaphor plays a central role in structuring how people think. But metaphor is only part of the story. Another part is image schemata, basic pre-conceptual structures that arise from embodied experience. Formed early in development [31], image schemata structure both metaphorical and non-metaphorical thought [8, 9, 29, 30].

Image Schemata

Daily life includes active physical motion toward objects or destinations (concrete or abstract): going to the door to let the cat out, walking or driving to work, and reaching out to grab a pencil or pick up the telephone. Life also includes abstract motion toward goals (abstract destinations): working to get a promotion, writing a dissertation to obtain a degree, and saving money for a trip. Each of these actions involves the image schema TRAJECTORY (also referred to as SOURCE-PATH-GOAL), comprised of a starting point, an end point, and a path between the two. Another image schema is CONTAINER, which arises out of bodily experience: swallowing things, entering and remaining in buildings, and so on. As we will see, these image schemata figure prominently in how people view obtaining information in the world and on the web.

In what follows, we explore the nature of people's metaphorical conception of the web. A specific goal of the study is to examine the presence and frequency of language reflecting underlying image schemata. We believe these elements structure much of users' conceptual structure of the web. Furthermore, as Lund and Waterworth [32] have claimed, an important step in the design process is identifying image schemata to provide a more experientially based environment for the user (see also [33]). In the current study, we investigate how users with varying levels of expertise talk about the web. We first describe a study that elicited verbatim reports from both experienced and inexperienced web. We next discuss reasons people use the metaphors they do, and finally, some implications of our results for the design of tools for navigating and collaborating in information spaces.

9.2 Study: Users Describe Web Experience

The purpose of this study was to further explore how people think about the web in natural settings. We looked specifically at how people conceive of the actions taken while using the web; for instance, to what extent users see themselves actively moving through space and to what extent they focus on the physical environment. We also wanted to observe differences between beginning and experienced users.

We hypothesised that beginners would talk about their experiences using the web in terms of the physical actions they performed more than experienced users would because beginners are likely to have only a partial understanding of the web domain. Along the same lines, we hypothesised that experienced users would generate more metaphorically consistent utterances than beginners would.

We analysed the data both quantitatively and qualitatively. In the quantitative analysis, we counted utterances of various types to compare beginning and experienced web users. In our qualitative analysis, we followed a method similar to that of Raubal et al. [34], who analysed the image schematic structure of talk about wayfinding in airports.

9.2.1 Method

Twenty-four undergraduates at the University of California at Santa Cruz took part, including 13 males and 11 females. All were native English speakers except five fluent bilinguals.

Participants first completed a questionnaire about their prior experience using computers and the web. They were asked about length of time using the web (e.g. one month or less) and hours per week used. Participants were then seated in front of a computer that was running the Netscape Navigator browser, which displayed the homepage for the University of California, Santa Cruz. They were then instructed to click on whatever icons or hyperlinks appeared interesting and to continue doing so for 5 minutes. The experimenter was extremely careful to avoid language that would bias the participant to think of the web metaphorically, such as, "Go to that page".

After each participant had spent sufficient time getting used to the task and experiencing the environment, he or she was instructed to look at a new domain: Yahoo!, a well-known catalogue in which information is organised hierarchically. The participant was again instructed to use the mouse to gain access to information that seemed interesting and to continue to do so for 5 minutes.

A tape-recorded interview followed the web session. To begin, the experimenter prompted the participant: "Tell me what you just did using as much detail as possible". If a response was not immediately forthcoming, the experimenter began, "Tell me what you did first", and so on.

9.2.2 Results

Participants were separated into two groups according to self-reported web experience: 12 beginners reported under 6 months of web use, and 12 experienced users reported over 6 months of web use.

Coding

In coding the data, we distinguished among seven kinds of verb phrases (verbs and conventional verb-preposition expressions) that correspond to seven kinds of web actions (see Table 9.1). For instance, the sorts of phrases coded included verbs such as "clicked" and verb-preposition combinations such as "went to". These were chosen based on discourse about the web that we had collected previously [18]. Only utterances that referred to what the participant did while using the web were assigned to one or more of the categories shown in the table. For example, statements such as "I'm on a tight budget", or "Using the web is pretty fun" were not included in our analysis.

In analysing utterances, we wanted to be careful not to confuse language referring to the information space of the web with language referring to the user interface of the web. Thus, in looking at verb phrases, we distinguished among three general types of action: (a) *outside actions,* which reflect the user's experience with things external to the web (such as typing on the keyboard, using the mouse, and clicking on browser icons); (b) *inside actions,* which reflect the user's experience conceptually within the web (such as going to a web page, and following a link); and (c) *miscellaneous actions,* which cannot be definitely classified as either outside or inside. Expressions such as "I *typed* something", "I *clicked on* the grapes icon", or "I *pressed* buttons" were coded as outside actions.

Expressions referring to inside actions were split into three types: TRAJECTORY, CONTAINER, and information actions. Motion of the user along a path in web space highlights the TRAJECTORY schema, such as "I *went into* this thing called Yahoo", "I couldn't *get back to* where I was", or "It *brought me to* the anthropology page". Transfer of information along a path from computer to user also highlights the TRAJECTORY schema, as in "It told me" and "It said". Sometimes a web site is talked about in terms of a container, instantiating the CONTAINER schema, as in "Yahoo! *contained* some cool stuff, or "Yahoo! *had* what I wanted". At other times, the web is described as a general information resource similar to a library or a phone book, as suggested by expressions such as "I *looked up* Chewbacca".

Table 9.1 Verb coding scheme

Category	Examples
Outside	click, press, type, scroll
Inside	go, follow, have, look up
TRAJECTORY	go, come, bring, follow
User Agent	go, follow
Web Agent	bring, come up, bring, show
CONTAINER	have, contain
Information Action	look for, lookup, search
Miscellaneous	look, see

The TRAJECTORY category was divided into utterances in which user is the agent and those in which web is the agent. Agency refers to who or what initiates and undertakes action. In some cases, the user is agent, as in "I *went* ...", whereas in others, the web is agent, as in "It *took me to* ...".

The miscellaneous category was used for verbs that could not be obviously classified into either of the other categories. This group mainly contained expressions beginning with "I *saw* ..." or "I *looked at* ..." because it is unclear whether these describe visual perception of the screen (an outside action) or visual perception of objects in web space (inside action).

Finally, note that we also could have examined use of prepositions to help code for TRAJECTORY and CONTAINER. For instance, *through* and *to* imply TRAJECTORY, and *in* suggests CONTAINER. For the present study, we looked specifically at verb phrases. A more thorough analysis would certainly include prepositions (for example, "I can't remember if I found information *in* Yahoo! or *inside* AltaVista") and nouns ("I took a direct *route* from the UCSC site to my homepage") as well (see [34]).

Qualitative Results

We first conducted a qualitative analysis of the data. To get a feel for the data and our coding scheme, consider the following utterance, which is fairly typical of beginning web users (participant 4):

> ... I clicked on uh grapes ... and it brought me to um ... this place where they had choices and then I clicked on bookstore ...

Note the presence of two outside actions ("I *clicked* on ..."), an instance of TRAJECTORY in which the web is agent ("it *brought me to* ..."), and an instance of CONTAINER ("place where they *had* choices"). In this utterance, the user clicks on an icon on the screen, is taken to a new location, and then she clicks again.

Now consider an utterance produced by an experienced web user (participant 14):

> ... I went to net search because that seemed like a good wholesome opportunity for going somewhere else ... I probably typed something and it told me I couldn't do it, so I dunno, I just went and clicked around a whole bunch ...

Here we see three instances of outside actions ("typed", "clicked", and "do"), two instances of TRAJECTORY with the user as agent ("went", "going") and one instance of TRAJECTORY with web as agent ("told"). (The verb "seem" and the second instance of "went" were not coded because they do not refer to actions taken while using the web. The use of "went" simply means "proceeded to"). In this case, the user's report blends different types of actions: metaphorically going somewhere, typing something, receiving information, and clicking.

These sorts of responses are representative of what experienced and beginning web users do: they both mix outside actions with actions inside the web's information domain. Nonetheless, we observed some interesting differences. For the beginner, the web can function as a kind of conveyance that moves the user ("brought me to"), but for the expert, the web is a kind of roadway on which the user moves ("I went"). In addition, for the beginner, the web passively contains information ("had choices"), but for the expert, the web actively provides information ("it told me").

Consider the report of another beginner (participant 2):

> ... I went into the um Brian's tattoo something or other, but when I clicked into it, it said that like it was gonna show tattoos of his body and like front, side, whatever ... it had objects to click on, and I clicked on em and there was no pictures ...

In this report, we see one instance of TRAJECTORY in which the user is agent ("went into"), and two in which the web is agent ("it said", "was gonna show"). We also see two outside actions ("click"), and one CONTAINER ("had"). As in both previous cases, outside actions are mixed with inside actions. Like the first beginner, this one refers to a web site as a container. Unlike the first beginner, however, this one also refers to the web as a kind of roadway along which people can travel ("went") rather than as a kind of conveyance ("brought me to"). For this beginner, as for the expert, the web actively provides information ("it said").

The utterance from participant 2 illustrates something our coding scheme does not recognise: the novel use of "click" in the verb phrase "click into". Whereas the verb "click" refers to an outside action, the preposition "into" specifies an inside location. Usually the verb "click" is followed by the preposition "on", and the construction refers to an icon or hyperlink visible on the screen. In this case, however, "click into" refers both to something visible on the screen and also to something contained in the information space of the web. We will return to this point in the discussion of conceptual blends.

Finally, consider a second expert's response (participant 23):

> ... I couldn't get through. I returned to the first page I started on and selected travel.

In this case, we note two instances of TRAJECTORY in which the user is agent ("get through", "returned") and one outside action ("selected"). The path is blocked ("couldn't get through"), and previous steps were retraced ("returned").

In summarising our qualitative results, we can see that both beginners and experts use the same sort of language overall. In reporting on their experience using the web, most participants mixed language about actions they did outside web space with those they did inside web space, especially actions reflecting the schemata TRAJECTORY and CONTAINER. In talking about the web, people also described the web as moving the user, or described the user as moving on the web. Their verbatim reports also suggest that the web can simply contain information, or it can actively convey or provide users with information. In any event, people seem to prefer to talk about their experience in using the web in more familiar terms, such as physical motion, physical actions, and physical containers.

Quantitative Results

The total number of verbs in each category was computed for beginners and for experts, as shown in Table 9.2.

Because we collected frequency data, χ^2 was used to compare beginners and experts along each of the seven action categories. As shown in Table 9.3, significant differences were obtained for TRAJECTORY versus outside actions, for user agent versus web agent, and for CONTAINER versus all other verbs. Thus, experts used the TRAJECTORY verb phrases rather than outside action verbs more often than beginners. Within the TRAJECTORY category, experts reported themselves as agent (i.e. actively moving through information space) instead of web as agent (i.e. information moving through web to user) more often than beginners did. By contrast, verbs phrases of the CONTAINER type were reported more by beginners than experts.

Overall, all web users reported a similar experience while using the web. Both beginners and experts talked about their experiences as if they had been moving from place to place though in fact they had not gone anywhere. The data also revealed noticeable differences between experts and beginners. Beginners more often mixed in their experiences using the keyboard, mouse, and other elements of the physical (non-web) domain (e.g. "I clicked on ..." or "I typed in ..."),

Table 9.2 Verb coding scheme

	Beginners ($n = 12$)	Experts ($n = 12$)
Outside	54	26
TRAJECTORY	56	87
User Agent	37	79
Web Agent	19	8
CONTAINER	22	11
Info Action	30	42
Miscellaneous	24	20
Total	186	186

Table 9.3 Percentage of verbs in each category for each group. The χ^2 statistic compares the difference between groups

	Beginners	Experts	χ^2
TRAJECTORY vs outside	51%	77%	16.49**
User agent vs web agent	66%	91%	13.60**
CONTAINER vs all others	12%	6%	4.02*
Info actions vs all others	16%	23%	2.48
Miscellaneous vs all others	13%	11%	0.41

*$p < 0.05$; **$p < 0.005$.

whereas experienced users did not. In addition, beginners were more likely to refer to the web as a container than were experienced web users.

9.3 Discussion

All web users in our study consistently used metaphorical language when talking about the WWW. In particular, they used verb phrases referring to physical motion to describe their experience using the web. However, there were differences between the language of beginning and experienced web users. In what follows, we discuss reasons why people use metaphorical language when talking about the web, and discuss implications for the design of tools for navigation and collaboration in information spaces.

9.3.1 Agency and Web Use

Our data suggest that web users – even those who had never used the web – view web activity as traversal along paths. In particular, participants most often see themselves as the agent, initiating and actively moving along these paths (even for beginners; see Table 9.3). According to the data, less often is the user viewed as the passive recipient of information or as a passenger being transported in some sort of web vehicle. This suggests that the semantic property of agency is primarily viewed as something inherent in the web user, rather than something inherent in the web.

One reason the user might view obtaining information on the web as actively moving through space toward objects is because of the ease of information access. The most common way of moving from one web page to another is by clicking on hyperlinks or using the browser's back button [35, 36]. Much less often do web users type in full addresses to obtain information. Simply clicking on links and instantly seeing new information creates a sense of fluidity and hence, the illusion of motion. One way to test this hypothesis may be to systematically vary the delay between clicking on a link and the subsequent presentation of information. Results from such a test, especially if conducted with novice web users, will tell whether longer delays result in fewer utterances in which the user is the agent.

It seems natural to talk about information access metaphorically in physical terms. After all, obtaining information in a library, in a reference book, or by telephone involves directed action. Thus, the reason why users talk about the web in terms of physical space most likely lies in human embodied experience [8, 30, 37]. The way people experience the web or other information spaces is shaped by human activities in the real world. A large part of human experience involves physical activities, such as standing up, walking toward a location, reaching out, and grasping what is desired. From these recurrent patterns of activity, people develop image schemata, as discussed previously. Thus, it is reasonable to assume that because directed motion toward goals is part of our embodied experience, it naturally structures how we think about and interact in information spaces, such as the web.

9.3.2 Conceptual Blends in Information Space

We now return to our finding that novice web users mixed talk about the outside domain with talk about the inside domain more than experienced users did (see Table 9.3). Recall the utterances of participants 2 and 4. These and all inexperienced web users often mixed inside and outside actions, seemingly unaware of the fact that they were switching between them. Sometimes this sort of blending happened at the sentence level, as in "I clicked on [outside] grapes ... and it brought me to [inside]...". At other times, it occurred at the phrase level, as in "I clicked into it", in which the participant created a novel verb–particle construction. These results indicate that in using the web, people naturally integrate two or more domains to create something more than simply the combination of its parts.

Such conceptual integration (also known as "blending") is not unique to web activity or even to language use, as Gilles Fauconnier and others have demonstrated [10–13]. In this framework, there are not just two domains, as in standard metaphor theory [21], but multiple domains. Through a complex interplay of mapping, or projection, from one domain to another, an emergent structure arises. This structure is to some extent independent of the meanings afforded by the domains on their own. A blend emerges from two or more input spaces, a generic space, and a blended space. The best way to show how the mapping works is through an example of how people create novel meaning by blending domains. The example comes from Coulson [38]: two college students are up late at night studying for an exam. One student grabs a piece of paper, crumbles it up, and throws it towards a wastepaper basket. The other student grabs the crumpled piece of paper and also throws it toward the basket. The actions of the students develop into a game in which the paper is a "ball" and the trashcan is a "basket". The students' understanding of this activity as "trashcan basketball" arises through integrating knowledge about different domains. In this blend, trash disposal is one input space and the conventional game of basketball is the other input space. The blended space combines elements from both the input domains. Importantly, though it involves the incorporation of elements from both domains, the emergent structure in blended space differs in many respects from the two input domains.

An example closer to home may be seen in Fauconnier's [10] discussion of the computer *desktop* metaphor. He argues that conceptual integration can account for the complexity of this familiar metaphor. According to Fauconnier, the desktop metaphor is constructed on the basis of two separate conceptual inputs: (a) traditional computer commands, such as saving a file, and listing a directory; and (b) work in an office, including a desk, files, folders, and trashcan. To create the desktop metaphor, a cross-mapping occurs whereby computer files are mapped to paper files, directories are mapped to folders, and so on. General

knowledge – such as image schematic notions of CONTAINER and TRAJEC-
TORY – mediate the mapping. Structure is selectively projected from the inputs,
yielding a coherent, well-integrated, emergent structure specific to the blend.
What emerges from these mappings is a "world" in which a trashcan can sit on
the desktop, in which double clicking opens files or applications, and in which
objects are routinely dragged from one location to another. The integration is
completely novel, but at the same time it is meaningful to the desktop interface
user. Note that if the mapping from the office domain to the computer domain
were simple (i.e. creating no new structure), the computer desktop could be no
better than a real desktop: such an interface could only selectively mirror the
world.

We believe conceptual integration provides a nice account of how web users
think about the web. It provides a plausible explanation for how novice users can
understand and use the web. For example, a person who has never seen the web
can sit down at a computer, browse for awhile, have the feeling of shifting
between inside actions and outside actions (e.g. "click into"). Conceptual
integration also provides some insight into how experts talk about the web less in
terms of outside actions than novice web users: Experts seem to rely on the input
from the abstract web domain to a greater extent than they rely on input from the
physical browser domain.

Conceptual blend theory also integrates web users' conceptual information
much more effectively than would a standard metaphorical approach, which
would be limited to a single source domain and a single target domain (e.g. [21]).
Of course, a traditional metaphorical account can explain the obvious metaphors:
(a) WEB SPACE IS PHYSICAL SPACE, which reflects to how users view the web
as a place; and (b) OBTAINING INFORMATION IS MOVING THROUGH
SPACE, which reflects how users view themselves as moving along paths to
information objects. However, it fails to say anything about how web users
naturally blend inside and outside actions, or about how this tendency interacts
with metaphorical thought.

Finally, as Rohrer [39] argues, blending can also explain how people can
understand and incorporate other, higher level metaphors of cyberspace,
including the popular *information super highway*. There are two parts to
understanding this metaphor. People understand it as highway upon which
movement occurs, much in the same way the beginners and experts in our study
understood the web, and as a road through time that allows travel into the future.
Rohrer provides nice examples from headlines and news reports to show the
dual, blended nature of this metaphor, for instance, "Prime Minister rides the
info-highway", "Congress suffers wreck on info highway", and "AT&T stalled on
the info-highway". In each case, there is the notion of movement through
physical space blended with the notion of "movement" into the future.

9.3.3 Designing Information Interfaces

If metaphorical language in fact reflects metaphorical thought, and people naturally think of the web as a kind of physical space in which they actively move along paths, what might be the consequences for the design of information navigation and collaboration tools?

Shum [40] points out many potential uses for the concepts of physical space in the structuring and presentation of information, such as Euclidean distance in two or three dimensions, direction, orientation, and depth. Nevertheless, Shum also notes that the key to adapting spatial metaphors to information presentation lies in understanding user tasks. Thus, adding a notion of distance to the information interface solely because physical space has distance would probably not be useful in all cases. For instance, distance in information space might reasonably be used to convey semantic relatedness (e.g. [41]) or expected download delay (see also [42, 43]).

Our data show that even novice web users conceive of themselves as actively moving on the web under their own control. Thus, we believe that the power of spatial metaphors for information presentation is not merely the result of people's *ability to use spatial metaphors*. Rather, its power lies in the fact that people *naturally use spatial metaphors* – that they cannot help but use them. It follows that interface designers should not construct virtual worlds that are merely consistent with ordinary experience and that merely use spatial attributes in task-relevant ways. Rather, the most useful information interfaces will target people's natural spatial understanding of information use and at the same time allow people flexibility to create an appropriate metaphorical understanding of the domain (see [44]).

Navigation in Information Spaces

Dieberger's [45] city metaphor for information navigation seems to be a good approach to information space design. In particular, Dieberger carefully balances spatially real interface elements with *magic features* that break the spatial metaphor. In a sense, magic features provide the user with known boundaries that can be used in guiding the conceptual blending process. For instance, because magic windows provide shortcuts between distal points in the information city, semantic-relatedness need not be determined solely by spatial proximity. Nevertheless, both sorts of connections can be understood spatially as TRAJECTORY, which provides a consistent basis for the mappings.

We also see much promise in Waterworth's experiential approach to information landscape design [33]. This approach offers an alternative to the traditional human–computer interface (HCI) approach, which is based on an objectivist cognitivism (e.g. mental models). One advantage of the experiential approach is that users are offered a more meaningful interface, one that affords metaphorical thought and action. An excellent example of such a design may be seen in Lund and Waterworth's SchemaSpace [32], which is grounded embodiment and which is structured – at least to some extent – in a way that reflects image schematic structure. For excellent discussion and compelling arguments against the traditional HCI approach and for details on the experiential approach, see Waterworth's chapter in this book (Ch. 8).

The key point is that people should not have to adapt to information space; rather, they should play an active role in determining how the space is used through their activities and practice [7]. As we have seen, people's conceptual experience of information space is largely structured metaphorically and based on embodied experience in physical space. We believe that web browsers or other tools for navigation in information space should be designed based on how people conceptualise and experience the environment.

Social Interaction in Information Spaces

Tools for collaboration in information space can likewise be informed by understanding how people conceptualise interaction in information spaces. Research on how people interact will likely reveal that people conceptualise virtual interactions with others much as they conceptualiase actual interactions. Nevertheless, differences between the two will undoubtedly arise, providing opportunities for creating interfaces that are both different from and possibly more effective than physical interaction.

For instance, consider Babble, a computer-mediated communication system meant to facilitate long-term, ongoing conversations [46]. One design goal of Babble was to enable those involved in a conversation to be made aware of many social cues, such as users' presence and actions with respect to a particular conversation. In addition to a text window that displays conversational content, Babble uses a very elegant graphical representation called a *social proxy*, which depicts a conversation as a large circle, individuals as small coloured dots within the circle, and chatting as movement of the dots toward the circle's centre. In this way, the Babble interface relies on a spatial metaphor in which an area of the screen represents a conversation, icons within the area represent individuals engaged in the conversation, and motion of the icons represents conversational action. This metaphor abstracts away many details of actual conversations, such as facial expressions and intonation, yet retains significant spatial relationships, such as proximity. In addition, unlike verbal conversations or other computer-mediated chat systems, Babble adds timestamps to each conversational action and can store the text of conversations indefinitely. This enables Babble users to retrieve previous interactions and to reconstruct all previous conversational contexts. Thus, the Babble interface is in some ways similar to and in other ways more effective than actual conversation.

Though our empirical data do not specifically concern social interactions in information space, it is reasonable to expect similar results. From our perspective, then, using space to depict conversations follows the principle that people conceptualise information activities in physical terms. Moreover, movement of the dots in Babble follows the TRAJECTORY schema, just as a chat action follows a TRAJECTORY. Being inside the circle instantiates CONTAINER, just as being in a conversation suggests containment.

In any event, Babble provides an environment in which users can create and participate in conversations. This communication system was not set up as part of a larger information space. But why not? Consider that the web is a fundamentally social structure – it enables users to publish and to read what others have published. Although web users interact through published documents, these interactions are asynchronous and lack the richness of ordinary communication. The web misses the people behind the documents. Users are invisible to each other because social affordances are not built into the web.

The WebPlaces system was constructed to enable social interaction on the web [47, 48]. The idea was to make interpersonal awareness and interaction an integral part of web activity by creating virtual places through which users can communicate. In this system, a *place* does not necessarily map to a location in web space, but might be automatically constructed based on the interests and activities of web users. To make users aware of one another, WebPlaces adapted Babble's social proxy: a circle represents the group or community of users, and small coloured dots represent individual users (see [48]). Motion of a dot toward the centre of the circle represents a group interaction (e.g. chat), motion of a dot toward another dot represents a user-user action (e.g. whisper), and motion of a dot around the circle represents an individual user action (e.g., browsing). In this way, Babble's social proxy was extended to maintain social awareness in a user community rather than in a conversation. Actions were included that are not specifically related to the ongoing conversation, but that are nonetheless relevant to the users who are gathered together. In coding various types of actions by these iconic motions, WebPlaces' proxy indicates both the state and activity of the users in a place. A glance at the social proxy tells a user how busy the place is, who is there, and what activity there is. Thus, by combining affordances of information space with affordances for interpersonal interaction, WebPlaces blends information activities with social activities, and in the process, WebPlaces creates a novel user interface that relies on TRAJECTORY and CONTAINER to structure user experience.

9.4 Conclusion

The way that people think about the WWW has implications for the way that they navigate it. The key to designing effective information navigation tools lies in discovering how people naturally conceive of information spaces, including the extent to which such spaces are thought of in terms of physical space. Likewise, to facilitate efficient collaboration in information space, it is critical that software be designed to reflect people's natural conceptualisation of the space. To discover how people think about the web, one type of information space, we studied how people talk about using it. In doing so, we found that people consistently refer to the experience in terms of user-directed motion through physical space toward information objects. That particular metaphorical language is used is no accident, even though there are many different ways to talk about the web. Such language is motivated by metaphorical thought, which is structured by the same basic image schemata that people rely on to mentally structure everyday life. Thus, the power of spatial metaphors for information presentation is not merely the result of people's ability to learn to use spatial metaphors. Rather, its power lies in people's tendency to naturally use spatial metaphors – they cannot help but use them. It follows that efficient interface design should go much deeper than constructing virtual worlds that merely include a few task-specific spatial attributes. The most useful information interfaces will target people's natural spatial understanding of information use, and at the same time allow people flexibility to create appropriate metaphorical and blended understanding of the domain. Because of the striking consistency in conceptualisation of information space across web users, collaboration would be well afforded by a user interface that makes explicit appropriate aspects of users' apparent common ground. The trick lies in discovering the conceptual differences between real space and information space, and then in using those differences to afford rich and effective interactions in information space.

Acknowledgements

Thanks to Rob Barrett, Seana Coulson, Gilles Fauconnier, Ray Gibbs, and Mark Turner for thoughtful discussions, to Chris Dryer for advice on statistical analyses, and to David Benyon, Alan Munro, Barbara Tversky, and John Waterworth for many helpful comments on a draft of this paper.

References

1. Thorndyke, P.W. and Hayes-Roth, B. Differences in spatial knowledge acquired from maps and navigation. *Cognitive Psychology*, 1982.
2. Tversky, B. Spatial perspective in descriptions. In *Language and Space. Language, Speech, and Communication*, P. Bloom, M. A. Peterson, L. Nadel, and M. F. Garrett, editors. Cambridge, MA: MIT Press, 1996.
3. Lynch, D. *Image of the City*. Cambridge, MA: MIT Press, 1960.
4. Passini , R. *Wayfinding in Architecture*. New York: Van Nostrand Reinhold, 1984.
5. Darken, R. and Sibert, J.L. Wayfinding strategies and behaviors in large virtual worlds. *Human Factors in Computing Systems: Proceedings of the chi '96 Conference*. New York: ACM, 1996.
6. Hirtle, S. Spatial knowledge and navigation in real and virtual environments. Position paper for the *CHI '97 Workshop on Navigation in Electronic Worlds*, 1997.
7. Benyon, D. Beyond navigation as metaphor. In *Exploring Navigation: Towards a Framework for Design and Evaluation of Navigation in Electronic Spaces*, N. Dahlback, editor. SICS Technical Report 98–01, 1998, pp. 31–43.
8. Johnson, M. *The Body in the Mind: The Bodily Basis of Meaning Imagination and Reason*. Chicago, IL: University of Chicago Press, 1987.
9. Lakoff, G. *Women, Fire, and Dangerous Things: What Categories Reveal about the Mind*. Chicago, IL: University of Chicago Press, 1987.
10. Fauconnier, G. *Mappings in Thought and Language*. Cambridge: Cambridge University Press, 1997.

11. Fauconnier, G. and Turner, M. Conceptual projection and middle spaces. Technical Report 9401, University of California, San Diego, Department of Cognitive Science, 1994.

12. Fauconnier, G. and Turner, M. Blending as a central process in grammar. In *Conceptual Structure, Discourse, and Language*, A. Goldberg, editor. Cambridge: Cambridge University Press, 1996.

13. Fauconnier, G. and Turner, M. Conceptual integration networks. *Cognitive Science* **22**: 133–187, 1998.

14. Maglio, P.P. and Barrett, R. On the trail of information searchers. *Proceedings of the Nineteenth Annual Conference of the Cognitive Science Society*. Mahwah, NJ: LEA, 1997.

15. Anderson, J.R. *Cognitive Psychology and its Implications*. San Francisco, CA: Freeman, 1980.

16. Coulclelis, H., Golledge, G., Gale, N. and Tobler, W. Exploring the anchor-point hypothesis of spatial cognition. *Journal of Environmental Psychology* **7**: 99–122, 1987.

17. Maglio, P.P. and Barrett, R. How to build modeling agents to support web searchers. *Proceedings of the Sixth International Conference on User Modeling*. New York: Springer, 1997.

18. Matlock, T. and Maglio, P.P. Apparent motion on the World Wide Web. *Proceedings of the Eighteenth Annual Conference of the Cognitive Science Society*. Mahwah, NJ: LEA, 1996.

19. Matlock, T. and Maglio, P.P. Untangling talk about the World-Wide Web. University of California, Santa Cruz, Psychology Department, 1997.

20. Benyon, D. and Höök, K. Navigation in information spaces: Supporting the individual. *Human Computer Interaction: INTERACT '97*. London: Chapman and Hall, 1997.

21. Lakoff, G. and Johnson, M. *Metaphors We Live By*. Chicago, IL: University of Chicago Press, 1980.

22. Gibbs, R.W. *The Poetics of Mind*. Cambridge: Cambridge University Press, 1994.

23. Sweetser, E. *From Etymology to Pragmatics: Metaphorical and Cultural Aspects of Semantic Structure*. Cambridge: Cambridge University Press, 1990.

24. Turner, M. *Death is the Mother of Beauty: Mind, Metaphor, Criticism*. Chicago, IL: University of Chicago Press, 1987.

25. Gentner, D. and Imai, M. Is the future always ahead? Evidence for system-mappings in understanding space–time metaphors. *Proceedings of the Fourteenth Annual Meeting of the Cognitive Science Society*. Hillsdale, NJ: LEA, 1992.

26. Coulson, S. The Menendez brothers virus. In *Conceptual Structure, Discourse, and Language*, A. Goldberg, editor. Cambridge: Cambridge University Press, 1996.

27. Turner, M. and Fauconnier, G. Conceptual integration and formal expression. *Metaphor and Symbolic Activity* **10**: 183–204, 1995.

28. Grady, J., Taub, S. and Morgan, P. Primitive and compound metaphors. In *Conceptual Structure, Discourse, and Language*, A. Goldberg, editor. Cambridge: Cambridge University Press, 1996.

29. Gibbs, R.W. and Colston, H.L. The cognitive psychological reality of image schemas and their transformations. *Cognitive Linguistics* **6**: 347–378, 1995.

30. Johnson, M. Philosophical implications of cognitive semantics. *Cognitive Linguistics* **3**: 345–366, 1992.

31. Mandler, J.M. How to build a baby: II. Conceptual primitives. *Psychological Review* **99**: 587–604, 1992.

32. Lund, A. and Waterworth, J.A. Experiential design: reflecting embodiment at the interface. *Proceedings of Computation for Metaphors, Analogy and Agents: An International Workshop*, University of Aizu, Japan, 1998.

33. Waterworth, J.A. Personal spaces: 3D spatial worlds for information exploration, organisation and communication. In *The Internet in 3D*, R. Earnshaw and J. Vince, editors. New York: Academic Press, 1997.

34. Raubal, M., Egenhofer, M.J., Pfoser, D. and Tryfona, N. Structuring space with image schemata: wayfinding in airports as a case study. In *Spatial Information Theory: A Theoretical Basis for GIS (COSIT '97)*, S.C. Hirtle and A.U. Frank, editors. Berlin: Springer, 1997.

35. Catledge, L. and Pitkow, J. Characterizing browsing in the World Wide Web. *Proceedings of the Third International World Wide Web Conference*, 1995.

36. Tauscher, L. and Greenberg, S. Revisitation patterns in World Wide Web navigation. *Proceedings of the Conference on Human Factors in Computing Systems (CHI '97)*. New York: ACM, 1997.

37. Lakoff, G. and Johnson, M. *Philosophy in the Flesh*. Chicago, IL: University of Chicago Press, 1998.

38. Coulson, S. Semantic leaps: the role of frame-shifting and conceptual blending in meaning construction. PhD dissertation. University of California, San Diego, 1996.

39. Rohrer, T. Conceptual blending on the information highway: How metaphorical inferences work. *Discourse and Perspective in Cognitive Linguistics*. Amsterdam: John Benjamins, 1997.

40. Shum, S.B. Real and virtual spaces: mapping from spatial cognition to hypertext. *Hypermedia* **2**: 133–158, 1990.

41. Chalmers, M. and Chitson, P. Bead: explorations in information visualization. *Proceedings of the Fifteenth Annual ACM SIGIR Conference on Research and Development in Information Retrieval*. New York: ACM, 1992.

42. Barrett, R., Maglio, P.P. and Kellem, D.C. How to personalize the web. *Proceedings of Human Factors in Computing Systems, CHI '97*. New York: ACM, 1997.

43. Campbell, C.S. and Maglio, P.P. Facilitating navigation in information spaces: road signs on the World Wide Web. *International Journal of Human–Computer Studies* (in press).

44. Kuhn, W. Metaphors create theories for users. In *Spatial Information Theory: A Theoretical Basis for GIS (COSIT '93)*, A.U. Frank and I. Campari, editors. Berlin: Springer, 1993.

45. Dieberger, A. A city metaphor to support navigation in complex information spaces. In *Spatial Information Theory: A Theoretical Basis for GIS (COSIT '97)*, S.C. Hirtle and A.U. Frank, editors. Berlin: Springer-Verlag, 1997.

46. Erickson, T., Smith, D.N., Kellogg, W.A., Laff, M., Richards, J.T. and Bradner, E. A sociotechnical approach to design: social proxies, persistent conversations, and the design of Babble. *Proceedings of Human Factors in Computing Systems, CHI '99*. New York: ACM, 1999.

47. Maglio, P.P. and Barrett, R. Adaptive communities and web places. In *Proceedings of Second Workshop on Adaptive Hypertext and Hypermedia (Hypertext '98)*, P. Brusilovsky and P. De Bra, chairs. Pittsburgh, PA, 1998.

48. Maglio, P.P. and Barrett, R. WebPlaces: adding people to the web. Paper presented at the Eighth International World Wide Web Conference, Toronto, Canada, 1999.

Chapter **10**

A Contrast Between Information Navigation and Social Navigation in Virtual Worlds

Paul Rankin and Robert Spence

Abstract

We begin by emphasising the need for clear definitions of terms such as browsing and navigation, which previously had been rather loosely described. Reasonably precise definitions are offered which have led to a new theoretical framework for navigation. The framework appears to be equally relevant to social as well as information navigation. An interesting contrast is drawn here between these two activities, speculating on some of the psychological processes involved. In both cases the navigational process comprises the four activities of browsing, modelling, interpretation and strategy formulation. These are set within a context determined by the user's intent and the constraints and biases of the search domain. Differences are found in the properties of interest, the types of conscious and unconscious activities in which the person engages, and the affordances which are presented to the participant as opportunities for interaction.

To illustrate the differences, two specific scenarios are considered, namely navigation of information while searching for a house and navigation of a social space at a cocktail party. The navigational framework is examined in the context of these two interactions – informational and social. Examples are given of the four main activities and the affordances on which they operate. Comparisons and contrasts between possible computer-assisted versions of both scenarios are drawn. Interesting challenges for the support of social navigation emerge, where remote users can encounter each other across the Internet, e.g. as participants in virtual worlds. This seems a ripe field for further studies involving experts from psychology, sociology and anthropology.

10.1 Navigation

Browsing has been defined by Spence [1] to be the elicitation of content. Often – though not necessarily – achieved visually and rapidly according to some formulated strategy of viewing displayed data, it answers the question "What's there?". There is no specific target being sought, though there may well be weights associated subjectively or objectively with item types. Browsing allows a *mental model* (or "cognitive map") to be created by sampling the information space which may facilitate beneficial movement.

The availability of a mental model allows *interpretation* of that model, typically leading to a decision to *formulate a new browsing strategy*, perhaps to determine the "best" direction in which to move in information space. Traversal of this sequence of activities constitutes the *navigational process*, a process which will be governed by some higher level *intention* (Fig. 10.1).

Browsing is distinguished from searching. Browsing is a single activity essentially asking "What's there?". Searching for a specific "thing" (for example, a house in two of the illustrations that follow), will generally depend on the navigational process to generate a mental model with sufficient detail. To allow a higher order decision process to deduce that, possibly within a consciously or unconsciously determined tolerance, what is being searched for has been found. At the opposite extreme, such as looking for a specific word in a list, feedback to the higher order decision process may well emanate from heavily weighted browsing.

10.2 Home Hunting

To illustrate the process of navigation in information space we consider a scenario in which a person must find a house to buy, and their aim is to identify three or four houses worth more detailed investigation at an estate agent's office. The person has no knowledge of the general area in which a house is to be sought, or of typical house prices.

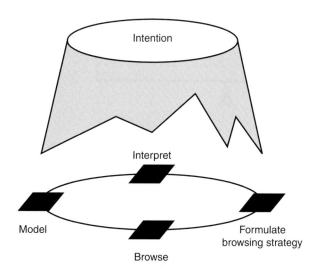

Figure 10.1 The navigational framework.

It is immediately apparent that the person will, by exploration, form a mental model not only of the population of candidate houses, but also of their environment. Proximity to a school, for example, may be just as important as an extra bedroom, and evidence of graffiti may be a strong determinant of choice of location. It is for this reason that we refer to *home-hunting* rather than house-hunting, and identify two aspects of the mental model that is built up during the navigation process (Fig. 10.2): one concerns the population of houses, the other the environment. Typically, both will concurrently be developed so that the final selection of candidate houses can be more effective.

There is, in fact, a third model, not part of the navigation process. It is the set of personal characteristics and assumptions (Fig. 10.3) of the home-hunter on the basis of which their Intention (Fig. 10.1) is formulated. The person's savings, for example, will determine the approximate price range of interest, and their expertise in do-it-yourself may suggest a low priority on state of repair. But as the models of houses and environment develop, it may be realised that different aspects of the home-seeker's personal characteristics and circumstances are more appropriate determinants of the intention: a scarcity of suitable houses may lead to a decision to accept grandmother's offer of financial help and hence to modify the price range of interest.

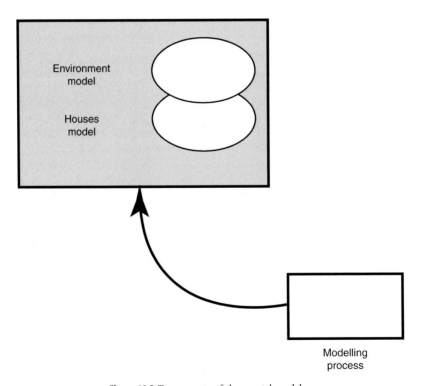

Figure 10.2 Two aspects of the mental model.

Personal Profile
Mental Model

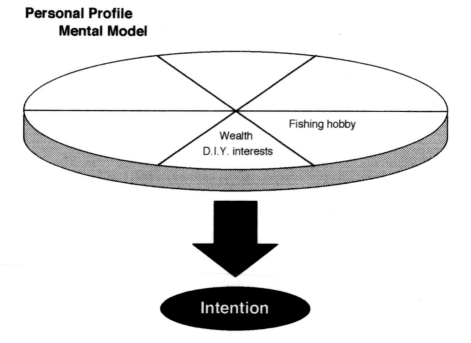

Figure 10.3 Parts of the personal profile form the intention.

To illustrate the concepts discussed above we provide below a description of a small fraction of the home-hunting scenario.

1. *Formulate an intention.* Based on part of the internal profile model of the home-seeker (see, for example, Figure 10.3), a three- or four-bedroom house with a medium to large garden, costing no more than about £70,000 and in any state of repair is assumed as the profile of an acceptable house.

2. *Browse.* The home-seeker drives to a small hill overlooking the area in which a house is to be found. As viewed from the hill, content elicited includes the nature of the area (more green than grey), presence of river, island, railway tracks, motorway, factory.

3. *Create a model.* (An *integration* of the above content, occurring virtually simultaneously with browsing, initiates a model of the environment.) Relative locations of the river, factory, railway, motorway. Model may unconsciously contain assumptions (not yet confirmed) regarding noise associated with a motorway, pollution with the factory and tranquillity of the river.

4. *Interpret the model.* In contrast to the integration leading from elicited content to a mental model, a form of *differentiation* is involved at this juncture. Within the environment model a perceived "gradient of desirability" points away from the dark factory and the railway towards an area focused on an island within the river. This gradient may lead to a refinement of the Intention to favour proximity to fishing, a particular interest of the home-finder.

Chapter 1

Footprints in the Snow

Alan Munro, Kristina Höök and David Benyon

1.1 Introduction

There are many changes happening in the world of computers and communication media. The Internet and the World Wide Web (WWW), of course, are various aspects of this vast network of interlinked machines. Computers are becoming increasingly ubiquitous; they are "disappearing" into everyday objects. They are becoming increasingly small, so much so that some are now wearable. They are increasingly able to communicate with each other.

These changes have had a major impact on our understanding of how computer systems should be designed and on what people can and want to do with them. The discipline of HCI (human–computer interaction) has not really kept pace with these changes in technology, and much HCI education still takes the view that there is *a* person interacting with *a* computer. Courses on HCI will emphasise the cognitive difficulties that people might have in using large software packages. They will present guidelines on how to design screens and human–computer "dialogues" and will discuss methods of evaluating computer systems in terms of the "tasks" that people undertake and in terms of the "errors" that people might make.

In the mid-1980s and early 1990s a new discipline emerged out of HCI and areas of study related to people working together: computer-supported co-operative work (CSCW). CSCW looked at how software systems and communication technologies could support the work that people were trying to do. It shared many of the assumptions of HCI, but also introduced more socially based ways of thinking. The importance of fitting technology into the workplace was emphasised, as well as *really* understanding what people wanted to do and of looking at what technologies were required in order to support and enable their activities.

During the 1990s a number of people were becoming dissatisfied with traditional approaches. People were using computers for many purposes other than work. People were not simply interacting with a computer but rather were interacting with other people using various combinations of computers and different media such as video and animations, touching, gesturing and so on. The notion that we could see people as existing in an information space, or in multiple information spaces grew, and was offered as a challenge to the predominant "people outside the information" view of HCI [1, 2]. Alongside this came the recognition that using computers needed to become a more enjoyable,

1

social activity. The development of the Internet, particularly for leisure activities, and the emergence of Internet service providers that bundled news, chat rooms, WWW access with remote game playing, shopping and so on resulted in designers looking outside of "traditional" CSCW and HCI for design principles and appropriate methodologies.

The chapters in this book capture much of the debate and ideas surrounding social navigation. The contributions deal with concepts of social navigation, ideas of the nature of information, the impact of people working with virtual environments, intelligent agents and wearable computers as well as incorporating design ideas coming from architecture, anthropology, experiential psychology and cinema. In this introductory chapter we aim to introduce many of the significant characteristics of social navigation and the backgrounds from which it has arisen.

1.2 Mapping Social Navigation

The concept of social navigation was introduced by Dourish and Chalmers in 1994. They saw social navigation as *navigation towards a cluster of people* or *navigation because other people have looked at something.* Computer systems known collectively as recommender systems [3] have been developed that implement these ideas. By collecting the likes and dislikes of a large number of people, an individual can specify one or two things that they like or dislike and the system recommends others based on the data collected from other people. Later, Dieberger widened this scope [4]. He also saw more direct recommendations of websites and bookmark collections, for example, as a form of social navigation. Since then the concept of social navigation has broadened to include a large family of methods, artefacts and techniques that capture some aspect of navigation.

Social navigation can be seen from several different perspectives and in several different domains, both in the "real" world of human activities and in the "virtual" worlds of information spaces of different kinds. Collaborative virtual environments (CVEs) provide an area where people may interact with each other in various ways. Sometimes these environments will be textual, such as a multi-user dimension (MUD) or a newsgroup; sometimes they include video and sound, and in others people are represented by avatars. Work on social navigation brings together and occasionally critiques aspects of CSCW, IUI (intelligent user interfaces), IR (information retrieval), and CVEs (collaborative virtual environments). Accordingly the underlying philosophies and disciplines come from work in cognitive and social psychology, anthropology, social theories of human action, HCI and artificial intelligence. The concept of social navigation brings some quite unique characteristics into our understanding of information, spaces, places, user interfaces, and the activities of those participating in these various worlds.

Social navigation does not have a narrow focus. Rather, its concerns range widely over "navigation" in different types of virtual worlds. Social navigation considers the creation of social settings and "places" in information space and behaviour in them, the sociality of information creation, people as members of groups and nature of information itself, its location, evaluation and use. We are seen not just as single users, but as members of many different types of information spaces, both real and virtual. As is suggested by the term "navigation", many writers on social navigation draw on work in architecture, urban planning, the visual arts and design.

1.2.1 A Design Scenario

Let us imagine that we are designing a online grocery store. This will include approximately 12,000 items that can be bought in the virtual store, paid for by credit card and then delivered to the doorstep.

If we look at this from a "traditional" HCI view we might think of this as an interaction problem: how can we design the system so that it becomes easy for users in front of their home computers to find the items they need, "pick them up", put them in their shopping basket, pay for the items, and make sure that they get delivered at the right time? Underlying our worries might be whether the interaction is designed optimally to be as efficient as possible from a user's point of view, down to the level of the number of actions taken to achieve a given task. We might also consider the aesthetic aspects: are the items displayed in a inviting way? Is the experience of walking through this information space a pleasurable experience? Is the right metaphor chosen, etc.? While all of these design considerations are highly relevant and should not be taken lightly, they are all based on a one-computer–one-user view of interaction. How can the ideas of social navigation be made central and be used to inform design?

The ideas of social navigation build on a more general concept that interacting with computers can be seen as "navigation" in information space [1, 5]. Whereas "traditional" HCI sees the person outside the information space, separate from it, trying to bridge the "gulfs" between themselves and information, this alternative view of HCI as navigation within the space sees people as inhabiting and moving through their information space. Just as we use social methods to find our way through geographical spaces, so we are interested in how social methods can be used in information spaces. How could the ideas of social navigation change our view on design in this particular context?

First of all, we would assume that other people would "be around" in the store. Instead of imagining a "dead" information space, we now see before us a lively place where (in some way) the user can see other shoppers moving about, can consult or instruct specialist agents and "talk to" the personnel of the grocery store. These are examples of *direct social navigation*. We also see the possibility of providing information pointing to what groceries one might buy based on what other people have bought, for example, if we want to help allergic users to find groceries and recipes that work for them, we could use the ideas of recommender systems; pointing people to products that, based on the preferences of other people, the system believes would be suitable. Sometimes we just like to peek into another's basket, or just take the most popular brand of some product. These are examples of *indirect social navigation*.

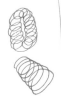

The form of social navigation chosen will depend on an understanding of the properties of the space (a large amount of grocery items that we can "see" but not smell or squeeze) and users' activities in the space. The activities in the space might resemble people's activities in real-world stores, but our design of the space and the possibilities for social navigation will change and shape users' activities. Some forms of social navigation might not "work" or be apposite in the case of the online grocery store scenario. For example, the *readware* idea (Dieberger, Ch. 3), where we can see from the texture of an item how it has been used or how often it is used, might not work if we take it literally: we would not like to see how many people have touched the tomatoes that we are about to buy. But if we think of readware as a way of overlaying the items in the space with behaviour patterns that manifest themselves in novel textures, we could imagine a system that introduces cream as a kind of texture that is placed on (or rather nearby) the strawberries. This would reflect the fact that many people buy strawberries and cream together.

Thus, social navigation will not be a set of ready-made algorithms and tools to be added to an existing space, and there might not be a direct one-to-one mapping from how social activities happen in the real world to how they take place in virtual worlds. Representation is a real and problematic issue (Dourish, Ch. 2). What we can do is to *enable* (make the world *afford*) social interactions and accumulate social trails. We might direct people or point them to areas "off stage" (Persson, Ch. 11). Social navigation will often be a *dynamic*, changing, interaction between the users in the space, the items in the space (whether grocery items, books or something else) and the activities in the space.

Grocery shopping is perhaps not a prototypical *information* space application, but it is something we all do and recognise, and this scenario shows the telling differences between approaches. This approach – designing for social navigation – is just as useful in a more "conventional" information retrieval context. Harper (Ch. 5) provides an analysis of what might be considered to be a prototypical example of an information space, the International Monetary Fund (IMF). In this situation the "desk officers" are concerned with obtaining information concerning different countries and their economies. Although such information may appear more "accurate" and "objective" and should therefore be handled under a more "traditional" information retrieval rubric, Harper argues that the "texture" of the information is much richer than can be assumed at first sight. Useful information might not necessarily be the most accurate, but what is known to be used by others, and what is able to be placed in a *context*.

1.2.2 Design Possibilities

Using the idea of social navigation will thus bring about a view on information spaces and users' activities in them that enables a whole range of design solutions that are by no means fully exploited in today's systems. With direct social navigation we shall be able to talk directly to other users and get their views on how to navigate, which will improve our navigation and understanding of the space. When we talk to someone else, the information we get back is often personalised to our needs. We are perhaps told a little bit more than exactly the information we asked for, or if the information provider knows us, the instructions may be adapted to fit our knowledge or assumed reasons for going to a particular place [5].

But direct social navigation can also be on the level of *interpreting* information based on how other people talk about or use information (Harper, Ch. 5). We can judge to what extent the information given can be trusted depending on the credibility of the information provider. Sometimes, even if it cannot be trusted, it still is of value as we know where it is from.

Social navigation is to an extent dependent on the modality in which one is able to operate; the affordances of it. A text-only system can afford fewer social cues than one using graphics, sound or video. Several authors in this book turn to three-dimensional information spaces as a metaphor and guide to ways in which we can show information, placing it in the space in some way, and using the position of the object to reflect its attributes.

We also talk to one another through body language or through the collective visible actions of a group of people. We follow the crowd of people who leave the plane in an airport assuming that they are all heading for the baggage claim. In a virtual environment avatars may be used to express some simple forms of body language that can invite others to join the group, or to indicate privacy (Jeffery and Mark, Ch. 7). The aura of the representation of a user can enable interaction with other people (Rankin and Spence, Ch. 10).

The possibilities for indirect social navigation also allow for aggregation of non-visible user behaviour. Designers can invent novel ways in which these things are displayed such as textures of items or trails of people, or it can be more direct instructions or recommendations: "others who bought this book also bought this one". We can also alter the way that data about people is collected, shown and mapped on the space in a dynamic way. The dynamics of this change is both good and bad: people shall have to adapt more quickly to new environments and the changed properties of the space. As seen in Ch. 6 by Buscher and Hughes, people are intrigued by spaces that do not behave as they expect them to, but they rapidly adapt and figure out new ways by which they can understand the space and the movements of people in them.

Both direct and indirect social navigation provide excellent opportunities to provide the seeker with a sense of security/safety: "since all these people have chosen this route, it must be the right one" or "if this is the way that my friend behaves in these sites, then I can probably do the same". Human beings are to a large extent behaving like flocks – we respond to other people's behaviours, body movements, use of tools, way of talking, and so on. We use dialects, clothing, rituals as means to make our group distinct from the rest of humanity. We create a feeling of belonging – it makes us feel good! These mechanisms are efficient ways of learning and still they are largely excluded from spaces like the web (Dieberger, Ch. 3).

1.3 Shifting Perspective

Social navigation does not just provide alternative design possibilities. It requires people to think differently about the nature of people and their interactions with computers and communication technologies. For example, Dourish (Ch. 2) encourages us to view social navigation as a phenomenon of interaction. If we populate spaces they will offer users "appropriation and appropriate behaviour framing, distinguishing them from simple spaces, which are characterised in terms of their dimensionality". Dieberger (Ch. 3) opens up the scope of seeing social navigation as a move away from the "dead" information spaces we see on the Internet today and in every way possible open up the spaces for seeing other users – both directly and indirectly. His claim is that "future information systems will be populated information spaces". Users will be able to point out and share information easily, guide each other and in general open up our eyes for various ways by which we can "see" other users.

Ideas of social navigation naturally lead us to consider the concept of navigation in real and virtual worlds. We also need to change our view of information from being decontextualised, objective portions of the world, to socially interwoven subjective views of the world. The relationship between space and how it can be turned into "place" where meetings can happen and where social connotations constrain and afford activities is also significant. Finally, we need to reconsider the relationship between the modality of the space and interactions that can take place in it.

Social navigation is not a concept that can be unproblematically translated into a set of particular tools. Instead it encourages understanding of human activities in space and place – both virtual and "real". People are active participants in reshaping the space. There is a dynamic relationship between people, their activities in space, and the space itself. All three are subject to change.

1.3.1 Navigation of Information Space

Social navigation both arises from and informs a novel way of thinking about how people interact with computers and communications technology. Within the discipline of HCI the typical view of how people interact with computers has been based, primarily, on a cognitive psychological analysis. This view sees the user as outside the computer or other device. People have to translate their intentions into the language of the computer and have to interpret the computer's response in terms of how successful they were in achieving their aims. This view of HCI leads to the famous "gulfs" of execution (the difficulty of translating human intentions into computer speak) and evaluation (trying to interpret the computer's response). Rankin and Spence (Ch. 10) provide a model of navigation that focuses on how people form intentions, browse and evaluate alternatives and then decide which "direction" (or course of action) to follow. In this sense Rankin and Spence are quite close to the work of urban planners and architects who have used this cognitive view of spatial awareness [7, 8]. Lynch beautiful entrance. Again, for any combination of trigger and action, the process may be *conscious* or *unconscious*. How do these concepts relate to navigation? The most likely explanation is that they influence the process of interpretation, thereby influencing the formulation of browsing strategy.

All actions will be directed generally towards enhancing the likelihood of intention satisfaction, a process supported locally by the creation of a helpful mental model. Artefacts such as specialised maps provided by an estate agent will be directed towards helping a mental model to be formed. Also, checklists provided by estate agents help the prospective buyer to crystallise their requirements through self-reflection by reference to their personal profile (Fig. 10.3).

10.3 Extension to Digital Representation and Social Navigation

The example above has illustrated the navigational process in the context of information in a physical world. In what follows immediately we address the same scenario – that of home-hunting – in the virtual information world. We then examine social navigation in both physical and imaginary worlds (Fig. 10.4).

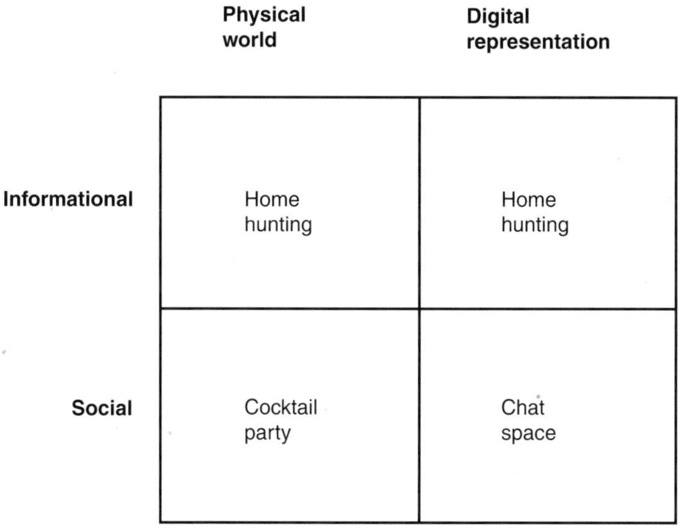

	Physical world	Digital representation
Informational	Home hunting	Home hunting
Social	Cocktail party	Chat space

Figure 10.4 The scenarios examined.

10.4 Computer-Assisted Home-Hunting

There is no doubt that the search for a suitable home can be supported electronically. One example is provided by the Attribute Explorer [4] and in the illustrative scenario which follows a suggested enhancement of the Attribute Explorer will be assumed. Here the home-hunter's overall aim is unchanged, as are the assumptions listed earlier. The personal model of Fig. 10.3 is still relevant to the formulation of intention. Selected episodes will suffice to illustrate the four activities comprising navigation, and are accompanied by views of the relevant computer-generated displays.

1. *Formulate the Intention.* As before.
2. *Browse.* The home-hunter views a displayed map (Fig. 10.5) and elicits content. Well-designed encoding (e.g. colour to denote price range) can enhance the browsing process considerably. Browsing can also be made effective through the efficient utilisation of available screen space to show a large number of houses and considerable environmental detail.
3. *Create a Model.* Here both models, of candidate houses and their environment, will be developed simultaneously. We note that a digital representation has the potential to provide a better overview than is possible in the physical world.
4. *Interpret the Model.* Interpretation of both models can take place, probably leading to a decision that the models need to be enhanced.
5. *Formulate Browsing Strategy, and Browse.* Mouse-click on houses randomly to cause a small image of the house to appear momentarily.
6. *Extend the Model.* Embody appearance information in the model of candidate houses and, if relevant detail is available in the images, in the environment model.
7. *Interpret the Model.* There are some attractive houses near the factory.
8. *Formulate Browsing Strategy.* Examine appearance of more houses near the factory.

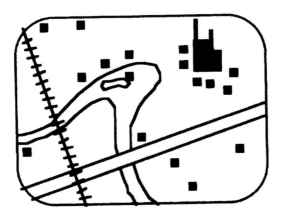

Figure 10.5 Computer-generated map showing houses for sale.

9. *Browse.* More houses near the factory examined

10. *Extend the Model.* Candidate houses model now has plenty of detail in the vicinity of the factory.

11. *Interpret the Model.* Model lacks information about prices and other house attributes

12. *Formulate Browsing Strategy.* Make use of available affordances to select the price histogram (Fig. 10.6) and explore the effect of setting upper and lower limits.

13. *Browse.* Elicited content includes spread and distribution of prices, and result of upper limit on the number of candidate houses.

14. *Extend the Model.* At this point the model may be more of a collage, with two distinct parts. One is "map-like" and influenced by Fig. 10.5. The other is based on Fig. 10.6. The provision of linking between simultaneously displayed figures can help to fuse the two components of the collage.

15. *Interpret the Model.* The model is still impoverished with regard to information about number of bedrooms.

16. *Formulate Browsing Strategy.* Select the "number of bedrooms" histogram (Fig. 10.7) and explore the effect of different requirements.

17. *Browse.* Vary upper limit. Note effect on availability. Note effect of price limits.

18. *Extend the model.* The existence of five-bedroom houses within the price limit is noted.

The effective facilitation of home-hunting in a digital representation clearly depends crucially on the provision of an appropriate combination of affordances, representations and presentations. There are no rules from which such a combination can be deduced, and much depends on the skill of the interaction designer.

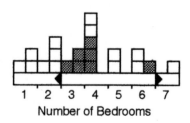

Figure 10.6 The price histogram of the Attribute Explorer.

Figure 10.7 Histogram of number of bedrooms, reflecting existing price limits.

10.5 Social Navigation: Cocktail Party Browsing

New private or business contacts always combine elements of chance encounter and deliberate search, seeded by some common interest ground. For example, a seemingly chance encounter in a music bar is never entirely random. People choosing to go there have some common tastes in music, style of establishment and probably even some commonality of mood. People waiting at a bus stop may be thrown into conversation by suffering the same situation of waiting for an overdue bus. The more specialised and intimate the discovered area of common interest, the stronger it seems as a seed for conversation, the more powerful it is as an "ice-breaker".

Contexts or *settings* (e.g. the place, topic or consensus social situation) bias and facilitate social interaction. The context focus, for example a specific topic of interest strongly promoted in a locale, will determine who is to be found there. Setting will affect and bias the communication processes between people's encounters, for example the aspects of personality and behaviour that people initially present to each other. In cinema or theatre the influences of various aspects of "setting" on character interactions are well known. It is easily appreciated that the aspects of self and one's interests that are disclosed during a discussion in a work context with a boss are quite different from those disclosed to a financial advisor, and again different from those disclosed on a first date.

The scenario in which we attempt to illustrate social navigation is that of a cocktail party, to which the "player", a young man, has been invited with only scanty knowledge of the other attendees or the reason for the gathering. This light-hearted description should be treated as speculative: expert psychologists can be far more precise about the social process and conversational signals involved.

1. *Form the Intention.* The party-goer wishes to have an enjoyable evening by talking with interesting people.

   The player brings with him a background profile of his personality, behaviour, roles and interests, what he finds enjoyable, the types of person he is attracted to and with whom he finds it satisfying to converse, etc. The player is partly conscious of this "back-history", but mainly unaware of his biases. Different aspects of this profile (Fig. 10.3) will be triggered according to the situations he enters and specific personal encounters. His intention is only very loosely defined.

2. *Browse.* On first entry to the party room he performs rapid audio, visual and olfactory sampling. Orient to the general setting and environment: a brightly-decorated hall. Note: noise level; nature of noise; typical ages of guests; style of clothing; stances; grouping, movements; gender, smells of barbecue food, etc.

3. *Model the Context.* Banners about a fund-raising event suggest the common reason for the participants' presence, thus a loose scope on the participants and common ground topic for conversation.

   Model the social atmosphere and general characteristics of the population: a rather noisy party, with much laughter, informal dress in general. The groups seem rather diverse, some with a welcoming open attitude and awareness of the player's presence, other groups seem closed, deep in conversation.

4. *Interpret the Model.* The player draws unconsciously on stereotypes of other people established over many years to simplify the model and make social inferences. A boy–girl pair is holding hands – lovers? A group of serious men discussing work, with backs to the others and who do not welcome intrusion – business associates? An attractive woman on her own looking rather bored – status? Approachability? A 20s group grasping beer mugs. Need for more content to refine the model of the context and its participants.

5. *Formulate Browsing Strategy.* Go slowly to drinks table, listening in passing to snatches of conversation, while appearing open for approach from others.

6. *Browse.* Boy–girl pair both mention "mum" and smile at you; men do not look around, mention "stocks ... city ... bank"; the single woman smiles and is wearing a brooch indicating an interest in folk music; 20s group mention "Arsenal team ... the cup"; sound of classical music emanating from a distant door. Much of this social browsing is done via subtle glance and gaze eye signals, where dwell time is critical. The same enquiry-and-response protocols being instantiated by different gestures, postures and glances in different cultures. For example, the mirroring in response to a smile or body posture is usually taken as a positive signal.

7. *Model the Context.* Boy–girl are brother–sister, possibly approachable. Group of men are discussing financial matters and do not wish to be disturbed. The 20s are discussing football. Single woman appears approachable.

8. *Interpretation.* Intention satisfaction is likely to be enhanced either by making the acquaintance of the single woman or by enquiring of the boy–girl pair where the host may be found.

9. *Refine the Intention.* Test the possibilities of a relationship with the single woman.

   The original, very loosely defined intention on arriving at the cocktail party now takes on a more specific target. Social goals may not be consciously framed, understood or even admitted by the individual.

10. *Formulate Browsing Strategy.* Approach single woman to enhance one's model of her. Perhaps ask her about the classical music ("Is it a concert? Do you like this music?").

11. *Browse* ... Observe facial signals and body language. Such signals may indicate gradients or directions for further fruitful investigation of common interests or attitudes. Proxemics (the role of body space) and are instinctively enacted, e.g. there is a cultural consensus on a comfortable interbody separation. People let in closer are treated as more trustworthy, likeable, etc.

Parallels might be drawn with the home-hunting scenario.

10.6 Computer-Assisted Remote Acquaintance

First, let us consider a simple digital representation for a social search for people (cf. wanted advertisements: "desperately seeking ...") by browsing through a

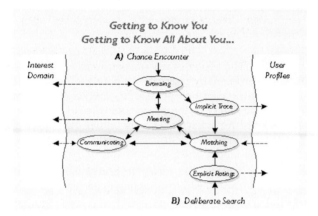

Figure 10.8 Using user data profiles in browsing or searching for people.

database of data profiles of candidates. Figure 10.8 schematically shows the processes that may be involved. Matching is done according to criteria in a general search profile (user profile), against those characteristics in a population of candidate individuals. Only certain facets of a general profile are examined during the search, those that are set by the current enquirer's interest or context.

A general representation of the user's own profile may be aggregated over time. This personal profile may be built explicitly by the individual, (e.g. by answering questions on their likes and dislikes) or may be built implicitly by a system that watches and traces the user's activity, making inferences about the user's characteristics (cf. web sites which trace user's actions). Such aggregate profiles may be very useful for the individual in order to personalise services or products, but are important commercial property for targeted advertising. Of course, privacy and trust issues with personal data are paramount.

In the next section we will take a richer digital representation than a database search, one in which remote presence and communication are supported, namely a virtual world.

Throughout it is useful to bear in mind that people instinctively interpret computers and media as if they were real people, unconsciously and inevitably attributing social qualities to any interface. These social attributions to an interface, or the social misinterpretations of remote communication with others through computer-mediated interfaces have been the subject of study [5]. Of course, it is exactly this tendency to anthropomorphise that virtual worlds try to exploit.

10.6.1 A Digital Representation of Remote Presence and Places

To explore a computer-rendered version of the social navigation scenario, a particular proposal for supporting remote encounter in virtual worlds will be taken. First however, some of the principles for constructing these virtual places and allowing participants to manipulate a representation for their remote presence, or "avatar" will need to be described.

In these virtual meeting places, new ways to support social navigation and the acquaintance processes are possible. These include searching for like minds in a crowd, stereotyping, grooming, presentation of self, disclosure, etc. [6]. One observation that can be exploited is that in "real life", acquaintances are rarely random, but are biased by meeting in a common context, interest or place. The contextualisation of human interaction by the locale is one that other developers of virtual environments (VEs) have also exploited [7]. A number of different locales within a VE can be provided for different types of remote activity or encounter.

One novel opportunity for VEs is the possibility of postponing the immediate biases that age, ethnic origin or gender introduce in normal face-to-face encounters. The initial appraisals of an acquaintance in a VE (like in an internet relay chat channel) might be based on matches of interests and other elements of personality or spirit, rather than gender, age, etc. For this reason, abstract metaphors for the environment and the user's avatar are investigated below.

Virtual Worlds Based on Abstract Metaphors

Virtual multi-user worlds modelled on a *literal,* photo-real representation of our surroundings and ourselves are currently popular on the web. However *abstract* representations offer more expressive, appropriate visualisations of many domains of human concern, from finance to music. With harmonious abstractions for users' presences we might turn such "information landscapes" into social places for communication and navigation. That is, if we can find ways of putting users into data visualisations, then "super-natural" mechanisms might then couple the users' embodiments, both with each other, and with the information, entertainment or education content of the VE, in novel responsive ways [8].

Before considering navigation processes however, the structure of the VE, affordances of presence and locale and the interaction within this virtual environment must be explained further.

A Model of the User's Profile

A vast aggregation of demographics, interests, skills, roles, aspirations, tastes and moods constitute a user's dossier. Trying to capture all these personal data in an understandable (for the user) model and provide an interface for the user to control context- and situation-dependent disclosure has hardly been addressed. A user interface has been proposed for managing this multi-faceted profile [8], exploiting the human's conscious awareness of distinct social contexts. Data disclosures are partitioned accordingly. Default masks over segments of data (also seen as "masques" of the alternative stereotypical role models) can be adopted,

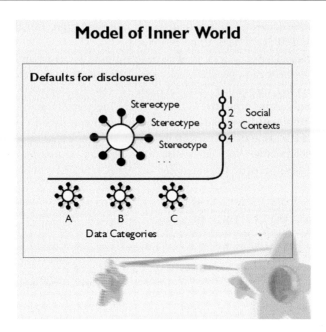

Figure 10.9 User profile model.

automatically triggered by the context of a VE locale and the company. These masks govern the style and pace of disclosures as well as what data is revealed or hidden during initial reciprocal revelations with other users (see Fig. 10.9).

StarCursor: Affordance of Remote Presence, Layered with Channels for Disclosures

Participants are represented inside this VE by their action point, an "avatar" based on a compound cursor. As well as being an instrument of action, the rudimentary heart, body, limbs, gaze and aura of this multi-part anthropomorphic cursor, the StarCursors [8] each form multimedia affordances for social browsing, personal disclosures and communicative cues (cf. [9]). For example, the audible aura of another cursor might signal its owner's mood.

Gaze direction, the focus of attention, is the prime user input control. In this concept, points of view for the user's virtual camera and microphones in the VE are automatically generated dependent on the user's direction of view and state of motion. However, their view (framing, first, second, third person point of view, etc.) is also dependent on external events and social interactions, for example, to attract attention. Meanwhile, the user's interest profile acts as a filter through which they perceive the space, for example, highlighting other users with common or complementary profile matches, or showing more detail in places that match the user's profile (see Fig. 10.10).

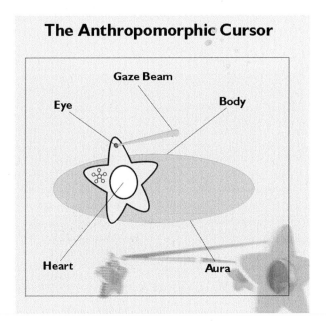

Figure 10.10 StarCursor parts.

Contextualised Social Interaction – the Outer Space

A three-dimensional VE , "ContentSpace" is proposed in this vision, linking different representations for content, from television programmes to games to web pages at a higher abstraction level [8]. In this space, walls project time-dependent audio-visual content and shape local property fields. These fields not only focus the local subject domain, but also influence the *social* properties of a zone such as its seclusion, ownership, access, the validation of participants, or the type of work or play activities. Both topic and social properties of zones refine, filter and mediate the interpersonal exchanges they enfold. Interactions between people and place should always be two-way. Content walls would respond personally to the user's gaze or proximity, while a user's profile would accumulate more detail from their activities in a zone. Both outer and inner spaces are therefore imagined as evolving through user activity and creativity (see Fig. 10.11).

Novel Navigation Mechanisms

Within such a VE, new types of social navigation are possible. For example, if forces of attraction and repulsion between StarCursors are implemented, based on the degree of match of their user profiles, then the system might allow the user to free-fall to a cluster of taste-mates. Alternatively, the user's StarCursor might

Stars in ContentSpace

Personally - responsive places

Figure 10.11 Cursors in ContentSpace.

be drawn to a particular place with particular social attributes within the VE, based on the user's current mood or social inclination.

Reciprocal Projection and Disclosure Mechanisms

Within this VE, distant auras of potentially interesting acquaintances might attract the participant's personalised eye and ear. However, observation of another user might carry a price, namely the leakage of some personal information.

Three mechanisms seem necessary to moderate the personal exchanges between users in such a VE:

- the zone sets a context, which in turn brings to the fore certain aspects and topics in the participant's profiles
- body space and the separation between two cursors moderates the intensity (depth of disclosures)
- eye gaze can be used to scan the items in another's profiles, which are related to different parts of their cursor body.

Additionally, to proffer new information to another user, by moving a part of the body forward, the gaze of the other's eye could be attracted to that aspect of self.

Separation between two cursors would automatically act, like body space, to moderate the projections of self. Profile disclosures and examinations would be

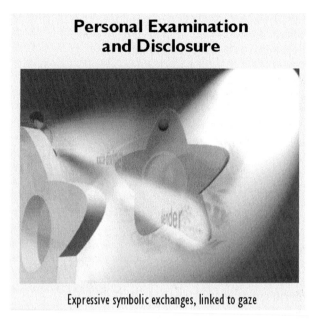

Figure 10.12 Two cursors examining each other.

conducted via an abstract body language where the gaze beam of one user's cursor triggers the emission of expressive multimedia objects and symbols associated with the other's profile from its cursor body. Longer gaze dwell times would explore deeper data in that aspect of the other's profile, down to their context-dependent mask. Body space could be expanded or contracted to suit the user's sociability to friends or enemies. Previous encounters with the same user in the same context are remembered, so personal barriers may grow or diminish on that segment of their profile (see Fig. 10.12).

10.6.2 Encounter in the Virtual World

A simple scenario of an encounter within the virtual environment imagined above will now be described. The intent is to show how some of the social cues of the cocktail party encounter might be supported in a digital representation and to facilitate comparisons with the digital support of home hunting.

1. *Formulate the Intention.* The male user connects to the StarCursor VE from his network-connected set-top box. He wants to find someone with a similar taste in classical music to exchange recommendations for music CD's and concerts.

2. *Formulate Browsing Strategy.* He looks at his own user profile, selecting a situation ("light entertainment with peers"), which in turn selects his default

stereotypical disclosure mask for that situation. The mask will be socially expansive about details of his leisure interests including musical tastes, without exposing his real name or gender.

3. *Browse.* The user turns from viewing his internal user profile, to scan the VE, using an input device that controls direction of view in the VE. Places appropriate for leisure interests within his gaze are highlighted in more detail.

4. *Create the model.* A mental map of the locations of zones in the VE concerned with leisure interests is formed.

5. *Interpret the model.* The top far right corner in the VE has an aggregation of zones about music, within which there is a classical music locale.

6. *Browse.* The user directs his gaze beam to the locale of interest and presses a button in order to be pulled along the gaze direction to that locale. On entering the locale, it responds to his cursor's presence and the selected aspect of his user profile by presenting images and music which lightly echo his music tastes.

7. *Extend the model.* Looking around the locale, the user sees four other StarCursors, indicating the presence of other remote users. Under his own eye gaze, the cursor on the left glows more than the others.

8. *Interpret the Model.* The cursor on the left is a better match in music tastes.

9. *Formulate Browsing Strategy.* Approach that user's cursor to find out more about the other user.

10. *Browse.* On entering the other user's rather expansive aura, sounds of light Mozart playing can be heard. The other cursor's gaze beam can be heard as it touches his own cursor, and the other cursor moves slightly towards the user's cursor.

11. *Interpret the Model.* The other user is in a bright mood and enjoys that composer. The other cursor has registered his presence and is open for acquaintance

12. *Formulate Browsing Strategy.* Exchange details of classical music tastes with the other user

13. *Browse.* Once cursors are close, the user moves his gaze over the body of the other cursor, his eye beam triggering the emission of images and sounds, corresponding to profile matches showing the other's deep awareness of Bach and Mozart, combined with an interest in American folk music. The user moves forward one limb of his cursor concerned with folk music. The gaze of the other is attracted to this movement, so the other cursor then becomes aware of their common interest in folk music ...

Eventually after getting to know more of each other, the two users agree to turn on their voice communication channel to speak with each other.

10.7 Comparisons: Information vs Social Navigation

10.7.1 Context and Setting

In both information and social navigation, setting is seen to be an essential element. In the home-hunting scenario, the setting is mainly that of the environment and the urban landscape containing the candidate homes for sale. Of course the population of houses and their disposition influences the environment and vice-versa. In a similar manner the place, its architecture and the current social setting in the cocktail party scenario also strongly influence the modes and aspects of human interactions between the participants. Again, the population of candidate individuals influences the social scene and vice-versa.

10.7.2 Search

The difference between search and browse can be illustrated in both scenarios. In home-hunting, the overall search for a few houses of interest specifically involves the four activities, of which only one is browsing, and involves many iterations between the four activities. The specific process of finding how little one needed to pay for a five-bedroom house, achieved by gradual reduction of the upper price limit until there are no green five-bedroom houses, involves a highly weighted browsing action

Termination of the search process will eventually be a very clear decision point in the home-hunting when sufficient weighted criteria are satisfied – that is on purchase of a property. In contrast, some social navigation scenarios may remain open-ended (until death!).

10.7.3 Browsing Mechanisms

Browsing mechanisms differ. Visual perception is the principal one employed in home-hunting. In the cocktail party the primary perceptions are aural and visual, but an additional mechanism is available; it is a form of social sonar. For example, smiling at a person and observing the response is a form of browsing to assess content, the latter being "readiness to engage in conversation". This browsing activity might be termed "personal".

10.7.4 Externalisation

Attributes of interest to the navigator are externalised in the home-hunting by appearance, location and abstract representation (in histogram form), and in the cocktail party by stance, clothing, jewellry, facial appearance, actions (e.g.

holding hands), drink (hand around beer mug, fingers holding champagne glass) and tone of voice.

10.7.5 Passive vs Active Affordances

Affordances in a digital representation are the ways in which the underlying properties of an object or space and the possibilities for interaction are communicated to the user. In the home-hunting scenario, most of the affordances of properties are passive: the user needs to interrogate a histogram or candidate for further information. In the StarCursor VE, many of the affordances (e.g. the sounds indicating type and activity of a zone, or the aura and emissions from another cursor) are proactive, without requiring prompting.

10.7.6 Conscious vs Instinctive Acts and Goals

Social and information navigation seem to differ considerably in the voicing of intent. During the browsing of information, goals may be rather precisely and consciously expressed, the search steps are taken consciously and deliberately (unless performed via a surrogate software search agent). In contrast, social goals are rarely well framed, or even admitted by the individual. Moreover, most of the communicative signals and responses during social interaction are instinctively transmitted and read. In trying to support these remotely, it is important to try to make such instinctive acts (e.g. like eye gaze direction) as easily manipulated as possible, so that the input controls for your avatar in a VE become "second nature".

10.7.7 Iterations between Models of Person and Models of Context/Setting/ Environment

It can be seen that there is a negotiation in both information and social situations between a model of self and a model of the setting, brokered by the intermediary, the formulation of intent, or search profile. During the browse and search process, both mental models (of the user to themselves, and the user's model of the context) are refined and detailed. For example, the existence of a feature in the environment may cause the participant to reflect on their attitudes and whether or not that aspect is acceptable.

10.7.8 Opportunism

The full nature of the available population of candidates, the complete setting and those aspects of the user's own personal nature which may be involved are only revealed during navigation. In both social and information navigation, all the constraints and desires of the user are never completely formulated to cover all possible matches or eventualities. Therefore an important aspect is that of opportunism, the reformulation of the overall goals and intent when new or unexpected aspects are revealed. The final result may therefore be quite different from the initial direction on which the user embarked.

10.8 Conclusions

The proposed framework for navigation has been explained by illustrations of conventional and computer-assisted home-hunting, and has also been shown to be relevant to social navigation by an illustration based on a cocktail party. Of course, there are large differences between houses and people, and which have not been covered. Houses are largely passive objects, there to be examined, while people, even remotely represented through an avatar, are very much proactive objects. Nevertheless, we hope that the contrast drawn is provocative.

Due to limitations on space, some issues directly relevant to the topic of this chapter have not been discussed, but deserve to – and will – receive separate and detailed attention. They include the design of the externalisation (i.e. display) associated with computer-assisted navigation, and which must ideally support all four navigational activities. The design of the externalisation must also take into account the possibility of perceptually triggering both situated and planned activities.

Social navigation has been illustrated in the context of the real world. Application of the framework to remote social interaction in imaginary worlds inhabited by anthropomorphic cursors poses interesting challenges and opportunities.

Acknowledgements

The StarCursor and VE concepts described here are the result of creative collaborations with many designers in Philips, especially C. v. Heerden. J. Mama , L. Nikolovska, R. den Otter and J. Rutgers.

References

1. Spence, R. A framework for navigation. *International Journal of Human Computer Studies* (in press)
2. Tversky, B. Cognitive maps, cognitive collages and spatial mental models. *Proceedings of European Conference COSIT'93, Spatial Information Theory – A Theoretical Basis for GIS. Lecture Notes on Computer Science*. London: Springer, 1993, pp. 14–24.
3. Tweedie, L. Interactive visualisation artifacts: how can abstractions inform design? In *People and Computers X, Proceedings of the HCI'95 Conference*, Kirby, Dix and Finlay, editors, 1995, pp. 247–265.
4. Spence, R. and Tweedie, L. *The Attribute Explorer: information synthesis via exploration, interacting with computers*, 11, pp. 137–146, 1998.

5. Reeves, B. and Nass, C. *The Media Equation: How People Treat Computers, Television and New Media like Real People and Places.* Cambridge: Cambridge University Press, 1996.
6. Goffman, E. *The Presentation of Self in Everyday Life.* New York: Anchor Books, 1959.
7. Fitzpatrick, G., Mansfield, T. and Kaplan, S.M. *Locales Framework: Exploring Foundations for Collaborative Support, Proceedings of OZCHI '96.* IEEE Computer Society Press, Hamilton, 1996, pp. 34–41 and pp. 332–333.
8. Rankin, P.J., v. Heerden, C., Mama, J., Nikolovska, L., Otter, R., Rutgers, J. *StarCursors in ContentSpace, Proceedings ACM Siggraph,* Orlando, FL, 1998, p. 250.
9. Greenhalgh, C. and Benford, S. MASSIVE: a collaborative virtual environment for tele-conferencing. *Transactions of ACM Computer–Human Interfaces* **2:** 239–261, 1995.

Chapter **11**

Understanding Representations of Space: A Comparison of Visualisation Techniques in Mainstream Cinema and Computer Interfaces

Per Persson

Abstract

In order to understand the characteristics of digital information spaces and the way they are being understood and manipulated by the user, it may be valuable to investigate how other forms of representations handle space. This chapter investigates six very common visualisation techniques used in mainstream cinema, and tries to find equivalent techniques in digital information spaces. Such a comparative study has to take into account the different functions cinema and computers are said to serve, as well as what kind of assumptions and knowledge the spectator/user brings into the situation. Throughout, implications for design are discussed.

11.1 Introduction and Some Distinctions

Space is a crucial dimension of human work and everyday life. In space, we perceive and identify objects and their relations *vis-à-vis* each other and ourselves (inside–outside, up–down, left–right). In space we can pick up and manipulate these objects and use them as tools for our own purposes. In space we can move around and get different views of things that can then trigger spatial reasoning and cognitive speculation about shortcuts, spatial layouts and possible routes. Human communication also often uses space to create a shared interpretative context in which verbal utterances become easy to understand (e.g. personal space behaviour, body language, referring to objects by way of pointing, glancing or diexis). In this perspective it is not surprising that spatiality is heavily grammaticalised in all natural languages [1].

Of course, space and objects have some form of objective existence "out there in reality" irrespective if someone perceives them or not (there is no reason to be a Berkeleyian), but it is also true that everyday experience of space is imbued with cognitive expectations and assumptions that we project on to that reality.

"Space" is always mediated through some form of *understanding* on the part of the observer, including his or her biological, cognitive and cultural constitution, as well as the pragmatics of the situation. For instance, Lakoff and Johnson [2] argue that *up–down* relations, which are extremely basic forms of spatial understanding, would be difficult to comprehend for a creature without a weighty body or for a creature living on a planet without gravitational force. Humans experience up–down relations as "objective and natural properties of reality" since bodies and gravitation are so fundamental features that we take them for granted and assume that we have a direct access to "objective reality". "Space" is thus not a static entity, but a constant activity and a negotiation between an external reality and cognitive/motoric work performed by the mind/ body complex (as well as socio-cultural expectations on space). Space is not objectively "out there" nor totally "in here", but in-between the two. It is this *spatial understanding* the present chapter will deal with.[1]

This discussion also applies to *representations* of space, or, to speak in computer-supported co-operative work (CSCW) terms, *information spaces.*[2] Whether we deal with text, hypertext, virtual reality (VR), computer games, cinema, paintings, photographs, icons, multi-user domains (MUDs), information bases or audio, the meaning and interpretation of these will not only be a matter of how that representation is formed and designed, but what competence the user-spectator-worker-player-reader is equipped with and the purposes of the situation. Represented space is not "in the representation", but is only meaningful through an active recipient on a specific occasion of use, thereby giving rise to different experiences of the same information space. The importance of the reader and pragmatics has long been known within the humanities (interpreting literature or cinema) and has lately also made it into the computer community (e.g. [4]). This does not imply that study of design and formal features of information space can be disposed with. On the contrary, the way information spaces are organised greatly affects the way they are perceived and used. But this has to be paired with a broader framework incorporating the user's understanding of the design and the functions it is said to serve.

The purpose of the chapter is to compare the understanding of mainstream cinematic representational techniques with those of computer interfaces. What techniques does cinema use in order to represent a space and its objects? What spectator competencies and expectations do these techniques exploit in order to work as well as they do and what purposes do they serve? Why is space in most mainstream films experienced as clear and coherent, while computer users customarily "get lost"? Can we find cinematic techniques in computer interfaces? Can computer interface design learn something from how film-makers and film industry "design" moving images?

---

[1]Compare this with Harrison and Dourish's distinction between space and place: "Our principle is: 'Space is the opportunity; place is the understood reality' " [3].

[2]I will use these two terms more or less interchangably. Also, note that *representation* refers to *external* representations and not cognitive–mental ones.

11.1.1 Different Beyond Compare?

Although cinema and computer interfaces are both "graphical" in a loose sense of the term, they seem to be poles apart in other respects. Before we examine the different techniques of the two, first we will discuss some of the apparent differences between the two media.

First and most obviously, cinematic and digital representations seem to serve fundamentally different purposes or functions. Computer interfaces are mainly used as *tools*, and thus designed with the purpose of maximising this feature. Computer interfaces are spatialised (buttons, menus, windows, pages, etc.) in order for the user to solve some problem or execute a task more efficiently, whether this be word-processing, layout, information seeking, etc. Mainstream cinema on the other hand is inhabited with objects, characters and actions, not to solve some problem, but in order to tell stories and to have emotional effects on the spectator. Spectators do not make the effort to go to the cinema (or go to the video store) because they have work to be done, but because they want to be entertained, surprised, frightened, excited or whatever. Obviously, the design choices of representations will be hugely influenced by these different purposes.

It should however be noted that there is nothing inherently necessary in such a division. There is no medium-specific feature of cinema that makes it better suited for narration and pleasure, and not as a tool. Documentaries and news are used by people to learn new things and gather information about the world in a "toolish" fashion. And computer games and MUDs are cardinal examples of computer space used for narrative and pleasurable experiences. In this way I think the apparent discrepancy between cinema and computer space is not about the representational carriers themselves, but the way they have come to be *used*. Media history teaches us that such functions are determined by the economical, societal and cultural factors involving the emergence and development of each media. For instance, the purpose of moving images was in the beginning of this century heatedly debated (science, documentaries, museums, circus, entertainment, poetry or what?), and although all of these functions do exist today, the entertainment function is perhaps the most central. This was not the outcome of some ingenious individual film-makers discovering the "essential" function of the cinematic medium, but an effect of the historical circumstances of early cinema. A good example of this is the narrativisation of cinema, which took place ten years after the actual invention of cameras and projectors. Up until then cinema was primarily concerned with singing, dancing and small sketches (1–3-minute reels), since it was exhibited in vaudeville, music halls, burlesques and other working-class entertainment forms. However, in order to attract a more stable and reliable audience, the industry turned to high-class art forms like theatre and narratives from literature, classical myths and religion. Thus the economical determinants of the industry pushed the way for an integration of moving images and narrative form, and it was in this wake the style and representational techniques of today's mainstream cinema developed [5–9]. Computer technology, on the other hand, emerged in the traditions of mathematics and engineering and was from its inception part of the academic and military complexes. Since computer technology was in the hands of scientists, their conception of the functions of computers determined its future. With the emergence of entertainment such as games, chat environments, interactive narratives and computer art, these functions are becoming more diverse and complex often embracing entertainment and social function as well as the narrative form [10, 11]. Such shifts typically place new and very different demands on design.

Secondly, and connected to the first point, mainstream cinema does not allow physical–manipulative interaction the way interfaces do. Games, databases, MUDs, hypermedia, the Web, word-processing programs and other kinds of software represent objects and space that the user is able to manipulate and rearrange. In word- or graphic-processing software, the user *creates* and *changes* documents and images. In cinema nothing of this nature seems to be evident. The spectator cannot modify fictive objects or warn or punish characters (even though he or she would like to). Of course, understanding and interpreting films and narratives requires that the spectator cognitively and emotionally interacts with the images on the screen, but this kind of interaction seems different if not in principle, at least in degree to the physical–manipulative one.

Thirdly, because of the different functions of the two, the space constructed in mainstream cinema will typically be *linear* in form, whereas the tool function demands some form of *non-linearity*. Hypertexts, the web and graphical software, in general, are constructed to encourage exploration and leave it up to the user to choose temporal order of actions and/or information. The fictive space of mainstream cinema is presented to the spectator in a particular order. First we get to see this and then that. This fits the function of narrative and emotional reactions very well, since suspense, surprise, humour, tragedy, horror and identification with characters are highly dependent on the fact that the right kind of information is presented to the spectator at the right moment. Of course, nobody *forces* the spectator to see the information in a particular order: you can choose to see scenes in any order you like (manipulating your video player), but then you will probably not experience those cognitive and emotional effects intended by the film-makers. (Non-) linearity is nothing *inherent* in the medium, but more of an agreement between sender and receiver: "if you read this in the order I specified, you will get all this good stuff!". Of course, that agreement can be broken by computer representations as well as cinematic ones. In the Taiwanese film *Blue Moon* (1998) it is up to the projectionist to choose the order of the reels (each one of them containing credits in one form or other). In these cases, just like in every hypertext or interactive narrative, the content has to be carefully designed in order to "fit" with any order of presentation. However, by handing over this liberty to the spectator or projectionist, the film-maker also hands over one of the most important means to achieve the emotional effects associated with the narrative function. The same is true for computer games and interactive narratives. In order to create some form of temporal linearity, these forms are seldom totally open (like the web), but the possible choices are restricted at every node. In this way the game "forces" the reader collect information or props that will be needed in future situations.

Finally, there is typically a difference between mainstream cinema and interfaces in the way that they adhere to a realistic or abstract representation form, or to put it in art theory vocabulary: to a figurative or non-figurative tradition. The nature of (non-) realism has been on the humanities' agenda since Plato, and I do not intend to explain it here. My distinction is a pragmatic one: with realistic representations I mean those that exploit and trigger everyday psychological competence of perceptual and cognitive nature. Mainstream cinema is realistic in this sense, because a great deal of the understanding of those images depends on our everyday knowledge of space, time and causality (and, of course, a host of other types of knowledge). Abstract representations, on the other hand, are formal structures of form, shape, colours and relationships, and often deal with higher cognitive functions, of course based in everyday understanding of the world, but not primarily dealing with it. A flow chart, a grapical-user interface (GUI) interface like the Macintosh or Windows, a Mondrian painting or a tree structure are good examples. They often fulfil different functions than the realist representations (see Fig. 11.1).

When I claim mainstream cinema to be prototypically realistic, I do not mean to imply that this *has* to be so. Abstract films have been around since cinema's origin, but since mainstream cinema took on the narrative task early on, it had to deal with depiction of characters and their actions and so came to fall into the realist furrow. And although abstract form seems to be the characteristic of the computer interface (again a product of its function as a tool), recently more realistic interfaces have been popping up, particularly on the "infotainment" sphere; (multimedia, games, collaborative virtual environments (CVEs)). Realism and abstraction can be seen as two poles dividing a spectrum, on which different types of representations fall. Modjeska's work on *textual* vs *spatial* structures investigates the effects of design choices falling on different points along that

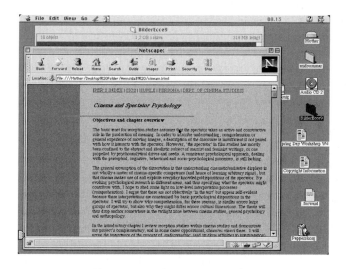

Figure 11.1 An abstract Mac interface.

spectrum [12]. In the more concrete discussions below I hope this distinction becomes somewhat clearer.

In summary, the different functions of cinema (narrative) and computers (tool) heavily influence the ways in which manipulative/non-manipulative, linear/non-linear and realist/abstract parameters are being handled and exploited in each medium. However, as long as we keep these distinctions in mind this does not hinder us from investigating and comparing more concrete levels of representational techniques and conventions. It is to this mission that we now turn.

11.2 Understanding Cinematic vs Digital Representations

11.2.1 *Mise-en-scène*

The way film-makers stage the scene *within* the frame is often referred to as *mise-en-scène* and can include composition and arrangement of objects, lighting, acting and character movements, colours, camera angle and so forth. Spectatorial understanding of this space includes many parameters.

First of all, the spectator uses his or her perceptual and classificatory knowledge to recognise and identify objects. The cinema goer does not have to learn that an image of a horse depicts a horse, since the photographic carrier is realistic enough to trigger the same recognition processes that the spectator uses in real life. If you can recognise a horse in everyday life, then you can do it in the theatre or in front of your television with no additional effort. This seems not to be the case with abstract space. In spite of the aspiration of personal computers to mimic real documents, files, folders and scroll bars, the appearance of, and mode of interaction with, digital objects are quite different from reality, and require some form of learning. Admittedly, for files, folders, links and menus, this transition period is very short for most people. We quickly integrate these as "everyday objects".

Some psychologists have suggested that some angles of a given object are more typical than others, i.e., views in which the recognition process is faster and more accurate [13–15]. In many orientations the object "hides" parts of itself. Such effects often occur when the viewpoint of the image is along the axis of the major component of the object (e.g. a body or other elongated forms). For instance, a cow drawn from above conceals all of the four legs. And there may be a frontal angle from which the two front legs hide the two rear legs and foreshorten the body. In these cases, important structural features of the "cow" are out of view, impeding or ambiguating the recognition process. This phenomenon can be exploited by the film-maker to create certain effects. In most mainstream cinema, easy recognition is desirable and typical views of objects are preferred. Cows and people are seldom shot from above or from behind. Some scholars have even suggested, on the basis of psychological research, that a three-quarter view of a face (cf. Fig. 11.2) is more typical than frontal (character looking into the camera) or profile views, which would explain why three-quarter angles are preferred in mainstream cinema [16]. Of course, in some contexts, easy recognition is not desirable. In *Psycho* (1960) a mentally disturbed Norman Bates dresses like his long-dead mother and randomly kills clients checking into his run-down motel. The spectator, however, is lead to believe that the mother is, in fact, alive and is the one who executes the killings. In order to keep up this suspense and withhold the real identity of the "mother" the dressed-up Norman is occasionally shot from straight above (e.g. when the detective gets killed). Here the atypical view serves important narrative and emotional purposes.

Comparing this with objects of a computer interface, icons of files, disks and folders and so on are presented mostly in typical views (cf. Fig. 11.1), which works well with the purpose of such environments. Disambiguous, fast and clear-cut object recognition is good if one wants to support efficient work with these objects. In games and other entertainment forms, the purpose might be different, thus exploring less canonical views.

Mise-en-scène also refers to the way objects are placed in relation to each other within the frame. Impression of three-dimensional space is an effect of such relations, and the cinematic image generally presents a wide range of depth cues:

- *Occlusion*: object A partially hides object B, therefore B is probably further away.
- *Relative size*: B appears smaller than A, therefore B is probably further away.
- *Height in visual field*: B is higher up than A in the picture plane, therefore B is probably further away.
- *Aerial perspective*: B is bluer or of lower contrast than A, therefore B is probably further away.
- *Motion perspective*: when camera moves, A moves faster than B in the opposite direction of the movement, therefore B is probably further away and so on (cf. [17]).

Depth cues are pictorial features as well as perceptual competence on part of the spectator. Again, if we can perceive depth in real space, then depth in images will not have to be "learned" (in fact it is difficult to *avoid* depth perception when looking at a photographic image). Impression of depth emerges in the clash between the depth cues of the image and the perceptual competence of the spectator, in spite of the fact that the image on the screen (and on the retina) is only in two dimensions. Three-dimensional space exists "in-between" the image and the mind of the spectator. This imaginary deep space is often exploited by film-makers to stage action in certain ways, serving narrative or aesthetic purposes.

In contrast, abstract digital representations use depth cues meagrely and inconsistently. In PC interfaces, occlusion is the major one (windows hide other windows or files – cf. Fig. 11.1). But occlusion only gives *ordinal* information, not *metric*; the user understands that an occluded object is further away, but on the basis of this cue only it is impossible to estimate distance. Perhaps some users notice the difference in size between a closed folder and an open one and interpret this as a depth cue. If the user assumes that the open and the closed folder is the same object with the same absolute size, then the difference in size on the screen between these two, may function as a metric depth cue. Such understanding may seem a bit far-fetched for some of us, but maybe this is how users out there make sense of their interfaces. Could the relative size cue be used in interfaces to support navigation? What if the user could arrange objects on his or her desktop, not only horizontally and vertically, but also in depth, having objects gain size when closer to the user, and loose size when placed "further away"? In what way should such a feature change our experience of the interface, and would it be helpful?

It might also be possible that some users interpret the vertical arrangements of icons on a given desktop as a height-in-visual-field cue. In such an understanding, the trashcan, often lingering at the bottom right of the screen, would be considered to be closer than icons placed higher up in the field, in spite of the fact that these icons do not get smaller (as they would in the real world). For modern westerners like us, this inconsistency between height-in-field and relative size cue seems unnatural, but in the depth cue system of post-twelfth-century Persian painting, this was the dominant mode of representation.

Besides object recognition and perception of depth the major function of *mise-en-scène* in cinema is to direct attention of the spectator to areas within the frame that contain narratively important information [7]. In order to present visual information it is necessary to create *hierarchies* of significance in the image. Lighting is important because it makes possible to subdue or even hide certain areas and emphasise others. Lighting from behind (*backlighting*) and thus creating a contrast of luminance, is an extremely common technique to make objects and characters stand out from the background (cf. Fig. 11.2). So is movement. Another strategy is foregrounding, which means to place more important objects in the foreground and less important things in the background. It is fascinating to imagine these attention directing techniques in abstract interactive information spaces (a lighted desktop?), and in games and

Figure 11.2 *Midsommarafton* (1995).

VR environments many of them are already being employed, bringing the trend of realist *mise-en-scène* and interactive information space together (cf. *Lighthouse* and *Riven*). The major problem in computer interfaces is, however, not the attention directing techniques themselves, but *when* to attract the user's attention and to *what*. In a linear form like narratives the film-maker knows what is important or not, but in interactive information spaces it is almost impossible to know what information is most relevant for the user and in what moment. Here, we are afraid, cinema can be of little help.

11.2.2 Close-ups and Object Permanence

Another crucial parameter of the human perceptual system is that it expects objects to have *permanence* or *continuity* even though they are partly or wholly occluded from view. Although I can only explicitly see the upper half of the chair in front of me (because the table blocks the view), I do not presume that it is cut in half and floats in the air, but rather I presume that it continues behind the table (for the infant's development of such expectations on the world, see [18]). The *close-up* of cinema and other pictorial arts exploits this spectatorial expectation. Although the frame ends at the neck of the woman in Fig. 11.2, the viewer does not interpret this as a picture of a decapitated head. The realism of the image triggers the expectation of the woman having a body continuing outside the frame into *offscreen space*. The spectator assumes the fictive space to possess the same features as the space of everyday life. The intuitive understanding and "naturalness" of this convention is due to the fact that this everyday-life assumption is so basic and fundamental. The same process is present whenever the film-makers indicate offscreen presences by having parts of objects bursting into the frame (shadows, shoulders, hands with guns, cigarette smoke, etc.).

The close-up convention was not a part of the early cinema style. Here, again thanks much to the vaudeville exhibition context, scenes were shot through long-shots, in an "overview-style", displaying objects and their spatial relations *within* the frame (Fig. 11.3). All objects and characters were displayed in total, and the camera "respected the wholeness" of the action. If we define off-screen space as that space not visible on the *screen*, but still part of the *scene* [19, p. 495], then off-screen space was almost irrelevant in early cinema style. The close-up convention, on the other hand, emphasises this suggested space and place; a somewhat higher demand on the cognition of the spectator. Of course, the introduction of this convention was pushed by cinema's move into narrative form (as was editing in general): close-up is a very effective means to direct spectator's attention to important parts of the scene (guns, facial reaction of characters, etc.)

The close-up device is put to use in graphical MUDs and in CVE where it is possible to zoom in and out and get closer views of people and objects. Close-ups in these worlds are understandable since the representation triggers real-world expectations: the graphics display interiors or exteriors with everyday-like

Figure 11.3 *The Life of Charles Peace* (1905).

objects, such as whiteboards, carpets, beds, trees. The space "stretches out" beyond the frame; concepts of left–right/up–down off-screen space are meaningful; objects look and behave more or less like everyday objects.

However, many (if not most) interfaces are not realistic in this sense. In *abstract* environments like the operating systems of Windows or Mac OS, the background is undefined (neither interior nor exterior, no horizon or walls), the objects (buttons, menus, document icons, windows) do not look like or behave like everyday objects (they are flat, float around, make windows pop up), and things might happen here that have no equivalents in everyday life (scrolling). The space off-screen (right–left or below–above) does not contain anything in particular and does not trigger any particular off-screen space expectations. Everything of interest is contained within the frame. The landscape does not "stretch out" into the distance in any direction.

Since this kind of representation fails to trigger user's everyday expectations on objects and space, it is difficult to exploit object permanence composition techniques. For instance, we have on several occasions observed fairly experienced computer users browsing a number of web sites and were struck by their inability to understand that information continued further down the "page" (accessible through scrolling). They considered the screen to display the whole situation (just like the early cinema spectator). Unlike real-world objects, web "pages" and "documents" are not everyday objects, and the user has no expectation as to what a "whole" document or page looks like (but see the hypertext documents in [20]). Would the problems be precluded if the design triggered such a conception in the user, for instance by designing web pages as human bodies?

Could we use partial views to suggest other users' presence in information space? Instead of having the ICQ indicator blinking, we might think of cigarette smoke or a shadow bursting into the frame. Could we perhaps turn our vision of field in direction of the source? That would be a considerable shift from abstract to realistic representation, suggesting the presence of another user in a particular direction (off-screen left–right, up–down), and so make the user apply his or her everyday expectations of space instead of abstract knowledge. Would that be helpful for some users and in some situations?

11.2.3 Gaze Direction – Point-of-View Editing

In everyday life following another person's gaze is a very basic and useful behaviour [21]. Joint visual attention creates a shared semantic space in which reference and meaning are naturally established, thus constituting an important proto-communicative setting for adults and infants. By looking at the same object as the gazer, we can also speculate about his or her mental disposition towards the object (beliefs, desires, feelings, intentions, etc.). This behaviour is not restricted to humans, but shows up in quite complex forms in animals which indicates its fundamental function [22, 23].

Gazing behaviour in cinema and cartoons is extremely pivotal in order to relate different shots and spaces as well as activating off-screen space. Consider Fig. 11.2 again. Following the gaze of the woman suggests to us that she is attending to some object/event off-screen, although we cannot see it for the moment. The image activates a "cognitive curiosity" on part of the spectator. If from this shot/image we cut or transfer to another shot/image of, say, a burning house, then we most likely infer that the woman is looking at the house. When the shot of the woman is shown we assume that there is something occupying off-screen space (left), and when the shot of house is shown, we infer that the woman is somewhere off-screen (behind/right?). This juxtaposition of spaces thus generates spectator inferences about the woman's *point-of-view* and to what she is attending, but also about the spatial closeness between the house and the woman (within sight's distance). This trope is prevalent in all genres of moving images.

It is interesting to note that "the woman looking at the house" is never explicitly present within the image frame. The spectator never explicitly *sees* the two together. The spatial relation between the shots only belongs to the imaginary, mental space that the spectator is *constructing* in her mind during viewing. In reality, these two spaces may never have met at all: the woman may be shot in Stockholm and the house in Edinburgh. But since our gaze competence is so basic and fundamental in everyday life, this understanding of the two shots happens habitually and unobtrusively [24].

Although recent digital environments are occasionally inhabited by agents, avatars, personal guides and assistants, gaze behaviour is seldom present. *Microsoft's Office Assistant*, for instance, is communicating textually/verbally and by bodily postures, but pointing and gazing at objects, within as well as outside the frame, would perhaps enhance the support. In graphical chat environments like *Palace*, the avatars are represented as smile-faces, but they cannot look at or display gaze behaviour like in social situations. In other collaborative environments like *DIVE* or *Active Worlds* the avatars are detailed enough to exhibit gazing behaviour. The user can turn around and change point-of-view (through keyboard and mouse commands), and other users can see in which direction the avatar is looking. Here off-screen space and gazing direction cues work in similar manners as in cinema (and the real world), and seem to be central for the understanding and experience of the space.

11.2.4 180° Convention

Another way to connect two shot spaces even more tightly with gazing behaviour is to substitute the burning house shot with *another* person looking, generating the impression that the two are looking at *each other*. Multiple gazes, however, make directions and spatial relations between shot spaces more complex and in order to support clarity mainstream cinema typically employs another rule of thumb. The 180° convention says that directions within the scene form "invisible" lines over which the camera is not to cross. Gazes and movement constitute such directions. Consider a scene from *Notorious* (Fig. 11.4), in which the gaze line between the two characters forms the 180° line. An establishing shot (camera position 1) informs the spectator of the overall spatial arrangements: A is to left-hand side of the image looking right, and B is on the right-hand side of the image looking left. These directions of gazes are maintained in the close-ups (camera positions 2 and 3), in spite of the fact that the angles of the camera have changed. However, if the camera would cross the 180° line to camera position 4, then suddenly A would be looking *left* (this shot is not included in the actual film – I have just inverted the shot from camera position 3). This might create some form of disorientation: the shot seems to suggest that A is looking *away* from B and not *towards* her. Crossing the 180° line endangers the spatial coherence

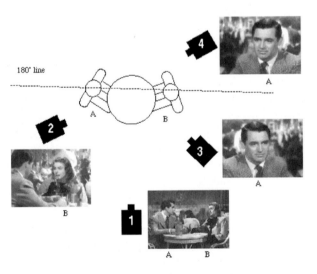

Figure 11.4

between shots and makes the spatial relations ambiguous. In a narrative form of art like mainstream cinema, which tries to convey story information as effectively and unambiguously as possible, such crossings are extremely rare. If they occur at all, the crossing is always explicitly marked in some form (like having the camera track over the line within the shot). These rare cases might be valuable for anyone who seeks to explain the 180° convention, because although it is frequently used and described within production and cinema studies, no one has, to my knowledge, tried to give a psychological explanation of why this effect arises.

Not many systems let the "camera" move around and change position in virtual space. In some games, and most CVE systems, the user's perspective is the camera's, and it follows where the user goes (either the "camera" is placed in the very eyes of the avatar with only hands visible (*Riven*), over the shoulder (*Tomb Raider*) or with the face of the avatar in full view). The user moves around, but the perspective stays with the user/player, which means that 180° convention is not applicable or necessary. In other systems the perspective of the camera is more "objective" and we get to see characters from an outsider's point of view. In games like *Full Throttle* (1994), the editing is carried out with cinema style consciously in mind, taking great care to follow the 180° convention.

Teleconferencing seems to be a suitable area for applying point-of-view (POV) and 180° conventions in CSCW. In a conference situation, gaze behaviour is crucial to enable effective communication, but the complex spatial relations between monitors, cameras and participants make it difficult to support this in an intuitive fashion. Here, the study of cinematic visualisation techniques could be fruitful. The 180° convention has been proven successful in cinema for the past 80 years and there is no reason to believe it will fail in other contexts.

In some dynamic 3D environments used for scientific and educational applications within molecular biology, computer engineering, and medicine, the user is freed from camera control since the tasks performed are complex and demand total attention [25]. In these systems the real-time camera planning is carried out by the system and entails selecting camera positions and viewing directions in response to the user's activities and object manipulations. Since the system has to make on-the-fly decisions about camera angles and distances, it is perhaps in this context that cinema editing techniques have their most useful application. In order to provide a continuous space and reduce the disorientation in the user and for him or her to execute the task as efficiently as possible, the 180° convention as well as other editing techniques would be worth considering [25].

11.2.5 Spatial overlaps

Early cinema's adoption of the narrative form around 1905–1915, placed a demand for editing and breaking up the long shot "overview" style deployed earlier (Fig. 11.3). This was partly in order to direct spectators' attention within the scene, but also to make spatial and temporal transferences (think of such crucial narrative features as ellipses and flashbacks). However, although closer-ups and cuts (of course in collaboration with acting and composition techniques) increased the articulation possibilities and hence the narrative potential, they also fragmented the long shot space and disoriented the spectator. The close-up's reliance on object permanence knowledge, the POV and the 180° convention all aimed at trying to subdue such fragmentation and support the spectator in establishing spatial relations between different slices of space. Another popular technique is to initiate a new scene with a wide *establishing shot* (Fig. 11.5). Then the camera typically cuts to closer view of the scene as the action or dialogue gets more intimate (Figs 11.6–11.8). Although shifting angle and position, the closer-ups stay within the same spatial segment as the establishing shot. This editing technique is called *spatial overlap* (Fig. 11.9).

Figure 11.5. *Notorious* (1946).

Figure 11.6

Figure 11.7

Figure 11.8

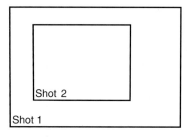

Figure 11.9 Spatial overlap.

Establishing shots undoubtedly have a number of functions, but an important one is to give the spatial setting for the following close-ups. If the film-maker would have cut from a preceding scene to the close-up of Fig. 11.8 the spectator would have had some orientational problems: where are we? What kind of situation is this? With whom is he talking? What is the spatial relation between this shot and the former scene?

As the scene is edited now, we first get an overview of the street, then the table with *both* characters. After that we cut closer to medium-shot, still with the other character's shoulder within the frame, until we reach the close-up of a face. Although this shot has no *explicit* clues to what goes on in the surrounding off-screen space, the former shots have presented the spectator with this information. Overview shots and spatial overlaps provide information and expectations on off-screen space in close-ups. If the film-maker wants to support the spectator's construction of the spatial layout, spatial overlap is a useful technique.

Computer interfaces often use overlapping strategies. Different sorts of *index pages* (on web sites, file systems, CD-ROMs, etc.) basically have the same function as the cinematic establishing shot. An initial overview provides the user with some conception of what is to be found outside the frame of the subfile he or she will be visiting. The index page sets expectations on the *off-screen information* of the upcoming subpage.

This comparison does halt slightly, however. Once again the computer space is often much more abstract than cinematic space. In contrast to the scene from *Notorious* above, in which the establishing shot provides the spatial surroundings and specify the off-screen space of the upcoming close-ups (in Fig. 11.8 we infer the woman is somewhere off-screen right), the index page does not establish *spatial* relations, but rather *semantic/structural*. When visiting a subpage, the user cannot tell whether other subfiles are to the left or right, and this is characteristic of abstract space.

Could we support navigation if we spatialised the index- and sub-nodes, and if the movement from the former to the latter were experienced as a zooming-in or a cut-in? Many three-dimensional browser environments seem to employ these techniques. However, we think spatialisation *per se* is to help navigation only if it is accompanied with a coupling structure-content. There has to be some relation between the spatial relation and the content ("the same kinds of things in the same surroundings"). How this could be applied in the large and constantly changing information environment such as the web will be a critical challenge.

11.2.6 Sound

Within cinema studies research on the sound track has been marginalised in comparison to studies of the image track.[3] The sound cannot be described and

---

[3]The same hierarchisation seems to be in place in perceptual psychology, with vision receiving far more attention than hearing or other sensory systems.

printed in books as easily as images. Many film scholars originated from an art study background and their preferences lay more on the pictorial side. Yet cinema was never silent. Music, singing, lecturers, live behind-screen sounds, and so on were, from the very beginning, pivotal elements in cinema exhibitions. In today's mainstream cinema, sound creates strong effects and yet remains quite unnoticeable by audiences. However, its relation to and reinforcement of the image track are many and complex and certainly worth studying. I will deal here only with its spatial functions.

Cinema scholars tend to divide sounds into *diegetic* or *non-diegetic* sounds [19]. In the latter case the sound comes from without the fictive world (*diegesis*). It could be a voice-over of a (non-visible) narrator or typically a musical accompaniment. The source of these sounds is not situated within the story universe. Voice-overs often have an authorial force, constraining the interpretation of the image and directing the attention of the spectator. Music typically has emotional functions.

Diegetic sounds, on the other hand, are issued from sources either *on-screen* or *off-screen* but still within the fictive world. The sound of a firing gun accompanied by an image of a gun firing, would be a good example of an on-screen sound. Mostly, the volume of the sound is regulated to the visual distance from the camera to the source, thus in practice to shot scale (long shot, close-up, etc.). Loud volume indicates closeness and this can be used to create certain effects (cf. the sudden bursts of loud sounds in horror films, often in combination with unidentified objects bursting into the foreground).

On many occasions, however, we hear sound whose source are not within the field of vision, i.e. either occluded by some parts of the *mise-en-scène* or wholly off-screen. In dialogue scenes, the camera may shift to a close-up of character A but we still hear character B talking. Or in a close-up of a character we hear a sound, triggering the character to look off frame, and thereby anticipating an upcoming shot of the source of the object. Or the spectator might never see the source of the sound at all, but have to infer the presence of objects and event on the basis of their sound only (in co-operation with other off-screen indicators). Off-screen sound in mainstream cinema has two major functions: first an overlapping sound from one shot to another *smoothens* the disruptive force of the cut. To speak in cognitive terms, it supports the spectator's construction of the fictive space and promotes the object and space permanence talked about earlier. By having character A talking although the camera cuts to character B, we know that character A is still somewhere in the vicinity although we cannot see him. The overlapping and consistent soundtrack makes the status of the second shot clearer: this new slice of fictive space is near the former one. Also, the volume and timbre of the off-screen sound often suggests the distance between the camera and source. A close-up of character A with B talking off-screen in a low voice, suggests to us that they are sitting quite close, although we might not see B at all in the shot (the scene might even lack an establishing shot). In another case the volume of A screaming, might decrease drastically as we cut from a close-up of A to a close-up of B, suggesting that A and B in fact are distant.

Another distinction within diegetic sounds is between *external* or *internal* sources. External sounds have a physical source, whereas internal sounds emerge from "inside" the minds of characters. A character's inner speech, while on the screen, would be a good example of internal or subjective sound. While external sounds are available to other characters nearby, the range of internal sounds are typically confined to one character.

Can we describe sound use in digital information spaces with these distinctions, and can they provide design suggestions? Well, although it is uncertain whether computer interfaces are *fictions* (and thus making the concept of *diegesis* problematic), they do have *action space* or *virtual space*. Generally, I think most sounds in conventional software have their sources in this space: files and folders being opened and closed (a zipper sound); pushing a button generates a click; *the Office Assistant* hopping around. In many cases the sound is related to the actions of the users and functions as feedback to confirm an action taken. It is also exclusively connected to some form of movement and change in the visual display (a button gets highlighted; a file swings open), which rhymes well with sounds in everyday life and cinema. The sound designer of abstract digital space also faces the problem of creating sounds for a source the user has never before encountered. Analogous to laser guns in science fiction films, folders and files or links are virtual objects and the user has no clear-cut expectation on how they are supposed to sound. The designer has to make some estimation of the credibility of sounds (are there some constraints on how a folder can sound?), but, as with laser guns, we quickly learn and accept sounds connected to fictional objects. The design of these sound is important because it can give the user suggestions on how to interact with a given object, in the same manner as visual metaphors are said to function.

Sometimes, however, the (non-) diegetic status of the sound is hard to determine. Is the source of an error message within or outside action space? Could not this type of sound be seen as an authorial voice-over, from the "system", not really related to the ground action (analogous to film music)? Or perhaps the only non-diegetic sound is the processor, which on many occasions can be a crucial source of information about what is going on in action space.

I have been making the argument that abstract interfaces do not generate expectations on off-screen space and in that respect it is rather natural that all sounds in these information spaces are on-screen sounds. When realism is introduced and space is understood to continue outside the frame, the use of off-screen sounds becomes much more customary. The ambient sounds of *Riven* and *Myst* are cardinal examples of environments that create a stark consciousness of off-screen space. Just like in cinematic editing, the cuts between different views are smoothened by the soundtrack, making it easier for the player to maintain space permanence and construct a cognitive map of the environment. Or in the realistic information space of *Active Worlds*, we can hear people talking in spite of the fact that they are not in the field of vision: it is not until we turn around that we see them.

This is of course not to say that abstract interfaces could not use off-screen sounds. The idea of having different "crowd sounds" accompanying a web document depending on how many visitors it has at a particular moment is a good example of an off-screen sound in an abstract information space. Here, however, the source of the sound does not have any *spatial* relation to the objects within the frame (the user does not conceptualise the other users as off-screen left–right or up–down), but seems itself to be abstract and seemingly coming from everywhere and nowhere. Its purpose is not to trigger spatial under-standing, but to give a general and abstract estimation about the number of visitors. Such sounds *could* give spatial information if we emphasised the spatial dimension of web browsing (three-dimensional browsers), and had each visitor represented in that space (*Active Worlds*). But then again, we must ask ourselves what spatiality and realism contribute in terms of navigability, usability and efficiency. The choice between abstraction and realism has to take into consideration the type of task, the type of user and the type of information presented. It is thus impossible to give any general design recommendations.[4]

11.3 Concluding Comments

CSCW in general, and ethnomethodological studies in particular, have shown that space and spatial understanding are central components of many work environments. The ways artefacts and tools are placed, manipulated and handled in space, thus constraining and enabling actors to behave and cognise the way they do, often provide important keys into the essence of a given work situation [27, 28]. When CSCW researchers try to support work or create totally new information spaces for workers to inhabit, it is crucial to take spatial understanding into account in order to make the right design decisions. What spatial expectations does the actor bring into the understanding of the system or situation? On the basis of this, what types of spatial understanding are triggered, enforced or subdued by a given representational technique? And how does this serve or hinder the purpose the system is said to serve?

I think we can better understand the affordances and properties of computer and digital information spaces if we compare them with the techniques employed in other visually and auditively rich "artificial environments" such as cinema. Understanding the *understanding* of cinematic representations may not only equip us with tools for how to analyse understanding of digital information spaces, but may even suggest new design. The present chapter is a first step towards such a comparative investigation, including not only descriptions of the techniques themselves, but also the *context in which they are used*, and foremost the knowledge and assumptions the spectator/user brings to the situation.

There is also, I believe, a lot to be learned from the history of other media. The comparison between information technology and cinema appears to be particularly instructive since there are an abundance of similarities. Both are forms of representation are deeply ingrained in technology (in contrast to

---

[4]See [26] for a more thorough and less cinema-studies-oriented view of sound in interface design. Many of their ideas gel well with those presented here.

literature), and in that sense children of modernity of the western world. Both are technologies that are imbued in commercialisation and consumption – they became industries fast. The emergence of both were intimately associated with juridical and commercial fights over patents and standards, effectively sorting out alternative technologies early on. The emergence of both raised questions about the functions of this new technology, with similar types of discussions and arguments. Both spurred the emergence of academic institutions (departments, publications, conferences). Both media are also surrounded by popular magazines providing a discussion forum for the "everyday person" (PC magazines and film-fan magazines). These publications often promote certain individuals as idols or stars, maintaining a cultural iconography (Harrison Ford, Lara Croft, Bill Gates, Steve Jobs). The censorship debate has always been present throughout the history of cinema and now information technology is entering this field of discourse – with remarkably similar arguments being made. With computer technology turning into media, questions of mass communication come into the fore, just like in cinema. These include most forcibly issues of ideology and the cultural effects of the medium. And finally, both are very young media, with a lot of historical evidence still easily available.

Information technology's deeply modern profile, with a strong emphasis on technology in general and *new* technology in particular, makes it rather unfashionable to look backwards. But perhaps the lack of sense of direction within this field is due to the fact that computer science has not yet written its history.

Finally, on several occasions I have hinted at and possibly also promoted a more "realistic" design of information spaces. Triggering expectations on space and objects is good because these can be used in the interaction process (cf. [3]). If the user can apply everyday knowledge we would get closer to *natural design* [29]. On the other hand, we must ask ourselves if realism really is the objective here. It seems that the very point of making and constructing virtual worlds is to provide the user with activities that are *not* possible to accomplish in the real world. On my virtual typewriter I am able to erase, copy and edit text in a much more efficient (although more artificial) way than on my real typewriter. And in abstract space I do not have to cover distance between information nodes, but just "teleport" myself there in a quick and efficient manner. Thus, designing information spaces seems to be a trade-off between easy of use and efficiency. It is precisely that point of intersection we have to strike in order to attract users with low abstraction and computer ability and still build forceful tools that will accomplish new and exciting things.

Acknowledgements

Figure 11.1 Screen shot reprinted by permission from Apple Computer, Inc. Portions © Netscape Communications Corporation, 1998. All Rights Reserved. Netscape, Netscape Navigator and the Netscape N Logo, are registered trademarks of Netscape in the United States and other countries. Figure 11.2

© the author. All the photographic material in Figure 11.4–11.8, © American Broadcasting Companies, Inc. (1974).

References

1. Frawley, W. *Linguistic Semantics*. Hillsdale, NJ: Erlbaum, 1991.
2. Lakoff, G. and Johnson, M. *Metaphors We Live By*. Chicago, IL: University of Chicago Press, 1980.
3. Harrison, S. and Dourish, P. Re-place-ing space: the roles of place and space in collaborative systems. *Proceedings ACM Conference on Computer-Supported Cooperative Work*, Boston, MA, 1996.
4. Schmidt, K. and Bannon, L. Taking CSCW seriously. supporting articulation work. *Computer Supported Cooperative Work (CSCW)* **1:** 7–40, 1992.
5. Gunning, T. *D.W. Griffith and the Origins of American Narrative Film*. University of Illinois Press, 1994.
6. Musser, C. *The Emergence of Cinema: The American Screen to 1907*. New York: Scribner, 1990.
7. Bordwell, D., Staiger, J. and Thompson, K. *The Classical Hollywood Cinema: Film Style and Mode of Production to 1960*. New York: Columbia University Press, 1985.
8. Burch, N. *Life to Those Shadows*. London: BFI Publishing, 1990.
9. Elsaesser, T., editor. *Early Cinema: Space – Frame – Narrative*. London: BFI Publishing, 1990.
10. Murray, J. *Hamlet on the Holodeck. The Future of Narrative in Cyberspace*. New York: Free Press, 1997.
11. Erickson, T. From interface to interplace: the spatial environment as a medium for interaction. *Proceedings of the European Conference on Spatial Information Theory*. Heidelberg: Springer, 1993.
12. Modjeska, D. Spatial and textual structures in virtual reality. In *Proceedings of Workshop on Personalised and Social Navigation in Information Space*, Stockholm, 1998.
13. Deregowski, J.B. *Distortion in Art*. London: Routledge & Kegan Paul, 1984.
14. Palmer, S., Rosch, E. and Chase, P. Canonical perspective and the perception of objects. In *Attention and Performance IX*, Long and Baddeley, editors. Hillsdale, NJ: Erlbaum, 1981, pp. 135–151.
15. Biederman, I. Recognition-by-components: a theory of human image understanding. *Psychological Review* **92:** 115–147, 1987.
16. Bordwell, D. Convention, construction, and cinematic vision. In *Post-theory. Reconstructing Film Studies*, Bordwell and Carroll, editors. Madison, WI: University of Wisconsin Press, pp 87–107.
17. Cutting, J. and Vishton, P. Perceiving layout and knowing distances: the integration, relative potency, and contextual use of different information about depth. In *Perception of Space and Motion*, Epstein and Rogers, editors. London: Academic Press, 1995, pp. 71–117.
18. Harris, P. Infant cognition. In *Handbook of Child Psychology*, Volume II, Haith and Campos, editors. New York: Wiley, 1983, pp. 689–782.
19. Bordwell, D. and Thompson, K. *Film Art. An Introduction*. New York: McGraw-Hill, 1993.
20. Páez, L., Bezerra da Silva-Fh., J. and Marchionini, G. Disorientation in electronic environments: a study of hypertext and continuous zooming interfaces. Paper presented at *ASIS'96, The Annual Meeting for the American Society for Information Science*, 1996.
21. Butterworth, G. The ontogeny and phylogeny of joint visual attention. In *Natural Theories of Mind. Evolution, Development and Simulation of Everyday Mindreading*, A. Whiten, editor. Oxford: Blackwell, 1991, pp. 223–232.
22. Gómez, J. Visual behavior as a window for reading the mind of others in primates. In *Natural Theories of Mind. Evolution, Development and Simulation of Everyday Mindreading*, A. Whiten, editor. Oxford: Blackwell, 1991, pp. 195–207.
23. Ristau, C.A. Before mindreading: attention, purposes and deception in birds? In *Natural Theories of Mind. Evolution, Development and Simulation of Everyday Mindreading*, A. Whiten, editor. Oxford: Blackwell, 1991, pp. 209–222.
24. Carroll, N. Toward a theory of point-of-view editing. *Poetics Today* **14:** 123–142, 1993.
25. Bares, W. and Lester, J. Cinematographic user models for automated realtime camera control in dynamic 3D environments. *Proceedings of User Modeling '97*, Sardinia, 1997.
26. Macaulay, C., Benyon, D. and Crerar, A. Voices in the forest: sounds, soundscapes and interface design. In *Towards a Framework for Design and Evaluation of Navigation in Electronic Spaces*,

PERSONA Deliverable, N. Dahlbäck, editor., 1998, pp. 159–173. (Available at http://www.sics.se/humle/projects/persona/web/publications.html)

27. Hughes, J., Prinz, W., Rodden, T. and Schmidt, K. editors. *Proceedings of the Fifth European Conference on Computer Supported Cooperative Work*. Dordrecht: Kluwer, 1997.

28. Green, T. editor. *Proceedings of the Ninth European Conference on Cognitive Ergonomics*, University of Limerick, 1998.

29. Norman, D. *The Design of Everyday Things*. New York: Doubleday, 1988.

Chapter **12**

The Role of Wearables in Social Navigation

Odd-Wiking Rahlff, Rolf Kenneth Rolfsen and Jo Herstad

Abstract

Wearable computers are worn on the body and mostly hands-free computer systems. These new devices will facilitate new ways of supporting collaboration and social navigation in what we call wearable-supported collaborative work (WSCW). WSCW systems make it possible to remain closely in touch with the work while still accessing shared information resources and collaborating with others in the physical space augmented with information. Such systems are contrasted to the PC-based computer-supported co-operative work (CSCW) systems where the windowed interface is the main interaction channel. We discuss modes of usage, implications for communication, and illustrate this with scenarios from our work with tourist applications and police officers fieldwork. Finally some interesting areas for future WSCW research are given, and we outline some useful design components.

12.1 Introduction

> *What if you could see better with an instrument on your nose, or could tell the time by looking at your bracelet.*

Eight hundred years ago this futuristic vision for extending the human senses would have been regarded as highly unlikely. Nowadays, we take spectacles and wristwatches for granted. The new research challenge in empowering the user is to create wearable computers, or just *wearables*[1] for short. These new computers are computational and communicational devices that are worn as articles of clothing, jewellery, or accessories [29].[2] These systems made possible by the

---

[1] This type of computer has its roots back to the late 1950s and early 1960s where some UCLA students built a computer that fitted into a shoe. With this clandestine device the students increased their winning chances in Las Vegas casinos in 1961 [1]. The term "wearable computer" seems to have been coined first by Dan Siewiorek when describing the Vuman-2 wearable computer [2].

[2] These devices might just as well be named *personal communicators*, or *comlogs*, giving more emphasis to their communicative and logging aspects (credits to Nokia and sci-fi writer Dan Simmons of *Hyperion* fame for inventing these names).

miniaturisation and merging of information and communication devices, open up new vistas for human–computer interaction and co-operation and navigation for mobile users.

For desktop-based computer-supported co-operative work (CSCW) systems the co-workers are stationary, sitting in front of their computers and interacting with shared material through a classic Windows–Icons–Menus–Pointer (WIMP) computer interface. The windowed interface becomes the main channel for the manipulation of virtual objects representing the actual tasks or physical objects, for example. Situations exist, however, where one needs to collaborate closely and be computer supported while still manipulating the physical objects, and where the gap between the actual work and the manipulation of an abstraction of such work needs to be minimised both in time and space. A user of a wearable can remain closely in touch with the actual, physical work while still accessing shared information resources and collaborating with his fellows. In this respect the *wearable-supported collaborative work* or, as we suggest, WSCW for short has the potential of bringing the collaborators *one step closer to reality* by allowing mobile users to be in contact while doing physical work at different locations.

We believe that small unobtrusive wearable computers, which are location-aware and always active, may have significant roles for supporting *social navigation*, especially in the case of fieldwork. A wearable is ideal for supporting co-operation in those cases where the user has to move because one or a mixture of the following two situations applies:

- asynchronous collaboration involving a given place or object (like *inspecting a common object*)
- synchronous collaboration involving potentially moving objects (like *hunting moving objects*).

Furthermore, a wearable can support the user by providing means of teleconversations, for example:

- A novice field-worker may need the assistance of a remote expert. (*Remote consultancy*).
- A mobile expert may be available for several field-workers. (*Enhanced availability*).

In this chapter we will look into how mechanisms of *direct* and *indirect communication* through wearables may provide personal navigation assistance as well as collaboration support. The mechanisms are illustrated by examples from our current work, which focusses more generally on mobile informatics. We have investigated the use of personal digital assitants (PDAs), cellular phones and different devices that are context-aware, for example, GPS/DGPS and biometrics sensors, but which nevertheless illustrate important aspects of wearables. Our observations point the way to new mechanisms and capabilities that these systems require in order to support social navigation.

12.2 Defining a Wearable

We can loosely define a wearable computer as:

> *A never-sleeping ever-present net-connected electronic butler that unobtrusively supports you in what you do wherever you may choose to do it.*

In this chapter we will somewhat narrow the definition of the term to mean a system consisting of a highly customised personal wirelessly networked computer with (semi-) transparent glasses and earphones for output and microphone for hands-free voice input. A wearable gives hands-free data- and communication-support with the ability of moving around carrying out tasks.

Table 12.1 describes what we shall assume that the typical wearable consists of.

Note that the current generation of wearables is as of hip-worn computers that mostly run Windows and show the standard desktop screen on a small eyepiece while using some limited speech recognition for input. However, these systems are in a fast development cycle, and we expect systems to close in on the already mentioned components in a few years.

A wearable has some important attributes:[3]

- it is personal
- it is (in principle) *always on*, that is, always worn at all places
- the user can *perform other tasks* while using the wearable
- the user can *move around* while still using it
- it is *connected* wirelessly to a network
- it is *context-aware* (especially location aware): it has one or more sensors, and the sensed values are used as filters for the information presented to the user.

By adding sensors to the wearable, such as a global positioning system, electronic compass, and various biometry sensors, the wearable can be constructed to support collaboration better by making it context-aware. The physical position of the user seems to be of key importance in the design of WSCW systems. As human beings in a physical world, objects that are near us are naturally perceived

Table 12.1 Components of a typical wearable

Main unit	Input units	Output units
● Main (belt)-worn computer ● Wireless network connection	● Microphone for spoken input ● A pointing mechanism ● Location-sensors (GPS-receiver outdoors, infrared indoors) ● Electronic compass for direction sensing ● Digital camera (optional) ● Other sensors	● Head-mounted (semi-transparent) display. ● May be built into glasses. ● Earphones

[3]Based on "The Wearable FAQ" from MIT: http://wearables.www.media.mit.edu/projects/wearables/FAQ/FAQ.txt

and interpreted by us. This three-dimensional natural interaction of ours may be exploited by letting positional data be utilised for the implicit partitioning of a large shared location-based information space supporting groups and communities. By such sharing and "grounding" of shared information artefacts, we believe the distance between our daily tasks and the shared information space will be reduced, and thus improve the chances of better co-operation.

12.3 Wearables and Social Navigation

The main reasons for wearables being interesting from a navigational point of view are that they may:

- bridge the gap between the physical work at a location and its documentation
- facilitate easy and contextual annotation of physical locations, and thereby inter-personal indirect communication
- enhance the availability window of users, and thereby simplify teleconversations
- allow individual time–space trajectories to be silently recorded, and possibly published within groups
- blur the border between professional work and free time usage, thus strongly enhancing the possible data basis of social traces

In the following we will use examples from current research projects we are involved in. These are:

- The European Union long-term research project, HIPS, *Hyper Interaction within Physical Space* (1997–2000), for a handheld and network-based location-aware and personalised tourist guide [3, 4].
- The *Wearable Media* project investigating bike messenger operations in Oslo and in New York [5].
- Other projects investigating police force bicycle operations, and a project involving installation and inspection workers in a large international company.
- Current internal strategic research projects on wearables.

12.3.1 Social Navigation

One way of viewing navigation is according to Downs and Steas' definition as an activity of: "orienting oneself in the environment, choosing the correct route, monitoring the route, and recognising that the destination has been reached" [6]. A wearable is designed to unobtrusively support the user while he or she moves in the physical world. Physical navigation is therefore a central issue for the user.

Dourish and Chalmers separate navigation mechanisms into *social*, *spatial* and *semantic* navigation, where an underlying semantic relationship in information is mapped on a spatial metaphor [7]. They use the term social navigation more similarly to what we will call *implicit social navigation*: navigation behaviour based on *sensing* what others are doing or what they have done. At one level of abstraction, we consider the activity of social navigation to be based on communication and interaction with others. This is similar to Dieberger's usage of the term within the context of navigation in information space [8].

Sørgård distinguishes between two kinds of communication within a co-operation: *explicit* and *implicit* communication [9]. He uses the process of moving a piano to illustrate his ideas. When moving a piano people co-ordinate their actions either through explicit communication (talking) or they communicate implicitly by sensing the movement of the piano and acting accordingly. Explicit communication is communication that is medium independent and explicitly generated, while implicit communication is medium dependent and passively generated by *the shared material*, see Table 12.2. An example of medium dependent and explicitly generated messages is the use of painted arrows at Pere-Lachaisse graveyard in Paris, showing the way to Jim Morrison's grave. Medium independent and passively collected communication is, for example, communicating through actions, expressions, or speech that is not explicitly directed.

Within all of these four dimensions we have a granularity of message targets, from directed person-to-person communication to the arrows at Pere-Lachaisse.

An enhanced availability window facilitates implicit communication of one's own context, for example, the position, current activity and physical surroundings. If the user opens such a window for members of a group, they can initiate teleconversations in a receiver-friendly way. This broadcasted context-awareness can support co-ordination of activities in general, not only human-to-human co-ordination [10–12].

Mentioning awareness, we find two kinds of social navigation in a co-operative setting, seen in the light of Dourish and Bellotti's use of the term awareness [10]:

- Navigation between sub-destinations or nodes, as process states. Awareness supports the co-workers in maintaining the inter-subjective comprehension of the state of the work.

- Navigation between and within co-operative activities. Awareness supports the co-workers in co-ordinating activities.

Table 12.2 Dimensions of social navigation through communication

Navigation	Medium independent	Medium dependent
Explicitly generated	Human-to-human communication, and other forms of *medium independent conversations.*	Communication through explicit generated information that is attached to a material. (*Object annotation*)
Passively generated	Peripheral awareness of others activity level. (*Enhanced availability window*)	Communication through passively generated information (*Object logging*)

These two kinds are not mutually exclusive. In co-operation, co-ordination is often based on *mutual adjustments* [13], which require a high level of awareness of activities and the state of the co-operative work, with respect to the co-operational goal(s). Hence, supporting co-ordination of activities can be viewed as supporting the co-worker in *navigation within the co-operative process.*

Another view of social navigation is whether the user navigates directly or indirectly [14], which is quite similar to navigation through explicit and implicit communication. In direct social navigation *users actively guide one another*, for example, by asking somebody for directions, whereas indirect social navigation *is indirectly influenced by other users activities*, for example when following the crowd to the baggage claim at an airport.

To support social navigation the system provides mechanisms for person-to-person interaction or interaction through some agent service.

In addition to such interaction mechanisms, wearables can also provide a set of different awareness mechanisms, which facilitate smooth synchronous interaction and communication with other users. One such mechanism can be called an *availability window*, in which the user chooses to broadcast his position, activities, and so on to a certain extent and to a personal choice of groups, and for a certain period of time. The trace can then be used by others more or less like "virtual pheromones".

Indirect social navigation requires knowledge of other people's activities. In the example above, the traveller wants to reach the baggage claim, he sees where the others are heading, and therefore chooses to follow the stream. Such navigation requires situational awareness of the surroundings and the actual task at hand.

Regarding asynchronous social navigation we will look closer into the cases of *object annotation* and *object traces*, within the light of the *shared augmented material* concept.

12.3.2 Using a Wearable

A wearable can be used for different purposes and in different user modes. Sometimes it is used as a cellular phone, sometimes it is used as a mobile PC, demanding concentration and inhibiting mobility, and some other times it is something totally different. We believe it useful to differentiate clearly between these modes of usage, as they are often conflated, leading to some confusion in the description of such systems.

One way of illustrating this point is to categorise the modes of usage of the wearable with respect to the levels of *user attention* and *user interaction*. These dimensions span an area where the wearable ranges from seemingly dormant to highly active. As the wearable is to give information support to the user, who is mobile, and in a physical context and situation, we cannot and should not assume that the user is always paying full attention to the system.

Our field studies show that PDA-users tend not to give the PDA the same amount of attention, compared to the use of a desktop computer. When working

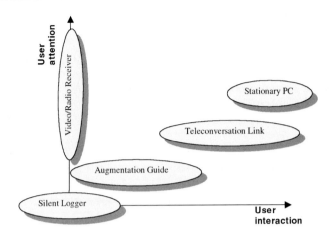

Figure 12.1 Usage modes of a wearable.

at a desktop, one usually uses both hands and pays attention to the screen. When working with a PDA, the attention is primarily at the work, and it is often inconvenient to use the PDA in the same way as a desktop. The same reasoning will apply even more to a wearable, and information on the head-mounted display must be designed carefully so as to not block the visual field more than necessary.

In Fig. 12.1 we have drawn what we believe to be useful usage modes of a wearable, with respect to the dimensions of user attention and user interaction.

Marshall McLuhan states that the content of any medium is an older medium, like the content of writing is the spoken word [15]. The wearable might actually serve as a candidate for a new medium encompassing older media as desktop PC's and cellular phones, but with a focus on enriching the user perception and memory with contextual information access and storage.

Stationary PC

The wearable might be used more or less like a traditional PC, although of a simpler kind due to factors like limited processing power, storage capacity, bandwidth constraints, etc. In this case the user enters stationary mode, i.e. he stops up completely or sits down while using it. The screen may be allowed to use all of the available visual field, as the user has chosen to step "out of context" anyway. Data input might be accomplished using a virtual keyboard like the one proposed by Goldstein et al. [16].

Our studies in the DART project have shown that people who use a PDA professionally and also have a PC at work or at home, tend to use the PDA instead of the desktop for certain tasks. Because of the easy and instant access to the PDA, several users tend to use the PDA to read mail, access the appointment schedule etc.

Teleconversation Link

One should be able to use the wearable computer for communication, like a cellular phone or a video-link. Wearables will generally be connected to a data network hosting several users simultaneously. Hence, it is natural to provide a medium for communication through the wearable.

Video/Radio Receiver

Using the wearable for video viewing or listening to radio broadcasts is an example of usage with low interaction and varying user attention.

Augmentation Guide

This aspect is used for presenting contextual or communicational information to the user in a non-obtrusive way. The Augmentation Guide annotates visually or audibly what the user sees, such as presenting information about buildings near the buildings themselves, or adding information icons or earcons to the user's field of vision or hearing.

This mode is similar to an awareness service in a collaborative environment. An awareness service captures and presents events that have taken place through the collaboration, based on the situation and different contexts like organisation and task contexts.

The Augmentation Guide is an agent which constantly filters basic sensor data. If there is a match between the captured data and some pre-selected criteria, the information may be presented directly to the user. The information can then be stored in a personal database, for later retrieval.

In the HIPS system, one functionality that we plan to implement is that of playing ambient sounds. For example, children laughing when the tourist-user is standing in front of a painting depicting children at play.

The visual experience can thus become audibly augmented.

Silent Logger

Using the wearable as a PC or a cellular demands user attention. In instances where the user pays no attention and does not interact with the system at all, we think there might be an unnoticed, yet important mode of a wearable system to identify. It is the role of the system when the user does not notice the system at all. We shall call this mode *The Silent Logger*. Here the system can silently take note of what is happening around the user. The system might detect familiar voices, knowing and noting the time and location, and may record video and audio trails for later use. The user can review these personal time–space trajectories later asking for more specific information.

The Silent Logger and the Augmentation Guide together comprise an information agent. The agent metaphor is an interesting abstraction when developing services for filtering information. An example is the "butler" metaphor – an agent that learns its owner's needs and performs accordingly [17]. These aspects are those where the system has the initiative [18].

12.3.3 Synchronous Communication

Generally, there are three settings in synchronous *human-to-human communication*. The co-worker with a wearable can communicate either with another co-worker with a wearable, a co-worker with a stationary computer, or a co-worker without computer.

Hence, direct human-to-human communication within WSCW is either conversation face-to-face or *teleconversation* through a wireless network (see Fig. 12.2).

In many cases the fieldworkers are working in the community and, naturally, at times interact with people face to face, in addition to remote teleconversation with the main office and colleagues. This indicates that the interaction channels are switching dynamically, which stresses the importance of interruption handing in such systems. The system should protect the user from *unnecessary* and *brutal* context switches, and instead, provide more harmonious awareness mechanisms for supporting informal interaction.

In the police bicycle field study described below we shall see that the police officersin the field should be in control of input and output mechanisms, and that the people nearby should somehow be able to detect whether the officer is addressing them or not.

Example: Police and Messengers on Bicycles

The following example illustrates some of the communication challenges involved in the use of wearables. It is based on a fieldwork study of the operations of bike messenger companies, the services delivered and the different

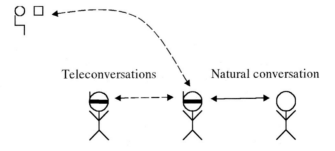

Figure 12.2 Two kinds of conversations.

type of users [19]. The main user group that has been studied are bike messengers who are currently equipped with communication terminals such as:

- cellular phones for voice communication
- private radio terminals for access to private radio networks
- pagers for messaging between the despatching centre and the messengers
- paper-based workflow system on clipboard.

The contextual enquiry [20] has shed light into the use of the existing communication terminals for this highly mobile activity. The terminals mounted on the body are at all times, during biking through the city, and while walking inside office buildings, on the pavement and so on. There are three findings from the field study that are relevant for the discussion of social navigation using wearables. These have implications for the design of wearable solutions in general. These three findings are grouped into the following areas of concern:

- awareness of others
- direct communication and communication at a distance.
- context variations.

Awareness of others

There is a need among the bike messengers to know where the other users are located. Awareness of the locations and the general situation of other users are achieved by visual contact, listening to the open radio channel and by placing calls on the cellular network.

With the police bike force this need for awareness is also seen. In the case of tracking down criminals, it is essential to know the whereabouts of other officers in the area.

Awareness is not limited to the information about the location of co-workers. It is seen that there is a need for awareness about what co-workers are doing, whether they are busy or not, what their plans are for the near future and so forth.

Confusing Switches Between Direct Conversation and Teleconversation

We have seen that switching back and forth from communication at a distance to direct communication is challenging to users. While the users are talking with customers picking up a packet, and at the same time using the radio network for communicating with the despatch centre there is a great deal of confusion. Often, it is seen that the customer is asking: "Are you talking to me?". There are no generally accepted signs or symbols to indicate "communication at a distance" to the customer in the near vicinity, as in Fig. 12.3 if the policeman suddenly starts responding to some question on the radio as they are giving directions to a member of the public.

Figure 12.3 Fast switching between conversation targets.

Context variations

Harmonious context switches are of high importance within WSCW and, in addition, the wearable must be aware of the context switches in the physical world, for example, if it is raining, if there is a time critical job going on, if there are changes of lighting and so on.

12.4 Augmentation

> The next care to be taken, in respect of the Senses, is a supplying of their infirmities with Instruments, and as it were, the *adding of artificial Organs* to the natural ... and as Glasses have highly promoted our seeing, so 'tis not improbable, but that *there may be found many mechanical inventions to improve our other senses* of hearing, smelling, tasting, and touching. Hooke, *Micrographica, 1665*

The physical world can be annotated with virtual data objects into what we call an *augmented world*. In such an augmented world, the connection between objects in the real world and those in the virtual world (cyberspace) can be one of *semantic* or *spatial* connection. By being physically near augmented places or objects, the user can implicitly navigate relative to the corresponding virtual objects using a wearable: browsing information content becomes a matter of moving. In a densely "populated" augmented world, the system will have to make clever choices as to what information should be presented to the user as he or she moves around.

An example of being in an augmented world is when the HIPS tourist-user is walking the picturesque Piazza del Campo in Siena, and the handheld HIPS system, by knowing the user's location and personal interests, tells her by earphones about the famous horse racing run there every year.

Figure 12.4 shows that a virtual artefact in information space may be related to a physical object, and the wearable computer can be seen as the *glue* between the physical world and the virtual information space. These relations may be used for implicit communication through the augmented material. Other researchers call this aspect *environmental-mediated collaboration* [21], or *augmentable reality* [22].

There are different approaches regarding infrastructure and technology for this: active badges and calm technology falls within the *ubiquitous computing* approach. Here, the use of computers is boosted by making many small computers available throughout the physical environment. Another approach, is used in the HIPS system, which encompasses a more traditional client–server architecture. There we use a thin "hippie" client to connect to a nearby server. The hippie senses its position, and requests the server for information, based on position and user context.

Regarding environmental-mediation it is essential to take into consideration that the user, not the environment, should be in control of the interaction channels. Cognitive overload arises if a user arrives at a museum and every picture she approaches leads to information being pushed at her. The wearable must filter the context for a suitable presentation to the user, acting more like a butler or agent on behalf of the user.

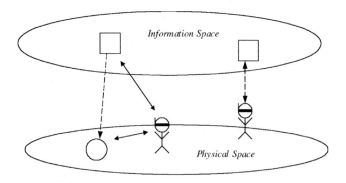

Figure 12.4 Accessing augmentations through wearables.

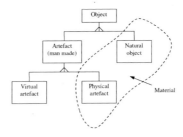

Figure 12.5 Object hierarchy of natural objects and artefacts.

12.4.1 Augmented Material

The main concept in a shared information space for co-operation, is access to the shared artefacts, or in Sørgård's terminology *shared material* [9]. The shared material is the focus of collaboration, for example, a document being edited by a group. One strength of wearables is the possibility, *in real-time*, to simultaneously be working on the virtual and the real material, that is working on the *shared augmented material*. Hence the wearable is reducing the gap between the physical objects and the virtual artefacts, which Baecker stated as one of the main challenges for CSCW back in 1992 [23].

We will be using the word *material* as being either a natural object or a physical artefact (see Fig. 12.5).

This definition of material differs to some degree compared to Sørgård's uses of the term. But it is suitable for our definition of a *shared augmented material* as an accessible material (as meant in Fig. 12.5) with one or several virtual artefacts attached to it.

Role of Physical Position

One of the characteristics of wearables is that they are meant to be carried and worn by users as they carry out different tasks. Although some of these tasks are performed while users are stationary, an example being industrial poultry quality checking, our focus is when users are on the move.

As the user moves, and thus changes position in space, the wearable should be able to detect this movement by comparing the user's position with a web of knowledge about objects and their locations, present the user with information that is relevant for the current location.

Tracking the user's position, both globally and relative to a given point, e.g. the stern of a ship at sea is an important challenge. Currently GPS/DGPS, radio or infrared localisation has some significant limitations and cannot be relied on to be present, for instance, in urban "canyons" where there is frequently no direct line-of-sight to the minimum of three satellites.

However, techniques for combining passive tracking of position with active user input for displaying positional hyperlinked information or use of dead-reconing may be used to somewhat alleviate this.

Position is therefore a key parameter for giving the user information within a geographical context. We might add this as characteristic of a wearable:

The wearable is location-aware; it senses the user's physical position.

12.4.2 Augmenting the Physical World

A special characteristic of wearables is that they constantly switch context as the user moves or the situation changes. The system senses the user's location and informs the user of available information in the surroundings. To attach information to places and objects, for later retrieval when interacting with the physical environment is a relatively new area of research. Generally speaking we can say that this is a matter of annotating the physical world like in the following example.

Annotation Example: Works of art

A tourist equipped with a HIPS unit enters a majestic room with paintings. Her hippie client detects her entry, and starts to download and play some background baroque music, as she has entered her preference for such effects previously. Walking to the north wall she hears a small beep in her headset, indicating that a nearby infrared zone has been detected. On her PDA, she can see outlines of several of the nearby paintings, and she clicks one representing "The Madonna". She now hears a story about the painting, adapted to her previous visit, and indicating special attention to an element in the picture that originates from her home country. She taps in a small message on her own and connects it to the picture hotspot for later retrieval, and gives her travelling group access to her "hyper-graffiti" annotation as well.

12.4.3 Social Navigation in an Augmented World

In a world of information, there can be several meaningful *information contexts* in which pieces of information are perceived to be *near* each other according to some nearness metrics. Often this metrics can simply be that of physical distance. Another metric could be that of semantic nearness of words and content.

Such different contexts can be used as a basis for filtering the (total world of) information, so that the right information is presented to the user. A well-known challenge in computer science is filtering non-structured information. Generally, one should filter information based on the *user's needs* and *current situation*. In this case the situation sets a context, for example, a work process in an

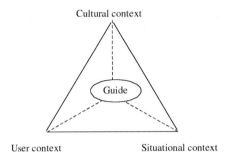

Figure 12.6 Guide agent exploiting contexts.

organisation or a position in physical space. User contexts may consist of user interests, current tasks, different user profiles, for example, age, nationality, and so on.

The triangle in Fig. 12.6 describes the contexts used for the activities of filtering information and for social navigation between places, in physical space as well as in information space. While the "pyramid" in the same figure illustrates a conceptual description of a mechanism that guides the user around in the augmented world. The *Guide* (see the section on the Augmentation Guide discussed earlier) allows cultural contexts, additional to user needs and situations, in order to decide what information is relevant to the user. The cultural context in this case is the footprints of others in similar situations and with similar user needs, for example, the well known virtual bookstore amazon.com and the use of popularity hotspots and annotations in location aware systems [24]. Popularity paths are cultural filtration contexts, because information that is often accessed by others might be of special interest to the user. By considering the footprints of others, the guide supports the user in navigating the augmented world.

Using the term *place* as a *space with meaning* [14] or *understood reality* [25], we could say that the orientation part of the navigation activity transforms the space into a place. In a wearable setting, where the users are operating in the physical world, the understanding of the reality is based on the characteristics of the situated environment, which, within a cultural context, is the knowledge of what others are doing or have done in it. A challenge of this concern is the construction of a common information space [26]. We look closer into this in section on wearables in context.

12.5 Time–Space Trajectories

A useful abstraction when annotating the physical world is what we call *time–space trajectories*. These are timelines that combine recorded and time-dated positions of objects. These traces function as personal or cultural remembrance mechanisms, laying the foundation or substrate for social navigation.

12.5.1 Personal Traces

Using a wearable a personal trace of your context can be recorded. This may contain time-stamped positions, activities, biometric data, digital images etc.

Example 1: Logging Data at the New York Marathon

Logging of data in context has been studied for a highly mobile activity, namely a marathon run [5, 27]. The research was done at the 1997 New York City marathon, where one of the authors ran it with a belt that logged contextual information such as: time, latitude, longitude, direction, speed, heart rate, temperature, running pace and voice (see Fig. 12.7). In addition to logging, the data was transmitted wirelessly to an Internet server. This case study is included to show a proof of concept about capturing and transmitting contextual information. This data may be logged as a time–space trajectory and may be used

Figure 12.7 Herstad logging his New York City marathon.

at the time of logging or at later times, by the person who does the logging, the network or other users.

12.5.2 Object Traces

An object-trace is the annotated time-line of an object. The object time-line (diary) becomes a mediator of communicating history of the objects working process, and handles the dynamics of planned and performed activities.

Object History Example: Inspecting Constructions

On approaching a large construction site an inspection worker is immediately given access to the repair history log of the object. This information can verbally be browsed while he is using both hands in a dirty environment.

12.6 Wearables in Context

Ensuring that a space becomes a place that reflects cultural and social understandings [25], is a matter of putting the information available in this location "in common" [26]. According to Bannon *et al.* a key concept within the field of CSCW is to construct, maintain, and use what they call a common information space (CIS). A challenge here is to facilitate asynchronous communication.

One perspective of a situation is that it refers to a point in the time and space dimension: something is happening at one time in one place. In order to communicate a particular situation, the sender of the message has to pack the information into a meaningful context for future use. Beside the fact that this is not a trivial requirement, it involves a "cost" to the sender. Dourish and Bellotti argue the an awareness mechanism that is based on co-workers actively informing each other of their activities has the potential to fail, with respect to the equality criteria of Grudin [28, 10]. The extra work of explicitly informing others may lead to work overheads which have no direct benefit to the provider.

On the other hand the recipient of an asynchronous message has to be able to unpack the information, in a way that re-creates the context. If the producer does not give the information a meaningful context, it results in an extra cognitive cost for the recipient.

Since a co-worker with wearables is "in the situation", i.e. in the augmented world where annotation and other articulation and co-ordination activities are done, the cost of such work may be reduced. The articulation work of synchronising and visualising the accomplishments in the actual work, is an extra "cost" for the worker, but less so if it can be done in context.

In a wearable setting it is therefore likely that the articulation cost may be reduced dramatically, because of the:

- *mobility of the system*: the user has the equipment at hand, or at the eye for that matter, so she doesn't necessarily have to go back into the office to articulate the work
- *silent logging*: a wearable has the ability to log the situation, take pictures, recording audio or video. The user does not have to fully describe the physical situation when the work took place, but can do with using spatial navigation instructions, like "look at the old stone ruin in the upper left area", or just record the situation visually, and then play the role of David Attenborough himself.

In an asynchronous working mode the contextual information about a past event may be of relevance to the user's current physical context, providing a wider contextual representation of it and thereby easing the interpretation.

In a WSCW setting the physical locations are the mediators for implicit communication, and give a natural (physical) context for the communication. In the HIPS system, for example, a user might be standing in front of the famous fountain Fonte Gaia, listening to a commentary about it. The same comment would not be meaningful somewhere else.

12.7 Future Work

We believe that wearables will have a great impact within the field of mobile informatics and the future use of information and communication technology in fieldwork situations. In all likelihood a wearable will be an important information and communication mediator for private use, as we will one day be living in an *augmented society*.

12.7.2 Some Useful Design Components

On the basis of the previous discussions we believe future wearables will need mechanisms for:

- Silently logging the interaction trace.
- Broadcasting its own context, i.e. position, heading, track, view, sound, etc. and availability window between members of a user-defined group.
- Interacting virtually with other users' display through a remote pointer mechanism.
- Selecting icons representing users, objects and places, supporting interaction.
- Annotating ("hotspotting") physical locations (onsite vs remote).
- Defining group ownership of information.
- Displaying an oriented map with its own position in centre, as an oriented arrow.
- Handling and displaying object timelines extending from the past (history) into the future (plans) with annotations on the timeline as pearls on a thread.
- Uploading/downloading wearable content to/from global network.

There are many issues arising from this use of wearables. We will only give short indications of some of these with possible corresponding problem situations:

- Organisational aspects – *my wearable is part of me, not of my corporation!*
- Social aspects – *Are you talking to me?*
- Cognitive aspects – **Are you talking to me?**
- Ethical issues – *Did you tape me just now?*
- Issues of equipment standards – *My new equipment does not communicate with the existing!*
- Issues of infrastructure standards – *I cannot access the American annotations.*
- Security issues – *I lost those darn glasses, and now all my latest files are gone!*
- Technical issues – *I am blind: My batteries are flat! The positional lock is too inaccurate!*

12.8 Summary – Creating Cultural Footprints

We have seen that wearables that contain the necessary functionality and mechanisms may be very useful for social navigation within the area of wearable-supported co-operative work.

The possibilities to create an interaction log silently or a strictly personal diary for the user as well as facilitating common access to augmentations of the physical world will certainly be some of the more interesting aspects of future wearables. Wearables will integrate personal *planning* (defining possible future events), *logging* (recording current events at the users location), and *remembering* (recalling past events) and have mechanisms for safely storing and optional sharing of such personal event logs with groups in a common shared information space.

In the long run, the use of wearables as our personal interweaving time–space trajectories are recorded, will profoundly change our society and our cultural footprints, as these information systems eventually grow commonplace in the next millennium.

References

1. Thorp, E.O. The invention of the first wearable computer. *Second International Symposium on Wearable Computers*, Pittsburgh, PA, 1998.
2. Siewiorek, D.P., Wearable computers: merging information space with the workspace. *IEEE International Conference on Computer Design: VLSI in Computers and Processors*, Cambridge, MA, 1993.
3. Not, E., Petrelli, D., Stock, O., Strapparava, C. and Zancanaro, M. Person-oriented guided visits in a physical museum. *ICHIM'97*, Paris, 1997.
4. Broadbent, J. and Marti, P. Location aware mobile interactive guides: usability issues. *ICHIM '97*, Paris, 1997.
5. Herstad, J., Thanh, D.v. and Audestad, J.A. A component based terminal – the invisible terminal. *The First Ericsson Conference on Usability Engineering, (ECUE '98)*, Stockholm, 1998.
6. Downs, R.M. and Stea, D. *Cognitive Maps and Spatial Behavior: Process and Product*. Chicago, IL: Aldine, 1973.

7.   Dourish, P. and Chalmers, M. Running out of space: models of information navigation. *HCI'94*, Glasgow, 1994.
8.   Dieberger, A. Supporting social navigation on the World-Wide Web. *International Journal of Human–Computer Studies* **46** [special issue on innovative applications of the web] 805–825, 1997.
9.   Sørgård, P. A cooperative work perspective on use and development of computer artifacts. *The 10th Information Systems Research Seminar in Scandinavia*, Vaskivesi, 1987.
10. Dourish, P. and Bellotti, V. Awareness and coordination in shared workspaces. *ACM Conference on Computer-supported Cooperative Work (CSCW'92)*, Toronto, 1992.
11. Gutwin, C., Greenberg, S. and Roseman, M. Workspace awareness in real-time distributed groupware: framework, widgets, and evaluation. *The HCI'96*, 1996.
12. Rolfsen, R.K., Distribuert samarbeid – hva skjer? [Distributed work – what's happening]. MSc thesis, University of Oslo, 1997.
13. Mintzberg, H. A typology of organizational structure. *Readings in Groupware and Computer-supported Cooperative Work*, 1993.
14. Dieberger, A. Social connotations of spatial metaphors and their influence on (direct) social navigation. *Workshop on Personalized and Social Navigation in Information Space*, Stockholm, 1998.
15. McLuhan, M. *Understanding Media: The Extensions of Man*. New York: McGraw-Hill, 1964.
16. Goldstein, Book, Alsio, and Tessa, Ubiquitous input for wearable computing: QWERTY keyboard without a board. *First Workshop on Human Computer Interaction with Mobile Device*, Glasgow, 1998.
17. Negroponte, N. Agents: from direct manipulation to delegation. *Software Agents*. Menlo Park, CA: Addison-Wesley, 1997.
18. Rahlff, O.W., Rolfsen, R.K. and Stegavik, H. "Where am I now, computer?" (Workshop presentation). *i3 Annual Conference*, Nyborg, 1998.
19. Herstad, J., Thanh, D.v. and Audestad, J.A. Human–human communication in context. *Interactive Applications of Mobile Computing (IMC 1998)*, Rostock, 1998.
20. Holtzblatt, K. and Beyer, H. Contextual design: using customer work models to drive systems design. *ACM CHI '98 Conference on Human Factors in Computing Systems*, 1998.
21. Schmidt, A., Lauff, M. and Beigl, M. Handheld CSCW. *Workshop on Handheld CSCW at CSCW '98*, Seattle, WA, 1998.
22. Rekimoto, J., Ayatsuka, Y. and Hayashi, K. Augment-able reality: situated communication through physical and digital space. *The Second International Symposium on Wearable Computers*, Pittsburgh, PA, 1998.
23. Baecker, R.M. *Readings in Groupware and Computer-supported Cooperative Work*. New York: Morgan Kaufmann, 1993.
24. Brown, P.J. Triggering information by context. *Personal Technologies* **2**: 1–9, 1998.
25. Harrison, S. and Dourish, P. Re-place-ing space: the roles of place and space in collaborative systems. *ACM Conference on Computer Supported Cooperative Work CSCW'96*, Boston, MA, 1996.
26. Bannon, L. and Bødker, S. Constructing common information spaces. *ECSCW'97: The Fifth European CSCW Conference*, Dordrecht, 1997.
27. Redin, M.S. Marathon man. Master's thesis in electronic engineering and computer science, MIT Media Lab, 1998.
28. Grudin, J. Why CSCW applications fail: problems in the design and evaluation of organisational interfaces. *ACM CSCW'88 Conference on Computer-supported Cooperative Work*, 1988.
29. Picard, R.W. *Affective Computing*. MIT Press, Cambridge, MA, 1997.

Chapter 13
Evaluating Adaptive Navigation Support

Kristina Höök and Martin Svensson

Abstract

From the few evaluations of adaptive navigation systems that have been performed, we see an emerging pattern where, depending on the domain, only certain types of adaptive navigation strategies work. The results indicate that adaptations should leave the interface somewhat predictable, they should not force users to interpret advanced annotations, and finally, the adaptation should not change the structure of the information space. Furthermore, evaluations of adaptive navigation support systems fail to recognise some of the more important aspects of why certain systems provide better support than others do. These studies typically measure task completion time, or how well the structure of the space is remembered. While these are among the important measurements that should be taken, other features, such as how much anxiety the system induces in users, how pleasant it is to navigate, or how much users actually learn of the information contained in the space, might be more crucial measurements.

13.1 Introduction

"Lost in hyperspace" is a feeling that is familiar to almost anyone using a computer. After a few actions, we do not know where we are, how we got there, or what our original goal was. *Adaptive navigation systems* have been proposed as a means to aid users in finding their way through information spaces. Several systems have been designed that adapts the navigation to users' knowledge [1–4], to users' preferences and goals [5], to users' tasks [6], or to users' spatial ability [7]. The hope is that if user characteristics are considered the cognitive workload can be reduced, or users' learning of the content may be improved, and so on.

The question we want to pose here is: exactly when do those systems in fact reduce workload, improve learning etc.? Can we find criteria that will give us better insight into how adaptive navigation should be designed to best assist users?

In order to investigate these issues further, we shall first review studies of adaptive navigation support systems, and then outline some criteria by which these systems should have been evaluated. Let us start by defining the concepts "navigation" and "information space".

13.1.1 Navigation and Information Space

We use the metaphor "navigating an information space", but what do we mean by "navigation" and "information space"? In Benyon and Höök [8] navigation is defined as both the more traditional *wayfinding* activities (when the destination is known), as well as *exploration* and *object identification*. According to Downs and Stea [9] wayfinding in the real world can be broken down into a four-step process:

- orienting oneself in the environment
- choosing the correct route
- monitoring this route, and
- recognising that the destination has been reached.

In an exploration activity, people are not trying to get anywhere, they are not trying to find their way; instead they are just interested in having a look around. So in exploration, the second step on Downs and Stea's four-step process is relaxed. Recognising that the destination has been reached must be interpreted as "having found out enough about the space to feel content".

When identifying objects, the user is not interested in the location of objects, nor is the user interested in finding a path or reaching a goal. Although object identification is somewhat akin to exploration, the purpose of the activity is different. While exploration focuses on understanding the contents of an environment and how the things are related, object identification is instead concerned with finding categories and clusters of objects spread across environments, with finding interesting configurations of objects and finding out information about the objects.

From this wide definition of navigation, it is obvious that the destination is something that is often negotiable and is altered as the user moves around. This is the challenge for designers of tools for navigation: they should support users in not only finding the destination, but actually improve the quality of their goal so that they find more information than requested, or perhaps rethink their goal and find a different destination to the one they originally aimed for. This should be done without failing to aid the pure wayfinding process.

We interpret the concept *information space* in a very wide sense, including real world spaces, augmented worlds, as well as virtual worlds. Real world spaces are often overlaid with artificial information such as street signs, symbols, or just paths that result from many people walking or using certain objects. Augmented worlds are, for example, museums with interactive information devices, route guidance systems in the car, or mobile phone systems connected to the Internet. Finally, virtual worlds include anything from your file system, databases, and so on to immersive three-dimensional virtual environments. What comes to mind first when discussing information spaces, are hypermedia spaces, in particular the World Wide Web (WWW). Most systems and evaluations we describe below are adaptive *hypermedia* systems, but we would like to make clear that adaptive navigation can also be concerned e.g. with the interface to a database system, the organisation of menus in a direct-manipulation interface, or a hierarchical file system.

13.2 Adaptive Navigation Systems

An adaptive system will try to infer an understanding of some user characteristics based on users' actions with the system. Based on this inferred understanding, the so-called *user model*, it will then adapt its behaviour to improve the interaction with the user. Many different aspects of user characteristics can be inferred. In adaptive hypermedia, the *user's knowledge*, for example, is used as a basis for educational hypermedia [1, 10]. The *user's familiarity with the structure of the hyperspace* is another factor that can help us limit the search for information. Vassileva [11] uses this distinction in her adaptive navigation techniques. The *user's goal or task* is mostly used to support navigation between nodes in the hypermedia structure [5, 11], but can also be used to decide what to show within a node [6].

As far as we know, there are no adaptive hypermedia systems that attempt to adapt to users' cognitive abilities, style, or personality traits (except in the sense that users' learning is a cognitive ability). We believe however that this is a fruitful direction of research since there are strong connections between cognitive abilities and ability to make use of hypermedia systems [12]. Benyon and Murray [7] made use of the fact that spatial ability together with experience of computers was related to how many errors users performed with certain database interfaces. Their system was adapted through selecting different interfaces for different groups of users.

13.2.1 Content Adaptation and Navigation Adaptation

Basically, there are two features of the hypermedia that can be affected by the adaptivity: the *content* of a page and the *navigation* between nodes.[1] According to Brusilovsky [13], we can distinguish five methods for content adaptation:

- *additional explanations* can be used for a special category of users
- *explanation variants* are used to present information in various ways depending on the user's knowledge of the subject
- *prerequisite explanations* and *comparative explanations* change the information presented about a concept depending on the user's knowledge of other related concepts

---

[1]Even if this pattern is most apparent in hypermedia systems, many other applications can be analysed in similar terms. For example, navigation in a direct-manipulation interface can be viewed as navigation between states (nodes), and adaptation can either be done on how to traverse the states or on the content of each state.

- *sorting* means that the information pieces about a concept that are most relevant to a particular user are placed in front.

Brusilovsky furthermore identifies four different adaptive techniques that affects the navigation between nodes:

- in *direct guidance*, the system decides which is the next "best" node for the user to visit according to the user's goal
- in *adaptive ordering*, links – on a particular page – are sorted according to the user model – the closer to the top of the list, the more relevant
- *in hiding*, parts of the navigation space are hidden or restricted by removing links to non-relevant pages
- *adaptive annotation* means that we augment the links with some form of comments which can tell users more about the current state of the nodes behind the annotated links (text or visual cues).

13.3 Evaluations of Adaptive Navigation Systems

There are few studies of adaptive systems in general and even fewer of adaptive navigation in hypermedia systems. When first interpreting the few studies of adaptive hypermedia systems which exist, it can be assumed that they are, in general, quite efficient in reaching their goals. In the second of two studies of HYPERFLEX [5], it was shown that the adaptive system could sometime decrease users' search time by 40%. In the study by Boyle and Encarnacion [4] on MetaDoc, it was shown that after using the adaptive system users solved a set of reading comprehension tasks in significantly less time, and they also had significantly more correct answers. Unfortunately, other studies give a more complex view.

Edward Carter [14], pre-structured a hyperspace in several different ways reflecting the domain content. During the experiment, the system would switch structure when the subject turned to a new question that would be more easily solved with the other structure. Users disliked this system, and they performed worse in terms of information seeking time, time spent on each node, remembrance of the hyperspace structure after the experiment, etc. Carter speculates that this may be due to the fact that a commonly used strategy when users get lost in a hyperspace is to return to a "landmark" node from where they know how to proceed. If the structure is changed, they may not be able to get back to their landmark, thereby loosing their bearings.

InterBook [15], ISIS-Tutor [1], and ELM-ART [16] are three educational hypermedia systems where the navigation between nodes is adapted through annotating the links. In InterBook, for example, depending on users' assumed knowledge of the domain, certain links are deemed as already known by the user (coloured in green), ready to be learnt (yellow), or too difficult (red). All three systems have been studied in order to determine their efficiency [1, 16–19]. These studies show that the adaptive navigation support will indeed aid users in traversing the space efficiently, avoiding nodes already visited or known to the user, but they fail to provide evidence that learning is increased.

In Weber and Specht's evaluation of ELM-ART II adaptive link annotation is compared to an adaptive NEXT button technique [16]. Two measurements are taken: how many exercise pages are visited, and how many navigational steps subjects used to solve the tasks. Regarding the first measurement, it is shown that the NEXT button is of use to novices with no previous experience of hypermedia nor of the material to be learnt, while adaptive link annotation is of use to more experienced users. Regarding the second measurement (number of navigational steps), no positive effects of the adaptive annotation can be shown, in fact, subjects perform worse with the adaptive annotation. With the NEXT button, they take slightly fewer steps. Unfortunately, this study did not measure whether users learnt more with the adaptive conditions.

In studies by Brusilovsky and Pesin [1, 17] on the ISIS-Tutor, it was found that the adaptive system reduced the number of steps, the number of concept repetitions, and the number of task repetitions.[2] Two conditions were studied: adaptive annotations and adaptive hiding. The authors conclude, "a significant difference between annotation and hiding techniques of navigation support was not found, however there is an evidence that unrestricted freedom of navigation is important for the user". Comprehension time was not affected by the adaptive conditions, and the authors say that:

> Adaptive navigation support can hardly improve the quality of learning and the comprehension time, but it can reduce the number of visited nodes – thus further reducing the overall learning time.

So in both the study by Weber and Specht and the study by Brusilovsky and Pesin, the adaptive navigation support has some positive effects on the traversal of the hyperspace. But, obviously, what should be measured for an educational hypermedia system is how much more students learnt with the adaptive system, as opposed to a non-adaptive variant. Something that is not done in these two studies.

In a study by Eklund and Brusilovsky [19] on InterBook it was found that when using the adaptive system subjects did not improve their learning, rather they actually seemed to learn less when using the adaptive system. However, further investigation revealed some interesting results. The more the subjects used the recommended links, the more they improved their learning. The subjects that did not use the adaptive feature or used it little, did not improve their learning. Eklund and Brusilovsky also found that the most widely used navigational aid was the non-adaptive "continue" button. Combining these two findings may imply that in order for an adaptive system to work it must be easy to use, such as, an adaptive "continue" button, and the adaptive feature must be used continuously throughout the system – i.e. trust in the system's adaptations.

So, in summary, studies of all the three educational hypermedia systems with adaptive navigation (InterBook, ELM-ART II, Isis-Tutor), show that even though they may reduce number of visited nodes, they do not promote learning – thus failing to reach their main goal. Eklund et al. [18] also showed that there are few

---

[2]Tasks and concepts are presented in separate pages in this hyperspace.

empirical studies in favour for adaptive annotation. Together with Carter's findings, should we conclude that adaptive navigation fails?

We mentioned above that Boyle and Encarnacion's study of MetaDoc did show that learning was improved by the adaptive parts of the system [4]. In the light of the studies we have just seen, this may seem strange. As it turns out, Boyle and Encarnacion do not adapt the navigation between nodes, but the information within a node The information space structure is thus stable and does not change, and each node will contain a description of the same concept each time it is visited – what is changed is *how* the concept is described. Boyle and Encarnacion have taken care to make sure the different explanations obey the same pattern of description, see for example, Fig. 13.1. As we can see in the figure, stability of presentation in the two conditions (novice and expert) in MetaDoc is maintained as much as possible. The node structure is the same when a user learns more and the only thing that changes are the explanations of the different concepts (e.g. kernel and shell).

In the HYPERFLEX system designed and evaluated by Kaplan and colleagues, users are actively restructuring a space themselves. The user's main task in this space is not exploration, but wayfinding. The space will be used over and over by users, and only when they have reused the space and adapted it to their needs often enough will it provide the users with some mileage for his or her tasks.

Still, in both the MetaDoc and HYPERFLEX studies it is not completely clear that they have measured the right aspect with respect to their users' tasks, when they measure time spent finding information.

Stability of presentation seems to be a crucial factor in the success of many adaptive systems. In a study by us [20], we showed that our system, that adapted the content of a node, did help users to find the most relevant information in a large online documentation system. It also reduced the number of actions. Even more importantly, users preferred the adaptive system to the non-adaptive variant. Our system also maintains a stable interface. What is adapted within the node is which headers are "opened" and which are "closed". An opened header is one where the text under the header is shown, while a closed one only shows

General System Structure (expert)

The AIX Operating System has three parts:

- The AIX Virtual Resource Manager (VRM)
- The AIX Operating System kernel
- The shell

General System Structure (novice)

The AIX Operating System (a group of programs that act as interface between the user and computer) has three parts:

- The AIX Virtual Resource Manager (VRM), a set of programs that manages the recources of the computer (main storage, disk storage, display stations, and printers)
- The AIX Operating System kernel, a set of program that send instructions to the VRM. It is a set of programs that control, using the VRM, the system hardware (the physical components of the system)
- A shell is often called an interface or a command interpreter. It is the part of the operating system that allows access to the kernel

Figure 13.1 Expert and novice explanations in MetaDoc.

the name of that section. By clicking on the header, the user can force a header to be closed or opened. The order of headers within a node is never restructured. What is perhaps more interesting with our study is that we measured how much *relevant* information users found with and without the adaptive system. In our particular domain, this was the crucial test of the system, not whether users could find any vaguely related information *faster* with the adaptive system, but whether the information that the adaptive system picked for them was the relevant information. If the system had adapted the navigation between pages, this may not at all have contributed to finding the most relevant information.

Meyer [21] and Debevc and colleagues [22] showed the same results. In both systems care is taken to make the adaptivity somewhat stable and predictable. In both studies of these two systems, users perform better and also like the adaptive system.

What is not clear in any of the studies mentioned above, is how well the system can adapt to the user. In the educational system examples, the system is supposed to keep track of the user's knowledge. But how much can be known about the user's knowledge from which links they click on? Unless we can be sure that the adaptations are "correct", these systems will be of no use. As pointed out by Kay [23] adaptations are always more or less stereotypical, and will therefore only be approximations of the individual users' characteristics. It is one of the crucial problems for the whole adaptive systems field to show that they can in fact predict user behaviour/knowledge/abilities more or less correctly. But those kinds of tests should be done separate from testing the usefulness of the adaptive behaviour as such.

13.3.1 Inferred Design Recommendations

From the few studies mentioned above we cannot claim that adaptive navigation in hypermedia is always bad, or that it is only efficient in reducing the number of nodes visited. The results may, of course, be due to bad design, bad adaptations, or the fact that users are quite unused to these kinds of interfaces. Once we have standardised ways by which adaptive navigation works, users may learn how to best utilise them. This is a matter of building a user interface culture known by users throughout the world. For example, on the WWW, links are coloured differently if the user has visited them. This very simple annotation strategy probably becomes useful to a large user population after a longer time of use. Also, since this behaviour is consistent throughout the web no matter which site is visited, it can become part of users strategies in general for keeping track of where they are and where they have been. (Remember though that this adaptation is not an example of proper user modelling as nothing is inferred by the user – the adaptation is mainly a reflection of user actions.)

So, further studies are needed in order to single out exactly what makes an adaptive design useful, both in the short term and in the long term. Interface design culture should not be underestimated when it comes to understanding why certain interfaces work while others do not (even when they are carefully designed!). Once we make certain adaptations into standard behaviour of many different information spaces, users will learn how they work and may well find ways of getting the most benefit out of them.

But what tentatively may be concluded from the studies mentioned above is that adaptations should:

- leave the interface somewhat predictable so that users do not feel lost (as in Carter's experiment)
- it should not force users to interpret advanced annotations, thus distracting them from their main tasks, and
- finally, the adaptive navigation support should not *change* the structure of the space (as in the adaptive hiding example or in Carter's example).

Of course, all systems should be evaluated with respect to the purpose of the system, the domain, and an understanding of the intended user population. From a domain and task analysis we may see if adaptation of navigation is at all feasible. For example, in a large domain that users seldom revisit and where there is no need for the user to learn the structure of the space, adaptive guidance might be very useful. Also, in a domain where the structure is of (nearly) no importance, as for example, in a collection of movies or food recipes, where any organisation can work, adaptation as a means of structuring the space according to preferences may work really well [24].

Finally, in a domain to which users frequently return (perhaps even daily), and where it is very important that they can create shortcuts through the space, adaptations based on interactions with users might be useful. This was shown in the system created and evaluated by Kaplan and his colleagues [5]. The system associated weights with the links between nodes. These weights were set initially depending on the relation between the topics in the nodes, as well as the relation between topics and users' goals. The users could also affect the weights themselves. The system presented a list of nodes, organised as a prioritised list that users could manipulate through dragging items higher up or lower down in the list. Users' thereby become active participants in reorganising the hyperspace.

In general, allowing users to understand (at some level) and/or influence the adaptations of the system is important [6, 25, 26].

13.3.2 Underlying Problems

Going back to the definition of wayfinding in terms of the four steps outlined by Downs and Stea [9], we find some clues to the underlying explanations as to why certain adaptations will only be of limited use and may have to be combined with other methods. It seems that the main problem with most existing adaptive navigational systems is not so much what they do, but rather what they do not do. Considering wayfinding as described above, it is safe to say that current systems usually address the problem of choosing the correct route. The four adaptation techniques presented by Brusilovsky: direct guidance, adaptive ordering, hiding, and adaptive annotation, all try – in one way or another – to minimise the number of steps that a user has to take to reach a specific goal in a hypermedia system. That is, the systems focus on finding the most effective (shortest) path through the information space. However, a good adaptive hypermedia system needs to support the other steps in wayfinding as well: orienting oneself in the environment, monitoring the route, and recognising when the destination has been reached.

First, users' abilities to orient themselves in the hypermedia system is often reduced by the current adaptive navigation techniques, especially adaptive ordering and hiding. These techniques alter the information space either by rearranging the links or by hiding certain links from users. The faster users get to know the whole of the hypermedia system, the easier it is for them to locate themselves in it, thus, increasing or decreasing the space by hiding links may delay this process. Instead of choosing the shortest path through the information space, one possibility is to adapt to the most logical and easy to remember path through the system, something which may increase users' abilities to remember the structure of the space. A strategy often employed by users is to identify one node as a "landmark" node. Whenever users are lost, they "walk back" to the landmark and reorient themselves from there. If the space is adapted, this strategy will be destroyed.

Second, a space that is dynamic (changing) can reduce the ability for users to monitor their given route. If the information space changes from time to time, for instance using hiding techniques, it will take a longer time for a user to get to know and understand the space, and thus, make it harder to monitor a route.

Lastly, adaptive navigation systems as described above, do not recognise that users may need help in reformulating their goal(s)/destination. On the contrary, the four techniques discussed above try to see to it that a user sticks to a given route. As indicated by our definition of navigation into wayfinding, exploration and identification of objects, users may not even have a particular destination in mind. In learning situations, it may in fact be important that users wander around in the space, revisiting certain concepts, turning to other information sources, such as textbooks, exercises, other students, etc. and then returning to the space. In general, in exploration situations measuring time spent will tell us nothing at all about whether users like the space and want to spend time in it!

13.4 Evaluation Criteria

From the studies discussed above, it also seems crucial to discuss how to measure the success of a navigational tool – what to measure in a study, and which method to employ.

13.4.1 Measurements

In the studies of adaptive navigation support systems above, in general, only certain measurements where taken, such as:

- number of visited nodes
- task completion time
- how well a user remembers the structure of the information space.

What do these measure in terms of successful navigation in hypermedia systems? Again, it is obvious that the first and second criteria are measures of shortest paths through a hypermedia system, but is that to say that the navigation was successful? We believe that visited links and task completion time are not good measures of successful navigation, or rather, they only show one aspect of navigation. Remembering the structure of a hyperspace can only be interesting if it is related to the structure of the domain and thus should be remembered. Alternatively, if the system is used often, and the user has to remember it in order to find relevant information. Remembering an arbitrary information space organisation just for the sake of it, seems a waste of time and energy.

In the introduction, we argued that navigation is a cognitively demanding activity. It increases the cognitive workload, it increases anxiety, and it puts demands on users' spatial ability, and the risk is that adaptivity can create an additional burden on users. We need to find measurements on users' anxiety level, how often they feel lost, how often they feel that they have to go back in order to find their bearings, etc. In addition to the three measurements above, we argue that four aspects of navigation need to be taken into account, in order to judge an adaptive navigational system:

- navigation should be a delightful experience that raises the users curiosity
- part of navigation is goal formulation
- measurements of a system's success must be related to users' goals/tasks/domain and not only efficiency
- adaptivity, unless carefully designed, may well introduce a certain amount of anxiety in the user.

Below we discuss these aspects and ways of measuring them.

Goal Formulation

In those cases where the user most often knows the destination, and the system is supposed to support wayfinding in its most traditional form, we can easily imagine study situations where we give subjects a particular task of navigating to a location. After having completed the task we can ask them their subjective feelings on how well they were supported by the tool. But as emphasised above, a large part of navigation is goal formulation: the goal or the quality of the goal should be influenced. In most educational systems the goals are implicit, i.e. there are several pre-defined goals in the form of lessons, however, in the same way that a tutor can help a student to formulate her goals, adaptive navigational systems need to be more flexible in this respect. So, we cannot put a particular destination or task in the hands of our subjects. We need to find innovative ways by which subjects can be given high-level tasks, and where we can test how much their perspective was changed (to the better) by the tool.

Task and Domain

Given a realistic scenario that we want to expose subjects to, we need to decide what it is that we want to measure? Different domains and tasks demand for different types of measurements. For example, in educational hypermedia we would probably have to combine, for instance, task completion time with various measurements on how much of the domain that was actually learnt. On the other hand, in a video-renting situation task completion time could be used alone, i.e. people usually want to find a good video as fast as possible, without needing to know anything about the domain.

Anxiety

One measurement that is often overlooked is anxiety. Even if a system adapts well, users may feel out of control or disoriented, in effect, creating an element of anxiety in the user. Woods [27] states: "flexibility, in the sense of autonomous changes by an interface mediator, creates uncertainty for the user – did something change?, Why did it do that?, What will it do next?" As discussed by Picard [28], computers need to be able to detect and act on emotions in order to be really useful. For example, when the educational hypermedia detects anxiety in the user it could change its adaptations.

Delight

There are other unorthodox measurements, apart from anxiety, that we need to consider. We believe that navigation should be fun, it should induce delight in the user. A system that is fun and pleasant to use will encourage learning, it will make users more willing to explore the space, thus supporting exploration and object identification. It is also our belief that if users feel that navigation is fun, the risk of feeling anxious is highly reduced. Connected to the delight aspect, navigation should (sometimes) engage users' curiosity. If they feel engaged and want to explore the space, they might not be as bothered about having to take many steps before reaching their destination [29].

13.5 Summary

Through re-evaluating a set of studies of adaptive navigation systems, mainly within the adaptive hypermedia area, we have identified a number of weaknesses of the systems and the evaluations of them. It seems as if reorganisation of space, or advanced adaptations that distract users will not reach the goal of the system be it learning or some other goal.

From our analysis of the studies, we then singled out some tentative design requirements on navigational tools, in particular, for *adaptive* navigation support tools. First, in the systems discussed we could see an overemphasis on wayfinding, i.e. situations where the destination is known, rather than exploration or object identification. This emphasis has lead the designers to focus mainly on aiding the user to choose the correct route, rather than helping them to orient themselves, monitor the route or improving the quality of the their overall activity. Secondly, if adaptive navigation is to be used, the domain has to lend itself to adaptation. It has to be a domain that users seldom visit or a domain where the structure of it is of minor importance. Otherwise, adaptation of the navigation between nodes will only mess up users' mental models of the space. Users should also be active or at least aware of the restructuring of the information space.

Finally, from our analysis, we concluded that there are some other measurements by which we believe that navigational tools should be evaluated. Those include studying how much the quality of the goal is improved, how much anxiety the tool induces in users, if their curiosity is raised, and whether the navigation is fun!

Acknowledgements

Thanks to John Eklund and Peter Brusilovsky for useful comments. This review has been done as part of the PERSONA project funded by the EEC, SICS and HMI research school. Thanks to all PERSONA members for good discussions on the ideas discussed here and related topics.

References

1. Brusilovsky, P. and Pesin, L. *ISIS-Tutor: An adaptive hypertext learning environment* In *Proceedings of JCKBSE'94, Japanese-CIS Symposium on Knowledge-based Software Engineering*, H. Ueono and V. Stefanuk, editors. Tokyo: EIC, 1994.
2. Brusilovsky, P., Schwartz, E. and Weber, G. ELM-ART: an intelligent tutoring system on World Wide Web. In *Proceedings of the Third International Conference on Intelligent Tutoring Systems*, Frasson, C., Gaulthier, G. and Lesgold, A. editors. ITS'96. Berlin: Springer, 1996, pp. 261–269.
3. Kobsa, A., Müller, D. and Nill, A. KN-AHS: an adaptive hypertext client of the user modeling system BGP-MS. In *Proceedings of the 4th International Conference on User Modeling*, A. Kobsa and D. Litman, editors. Hyannis, MA: Mitre Corp, 1994, pp. 73–78.
4. Boyle, C. and Encarnacion, A.O. MetaDoc: An adaptive hypertext reading system. *User Models and User Adapted Interaction, (UMUAI)* **4**: 1–19, 1994.
5. Kaplan, C., Fenwick, J. and Chen, J. Adaptive hypertext navigation based on user goals and context. *User Modeling and User-adapted Interaction* **3**, 193–220, 1993.
6. Höök, K., Karlgren, J., Wærn, A., Dahlbäck, N., Jansson, C.-G., Karlgren, K. and Lemaire, B. A glass box approach to adaptive hypermedia. *Journal of User Modeling and User-adapted Interaction* **6** [special issue on adaptive hypermedia], 1996.
7. Benyon, D.R. and Murray, D. Developing adaptive systems to fit individual aptitudes. In *Proceedings of the International Workshop on Intelligent User Interfaces*, W.D. Gray, W.E. Helfley and D. Murray, editors. Orlando, FL. New York: ACM, 1993, pp. 115–122.

8. Benyon, D. and Höök, K. Navigation in information spaces: supporting the individual, In *Human-Computer Interaction, INTERACT'97*, S. Howard, J. Hammond, and G. Lindegaard, editors. London: Chapman and Hall, 1997.

9. Downs, R. and Stea, D Cognitive representations. In *Image and Environment*, R. Downs and D. Stea, editors. Chicago, IL: Aldine, 1973, pp. 79–86.

10. Kay, J. and Kummerfeld, R.J. An individual course for the C programming language. *Proceedings of the Second International WWW Conference '94 Mosaic and the Web*, 1994.

11. Vassileva, J. User modeling in information retrieval systems. In *Proceedings of the 4th International Conference on User Modeling*, A. Kobsa and D. Litman, editors. Hyannis, MA: Mitre Corp, 1994, pp. 73–78.

12. Dahlbäck, N., Höök, K. and Sjölinder, M. Spatial cognition in the mind and in the world: the case of hypermedia navigation. *The Eighteenth Annual Meeting of the Cognitive Science Society, CogSci'96*, University of California, San Diego, 1996.

13. Brusilovsky, P. Methods and techniques of adaptive hypermedia. *Journal of User Modeling and User-adapted Interaction* [special issue on adaptive hypermedia] **UMUAI 6**.

14. Carter, E. Quantitative evaluation of hypertext: generation and organisation techniques. PhD thesis, University of Edinburgh, 1996.

15. Brusilovsky, P. and Schwartz, E. User as student: towards an adaptive interface for advanced web-based applications, user modeling. *Proceedings of the Sixth International Conference, UM97*, CISM 383, A. Jameson, C. Paris, and C. Tasso, editors. London: Springer, 1997.

16. Weber, G. and Specht, M. User modeling and adaptive navigation support in WWW-based tutoring systems. *User Modeling, Proceedings of the Sixth International Conference, UM97*, CISM 383. A. Jameson, C. Paris, and C. Tasso, editors. London: Springer, 1997.

17. Brusilovsky, P. and Pesin, L. Adaptive navigation support in educational hypermedia: an evaluation of the ISIS-Tutor (in press).

18. Eklund, J., Brusilovsky, P. and Schwartz, E. A study of adaptive link annotation in educational hypermedia. *EdMedia98* (http://www.education.uts.edu.au/projects/ah/edmedia.htm), Freiburg, 1998.

19. Eklund, J. and Brusilovsky, P. The value of adaptivity in hypermedia learning environments: a short review of empirical evidence. *Second Workshop on Adaptive Hypertext and Hypermedia HYPERTEXT'98*, Pittsburgh, PA, 1998.

20. Höök, K. Evaluating the utility and usability of an adaptive hypermedia system, *In Intelligent User Interfaces (IUI'97)*, Orlando, Florida. Also in *Journal of Knowledge-based Systems* **10**(5), 1998.

21. Meyer, B. Adaptive performance support: user acceptance of a self-adapting system. *Fourth International Conference on UM*, Hyannis, MA, 1994.

22. Debevc, M., Meyer, B., Donlagic, D. and Svecko, R. Design and evaluation of an adaptive icon toolbar. *User Modelling and User-adapted Interaction* **6**: 1–21, 1996.

23. Kay, J. Lies, damned lies, and stereotypes: pragmatic approximations of users. In *Proceedings of the 4th International Conference on User Modeling*, A. Kobsa and D. Litman, editors. Hyannis, MA: Mitre Corp, 1994, pp. 73–78.

24. Shardanand, U. and Maes, P. Social information filtering: algorithms for automating 'word of mouth'. *Proceedings of the Human Factors in Computing Systems '95*, Denver, CO. New York: ACM, 1995, pp. 210–217.

25. Höök, K. A glass box approach to adaptive hypermedia. Doctorate thesis, Department of Computer and Systems Sciences, Stockholm University, 1996.

26. Cook, R. and Kay. J. The justified user model: a viewable, explained user model. In *Proceedings of the 4th International Conference on User Modeling*, A. Kobsa and D. Litman, editors. Hyannis, MA: Mitre Corp, 1994, pp. 73–78.

27. Woods, D. The price of flexibility. *IUI'93, Intelligent User Interfaces*, 1993.

28. Picard, R. *Affective Computing*. Cambridge, MA: MIT Press, 1997.

29. Morkes, J., Kernal, H.K. and Naess, C. Humor in task-oriented computer-mediated communication and human-computer interaction. *Human Factors in Computing Systems '98*, ACM, Los Angeles, CA, 1998, pp. 215–216.

Chapter 14

Footsteps from the Garden – Arcadian Knowledge Spaces

Andrew McGrath and Alan Munro

Abstract

This chapter describes work in progress on a new way of approaching *social navigation* (e.g. [1, 2]) through the use of populated, growing, *knowledge gardens*. These shared virtual landscapes provide an online space where people communicate and information can be "tended" through the affordances of an ecological metaphor. If we define social navigation as "finding things or going to places via, or with, other people", and take that the whole process of categorising and finding information is a largely social process, then these Arcadian landscapes can provide a useful approach to social navigation in co-operative information applications.

14.0 Introduction: Personal and Social Navigation in Shared "Organic" Landscapes

> All our progress is an unfolding, like a vegetable bud. You have first an instinct, then an opinion, then a knowledge as the plant has root, bud, and fruit. Trust the instinct to the end, though you can render no reason.

Ralph Waldo Emerson (1803–1882)

In the century since Emerson wrote the lines of our opening quote we have seen a change in the relationship between humankind and nature. We are now entering the biological age. Our leading "public" technologies are biochemical, our sympathies are with organic products and our relationship with nature is becoming more complex; one of custodian rather than exploiter. This profound change in our perception of the world could change the nature of the systems we build to make sense of our information rich society, and the work we do. Organic metaphors are now being used to *explain* technology to us, the ecological view is one we are beginning to use to examine our processes and their effectiveness. What if such a view was used to make sense of our information, of our collective knowledge, what if such a view became the backdrop to our social navigation systems? In the following pages we attempt to show how the rich, mysterious nature of spatio-social collaboration/navigation might be mapped through the affordances of ecological information spaces.

14.1.1 Helping People Who Cannot be Together, Work Together

Chance meetings, recruited conversations, seeing what people are doing or reading, seeing when people are available, unplanned, unforced, social interaction, the ability to ignore people politely when passing them in the corridor – our work-lives are made up of these things. We might view these things as "by-products" of co-location. They are often difficult to support, or even absent in systems given to flexible workers. There is a whole genre of papers on this theme, of which we will mention but a few. In many cases these "by-products" would not be described as the actual *work* of the people concerned. The "desktop" metaphor of the computer is pervasive. It may reflect an element of our understanding of work as well; that it is something done at the desk, isolated from sociality, from others; planned, and done according to procedures and schedules. The field of computer-supported co-operative work (CSCW) has been concerned to look closely at work, the artefacts, practices and everydayness of it, looking at the things which slip by the way-side in our normal understanding.

The desktop metaphor is, of course, not only pervasive in the work place. We can look at the normal desktop and its effect in our early ideas of what an "Internet leisure environment" might be, the cybercafé. As one commentator points out, often what we here is very little "clatter and din" which is normally associated with café-society, rather "there is a hushed silence broken only by the mouse-clicks of customers silently roaming their own private corners of cyberspace" [3]. This is not to say that there are no applications which support sociality and which do not help us work. What we are arguing is that often these applications do so in a way which supports either the "meeting" or the "social chat" but does not have the multiple possibilities for action of real space. It is hard to "bump into" appropriate people in these spaces. There have been various attempts to do this, but often they rather violate the methods of approach which we have in the real world (cf. CRUISER). What is interesting is the possibility to bump into others, have a social chat, then be able to act on the discussion in the space, sharing information or navigation or indeed simply ending the conversation and continuing with our activity.

How we work is still quite a mysterious business [4, 5]. There has been a significant body of research over the past ten or so years in the field of CSCW and general sociology looking at the ways in which work happens [6], in particular at the ad hoc and situated nature of work. Moran and Anderson note the importance of the informal, the meetings in the corridor, around the coffee machine, which might not normally be termed as "work". They make a persuasive argument for these to be considered, despite their informal and loose nature, despite being unplanned and "just happening", as somewhat central to what goes on in an organisation. They are ways in which organisational issues get raised, where information is disseminated, where things are brought to one's attention. This sort of activity has been increasingly studied in this field and a rich picture emerges of its usefulness in the work environment.

Button and Harper [7] have demonstrated how we can easily be misled by a cursory look at what people do in the workplace. The partial picture we get is likely to be one which relies too much on organisational ideas of what people do (as well as their own "glosses" of what they do) rather than seeing the ad hoc, everyday work-arounds which make up real work. These work-arounds can be often seen as messy and inelegant. For example, they may involve the everyday flaunting of accountancy procedure in order that a thing gets done quickly and efficiently, rather than in the "right" way. In the case of a furniture factory studied by Button and Harper, the way in which we might be lead to believe that the work is done can often be far from what we find after a more contextual analysis. It seems as if the work goes as follows: an order comes in to the office, a price is then negotiated, and work proceeds through the various parts of the factory. It is then delivered with an invoice. A computer system is designed to support this procedure, and it does so, following through the various stages, one after the other, and quite defensibly from this analysis, it does not allow the stages to be done any way around. This seems entirely natural; the stages being done any other way would not make sense. For example, it would be odd to start devoting work and material resources to an item without negotiating a price and assigning an order number. But a close look at the every day "messy" reality of the factory reveals that this sort of thing actually happens. A job is begun for certain established clients without a formal contract or agreed price. The actual negotiation and paperwork is done retrospectively. These everyday work-arounds can be time saving, and generally useful, but they are also "messy" and often to outsiders made invisible. They often rely on informal, personal contact and informal practice. In this case, the new system, embedding in it an "accounting" view of the work, becomes a problem which has to be overcome, rather than a resource.

It is easy, therefore, to miss these informal elements of work practice, and discount them as legitimate parts of the work, though they have been demonstrated to be effective and useful. It is precisely these elements which may be missed in a more "mediated" encounter. Thus it is vitally important that these elements are not overlooked in an account of work which is incorporated into the design of any given system for remote working, that the parameters of a given set of design possibilities are sufficiently wide enough to include these aspects. It is not just enough to include issues of usability; it may be in a much wider dimension that the system succeeds or fails, for example in the arena of organisational politics, cf. [8]. For example, much work is not just about doing. It is also about being seen to be done. Some recent work e.g.[5, 9] discusses this. Hinrichs and Robinson [9] discuss the "work done" in a phone call from one small organisation to another larger organisation. The larger organisation acts in a number of roles: as a funding body, as an information provider. The phone call, they argue, does not only ask for information (which could perhaps be done more "efficiently" via the web). A cursory analysis might assert this. Rather, the phone call also can work as a "marker" that organisation X is interested in the area of Y. It can also re-establish relationships, taking them up from where one left off, allow for gossip to happen, etc. These "by-products", are as much part of the "work" of that call. Coming back to copresence in working environments, Boden says:

> The flexibility and multidirectionality of copresence allows seemingly irrelevant (but actually important) talk to occur on occasions otherwise dedicated to prescribed topics ... in ways that provide formative updates on colleagues' activities and moods ... These strips of talk provide interactional biographies as they provide organizationally relevant information [10].

This chapter describes one attempt to create a system which attempts to embody some of the aspects of work as we do it today; our various and many encounters with others whether planned or unplanned, with informational artefacts. Essentially, it could be viewed as one of a number of attempts to bring sociality into our encounters with different types of informaton artefacts (cf. populated information terrains). It is at an early stage, but is heuristic in generating some new ways to think about work and information representation, which we will discuss further below. Principal of these are *organic metaphors* for these represented virtual spaces. In particular, we utilise a set of *arcadian* metaphors, that is, metaphors utilising various ideas of the garden. In the next section, we will go on to look at information as it is used by people, and argue for the centrality of a social navigation approach. We will then discuss aspects of these organic metaphors, ending with an illustration of some of these concepts in a system which has been constructed at BT Labs.

14.2 Information Finding, Memory and Social Navigation

In our information world, we might not understand the need for sociality. Might we not be better with decent search tools, information interfaces? With the volume of information in the world, there is surely little need for consulting others. There is a cornucopia of information from all corners, at which we can look, or perhaps get some device to filter it. Perhaps this is part of the answer, but we would like to argue it is most certainly not all. We will now go on to look at the nature of information as it is used by people. We argue that information searching and use in our everyday worlds can be profoundly social in a number of different ways. This for us is the crux of the *social navigation* approach.

"Recorded" information in the work place once consisted simply of hard copy "paper" document in physically located storage systems. We are now seeing more and more of this "recorded" information being kept on the intranet or the web. In addition the quantity of information is growing and is beginning to appear in a number of media formats. The need to find information on these networks has given rise to a number of tools. These range from the "finder" on the desktop and the "recent documents" menu to the search engine or the portal. The effectiveness of these in particular circumstances is clear. These tools do help us to find information. What is not clear is that we have always found the *right* information or indeed where we might start to search for it. The concept of social navigation acknowledges that we may begin to look for information by using others to help us, either directly or indirectly. Information may be found directly, by asking other people, and indirectly by perhaps watching where others go, following trails like footprints in the snow. Information can be "actively" sought, or "passively", by information being sent by others because they know one is interested. The notion of "active" and "passive" is a problematical one, as we will see later.

Benyon and Höök [2] discuss the kind of information we tend to get from recourse to other humans. The information tends to be personalised (to the other's situation), for example, route navigation might be radically different whether one is a pedestrian or road user. It tends also to be embedded in social interaction and perhaps even further in a weave of personal relationships.

Star, Bowker and Neumann [11] discuss the relationship between our social worlds and the information artefacts we use. In particular they focus on the differences between the ways that professors "keep up with their field" and the way that undergraduates find information for their various types of assignment. There are telling differences between the two groups. They particularly focus on the concepts of "transparency" and "ease of use": "transparency and ease of use are products of an alignment of facets of information resources and social practices; each of these facets are interrelated and in motion". Notice here there is an explicit connection made between the social practices of individuals and their information resources. This is very evident in the two groups' differing ways of getting and "accessing" information. Note here, that we use "accessing" in inverted commas. In fact in the case of the senior academics, what is often happening is that information is often "at hand". Senior academics are often very much part of the community from whom they might wish to seek information. Papers, journals, preprints often come to them unasked for, as the result of membership of editorial committees, conference programme committees, etc. Here the information is often quite inextricably linked to the social worlds in which they work. Here we come back to what is problematic about the terms "active" and "passive". It may well be the case that an academic, being sent information could be seen as being a "passive" recipient of the information. However, one can look at the "active" way in which the academic maintains a manageable weave of social relationships, friendships and acquaintances, "tending" his or her social world.

As Star, Bowker and Neumann go on to say:

> the sharing of information resources is a dimension of any coherent social world – be it the world of homeless people in Los Angeles sharing survival knowledge via street gossip, or the world of high energy physicists sharing electronic preprints via the Los Alamos archive. On the other hand, any given social world itself generates many interlinked information artefacts. The social world creates through bricolage (a loosely coupled yet relatively coherent) set of information resources and tools ... put briefly, information artefacts undergrid social worlds, and social worlds undergrid these same information resources [11].

We see that in the case of the senior academics, the concept of information access is often a complete misnomer. Rather we could talk about the information world "at hand". This is because the information artefacts that the senior academic uses are almost in complete convergence with their social world. Latour [12] states quite categorically that "an isolated specialist is a contradiction in terms". Specialism is very much bound up with membership of communities, and the resources that they reflect. In the case of undergraduates, it is very much a case of "information access" of having to actively "find" information. Their concerns and informational resources may be more to do with organisational realities, the paucity of books and journals, the everyday problems of finding these articles. In their case, this reflects themselves in the information they use, taking perhaps local irrelevant articles over relevant ones. As Star and Bowker point out, this is not the result of apathy. It is a reflection of the lack of any convergence between social and academic worlds.[1]

Further, the undergraduate may differ from the academic in ways other than just the getting of information. We can consider what the information *means* to either group, or, in terms of Harper (Ch. 5 of this book), how to *place* it. The information world of an undergraduate is often devoid of, for example, any understanding of the *rhetoric* of a particular paper or perspective. Advancing in any field may be a lot to do with going from seeing just a bunch of alien names, research groups, etc., to seeing trajectories of argument and rhetorical frameworks; for example, McClosky [13] investigates and discusses the rhetoric of economics, and Latour discusses the rhetorical structure of scientific papers. See also Gilbert and Mulkay [14] and Woolgar [15].

We can also see this in the informational world of the desk officer at the International Monetary Fund (Harper, Ch. 5). Not only is the information content important, but so too is the political and organisational meaning of this content. In the case of the desk officer's placing of information from the economics institute of Arcadia, perfectly "good" informational resources are not utilised by the desk officer, because he or she knows that the hierarchy of the particular country does not use them either. What is far more informing is the biased "official" press, if one can see what might be behind what it says.

14.2.1 Casual Navigation: Of Hanging Out, Gossip and Bricolage

Searching or navigation varies from very directed to very casual. In the most directed sense we go to a web resource and search on keywords or we contact someone we know to be an expert. Casual searching presents a broad range of interesting scenarios. The academics in the scenario above "navigate" information through the colleagues they have and the positions they occupy.

---

[1]However, we must be careful to remember that the difficulty here lies in the non-confluence of academic informational resources and the social worlds of the undergraduate. Their social worlds may be rich in informational resources where the academic's are not; finding cheap places to drink, parties after the pubs have shut. Not all academics, sadly, will have this kind of information "to hand".

We argue, in line with this, that sitting at our desks in an office can be a form of navigation. We put ourselves in the middle of a dynamic mix of information and people and we interact with both these entities as part of our working day, not to mention the "corpuses of information" (cf. Harper, Ch. 5) we find on our desk, signifying our current projects. In some flexible working trials it has been noticed that people will come in to their desks when they want to catch the mood of the "office". Sometimes we also move through the space and find information or people serendipitously.

This type of "navigation" is very useful when what we need to know is not clearly defined. In an office we are among a resource that can help us with information in a general area even if we do not know what the question is exactly, nor exactly when it will arise. Further, we argue that searching for what we cannot define could be addressed in a more "social" way. We need to know more about CSCW so we go to a conference as well as go to a library. In fact with the speed of change nowadays it is often felt that if it is in a book it is too outdated (though often publishing lends information "authority"). We therefore gravitate towards a "place" where we can search by *being there*, by experiencing a rich flow of information and conversation. We find what we are looking for quicker, more effectively and more usefully for us. This is coming on information through being who we are talking to some one who may just be the right person. It is, through this social interaction that information may come to us through them becoming part of our social network.

But this is not all. There is an interesting thing about social navigation, of knowing "who" rather than often knowing "what" (or to be a little more realistic, of knowing "kind of what but are not sure" but knowing "who may know this kind of thing, perhaps"), which seems a profound advantage over other types of information browsing or search. It is often the case that what is needed is not written down anywhere but is *"in someone's head"*, at least to an extent. It may be that in talking to others, that conversation and the whole business of asking the question may actually provide part of the answer. It may be that the current set of concerns of the other means that a *creative bricolage* is created by which an "answer" is seen which grows out of the interaction. In fact, to go further, the idea of "questions" and "answers" rather breaks down, and might be seen as *post hoc* categories imposed on the interaction, cf. [16]. It may for instance take a long time to see just what the "answer" actually is, and when we find it, it might help us phrase our question.

Thus, there is a whole complex of issues surrounding information gathering, which makes it a profoundly social process. Taking into account this kind of rubric, we start looking at quite different approaches to information "seeking" than traditional information retrieval models. We need to see others, what they are doing with information, so that we can "use" them for our own informational needs, and for them to use us whether by asking or by looking at what we do with it.

We have also hinted that there are a number of more "societal" concerns in terms of the efficacy and usefulness of this kind of system, in terms of what one gets out and what one puts in (cf. Dieberger, Ch. 3 of this book). There has to be an equal spread of benefit to all users, otherwise such systems are likely to gather dust. Other "political" aspects of such a system should not be ignored either. An information sharing system that is seen to be a place where we are peered at by our hierarchy rather than a place to share with our peers may also fail. This failure will be directly attributable to the culture of the organisation in which it is trialled (cf. [17]). We would argue that people have to be able to negotiate and control their usefulness to others as well as arguing that a system must balance the often conflicting concerns of the management and the managed.

14.3 From Applications to Agents - Ecologies in the Interface

We will now begin to look selectively at some new possibilities for looking at our information worlds, particularly tools which might help us manage information which occurs virtually.

New types of computer tools are beginning to encroach on the hegemony of the user driven application. These tools are characterised by the fact that the user only partially drives them, only occasionally is manipulation direct. These new computer tools are commonly referred to as agents. Whereas applications need to be driven or constantly controlled via an on-screen window, agents do things for you even when you are not directly in control of them. Agents such as BT's ProSearch and Jasper, or GMD's BSCW can be configured to send you information as and when it "happens". They make us aware of things others are doing and things that have happened "elsewhere". The kinds of information that these agents are beginning to deliver is also not straight "readable data" for instance web pages and documents. These agents are also starting to deliver information about user activity, newness, relevance, interest and awareness. The information delivered is often meta-information.

As they do things for you we argue that these agents will need a place to show us what they are up to. They need a place to be seen to be working, a place to deliver the "bricolage" of results. The typical place for this information to "happen" is the e-mail inbox. While this undoubtedly brings information to one's attention it can also clutter up one's inbox with *too much* information. Instead we argue that the place where some of one's information "happens" cannot demand one's full attention all the time, it has to use one's powers of peripheral awareness to help you manage your communication and information needs. In the way we use our peripheral awareness to manage the information being streamed to us in a real space, so we have to do the same with online information. In particular, the information that is not exact, which is more about change and flow rather than specific fact. For example, the fact that someone is looking at my document or that someone is looking for me, I might want to know, but do not need to know in a way that requires my full attention. This kind information has to be peripheralised as similarly oblique information is peripheralised in the real world. Of course, the real world is rife with the possibility of information overload, it has super-high fidelity resolution, infinitely variable views, chaos, people, sounds, seasons, objects, etc. Add to this the complexity we add ourselves through documentation and physical architecture and you could argue that we need protection from life. But on the whole, people manage with life and with copresent communication. "Copresence is 'thick' with information. Under any media condition words derive their meanings only from contexts; copresence delivers far more context than any other form of human exchange." [10].

How are we able to survive this deluge of information in the real world? Well, some of it is down to this use of "peripheral awareness" cf. [18, 19]. The only way to put the equivalent of those useful by-products of co-location into a shared space is to avoid cramming it all into the same channel. Otherwise there is a danger that information manifested in the real world through a multitude of audio, visual, temporal and spatial channels now gets diverted into a single, often inappropriate, channel; usually e-mail. We need a space where those channels of information can be "peripheralised".

It is important to realise that this type of peripheral information space should not be an application in which one does one's work. Rather, it is a space one inhabits. So the information space might, in fact, be space for "hanging out" in, a place to manageably meet our colleagues or people you "ought" to interact with if only you were co-located. It is also a place where your information resources, as generated by your agents, "happen" and can be interacted with in a manageable way. The peripheralising of some of the information may help to reduce the sense of information overload.

Tools for dealing with information overload are talismanic in today's business world. In the past, there was too little information delivered too slowly. Now, there is too much, arriving all too fast. There is also a realisation that a key issue is also not so much dealing with information as generating knowledge and understanding the appropriate information. Information is arriving so quickly and in such quantities that it can be helpful to view it as if it was organic in nature. We can combine a peripheral information space with that of a growing, changing, organic information metaphor to help users cope with the quantity of information, cope with their inability to ever master all of it, and cope with the tools for manipulating and "training" search strategies.

We will now go on to talk about an organic information space which has been developed at BT Labs. This information space make use of the various elements discussed above, albeit in a number of different ways. It is called the Knowledge Garden. It is a prototype taken to "proof of concept" stage and was conceived to be the type of interface we might be using in the next three to five years. In addition, the effort required to build a space that truly "grows" has meant that some of the navigation and design assumptions of the work had to be coded into the "growing" algorithms. We have a garden that grows but our thinking about interacting with that garden has moved on somewhat. The spaces cannot fully be "used" and so true user feedback has been difficult to get, however, any feedback has been generally favourable and has led to the development of new interfaces being prototyped as we go to print.

14.4 Organic Metaphors

> Like Cybernetic space, the garden is an invented place, clearly and always a fiction, a contrivance [20].

It could be said that the driving technologies of the age give rise to the figures of speech that we use to describe the world around us. From the advent of the mechanistic age the way we looked at the world became mechanistic. Ideas that worked for engineered devices were assumed to work for other areas of human endeavour like economic theory and the study of human behaviour. Now we see the rise to prominence of the biochemical technologies. As a result we notice the development of more organic ways of looking at the world. The recent UK Renault Laguna advert describes it as an organically developed automobile, the new 3 series BMW is described using imagery of DNA strands.[2]

Another, older, ecological metaphor that has a bearing on our chapter is the cultural image of the garden or the Arcadian landscape. The garden as a strong cultural image has a very long history and a strong cross-cultural identity. From the past we can look at the epic of Gilgamesh and see the Garden of the Gods. In Christian theology the earliest scenes are played out against the Garden of Eden and the key moments set around the Tree of Knowledge while in Muslim culture the afterlife is seen as a garden paradise.

In Japanese culture we are familiar with the Zen garden or the Shinto shrines and similarly the Celtic tradition is closely associated with nature. We have a strong Arcadian folk memory. The Knowledge Garden acknowledges this memory and takes advantage of our architectural and social preponderance to "get a view", to diminish the claustrophobia of our environment. The Windows desktop of today is, in truth, a claustrophobic windowless environment. In the real world we don't tend to want to sit in a windowless room. With the Knowledge Garden users can look into the organic social world while the very "life" of the space encourages us to "stay connected to it" because like copresent interaction it "both engages and entraps us" [10]. We want to be connected to the life *in* the screen. Connected to the information and the people we find in there, in our shared garden. When the video-phone was first marketed in the UK in the early 1990s it was sold in pairs, because communication devices rely on a certain market momentum. It was vital that people buying them had someone to phone! Similarly the design of the Knowledge Garden relies on people being present online, at least through an agent intermediary. The physical, financial and psychological cost of staying online all the time is too high. Thus, in the Knowledge Garden, agents give an indication of the gardens inhabitants degree of online presence. In the garden we might "see" someone who is not online at all but representation in a way to indicate this to others by the agent.

---

[2]http://www.bmw.co.uk/new3series/home.html

14.4.1 Tacit Properties of an Organic Metaphor

The name "Knowledge Garden" alone can elicit strong responses from people even without any graphical images to reinforce the concept. We have found that people's responses are about what they expect to be able to do or to find. They expect that the information will be growing with both wild and controlled qualities. Even when cultivated there is a perception that the garden is not (need not) controlled completely. People understand that gardens have to be tended (not controlled) to get the best out of them.

> Gradually, as my awareness grew, I realized the garden was not my problem. The yellow roses and the weeds were still there, always had been. My job was to cultivate the roses and get rid of the weeds, you know, keep the garden neat and orderly. Gradually I'm supposed to build a path through it, winding past a fountain and a stream, through mazes and by benches where I can rest and smell the roses; the possibilities are endless.[3]

We have found that people seem quite happy to consider information in an organic way. Viewed in an organic way they realise too much information is "not their problem" their job is to cultivate the information and get rid of the "weeds". They can imagine information growing, dying (in importance) and are happy to engage with the metaphor. In fact other systems using the "language" of organic metaphor already exist [21], although none until now have combined the language of metaphor with the visual imagery and tools to allow the user to fulfil the "purpose" of the garden.

The garden metaphor allows the user a place to "hang out in" that does not need to be driven. It can happily sit in the corner of their screen as it is a space not an application.

> A garden ... is a place that offers a sense of separation from the outside world. That separation must be created by making threshold – whether real or implied. This threshold creates the ability to leave one world and enter another. That is the fundamental function of garden: to allow a person the psychological space to dream, think, rest, or disengage from the world ... It is meant to describe another type of psychological space, to help you touch real self [22]

We are using this ability of the garden to separate us a little from the real world and allow us to enter into a shared online Arcadian space.

The organic world is mutable and chaotic, we know this and expect it. In fact without a certain amount of chaos, of variance based on recognisable laws, the natural world does not seem natural at all. We know that there is a largely unchanging part that varies only in small understandable details. We also know there is gradual change and there is a small amount of radical change. The world generally remains the same but the gradual change serves as a backdrop to our actions and allows us to talk about the changes around us as a way of beginning conversation. The chaotic events can be used to remember things or again to help us to communicate. The organic and chaotic nature of such a metaphor can be useful when searching for information by remembering exactly the "incidentals"

---

[3]http://www.hevanet.com/vanrees/god/stories/garden.html

of a space, the things that we could never search on before without recourse to asking another sentient being.

The mutability inherent in the organic metaphor can be used to meet our identified requirements for a social navigation system. Now changes in the real world happen, however, changes in a virtual world have to be purposely built in. Because of the high overhead of content production this means that in practice virtual worlds do not change once they have been created. There are a number of ways to get affordable and useful mutability into virtual worlds. These are detailed in the paper "Strategies for mutability in virtual environments" [23]. The strategies range from letting users generate content through simple chat or even geometry creation to worlds where the geometry and animation (the "incidentals") of the space are part of an evolving environment generated by agents. In the Knowledge Garden the mutability occurs by linking information to the "incidentals" of the space and therefore provides an information rich backdrop to human communication. It is

> ... a composition of man made or man-modified space to serve as infrastructure or background for our collective existence; and if background seems inappropriately modest we should remember that in our modern use of the word it means that which underscores not only our identity and presence, but also our history [24].

14.4.2 Arcadian Interfaces: Some Examples

> The garden has always possessed the potential to offer its visitors the same sort of experience as is nowadays available in forms we call virtual reality [20].

John Dixon Hunt argues that the garden was the first virtual reality environment and so it is no surprise to know that the concept we have developed takes visual cues and underlying structure from real gardens. There is no slavish copying of reality though, "the ordering devices of abstract art and minimal art are appropriated and applied to the landscape realm" [22]. The metaphor is used to the point at which it stops becoming useful to aid the understanding of the space. There is no need to take the metaphor too far, there is no "Privileging internal coherence over empirical validity" [25]. There are no worms, there are no fallen leaves, there are no picket fences. Our garden is an abstraction of elements of real gardens, and as with any abstraction, there is always a moot point over what is abstracted, what falls by the wayside. Like any process of abstraction, what we are leaving out could be in some ways a political decision (see Fig. 14.1).

Many contemporary garden images are becoming idealised and increasingly geometric. We can take the example of the garden that the Teletubbies on BBC TV live in. Their environment is a cross between a garden and a geometric play space. The emerging use of this kind of garden imagery could be a reflection of our increasing ability to dominate nature through our biological knowledge. The garden becomes a place we can bend to our will. It has being product-ised. If we look also at the work of the Landscape artist Martha Schwartz we see this notion of geometricised nature again. Her Splice Garden combines the "natural" and the

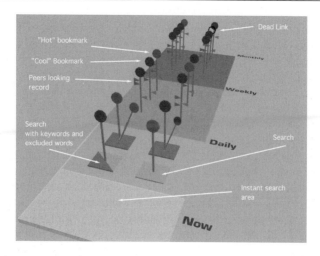

Figure 14.1 An early concept design of the Knowledge Garden showing what the geometric "plants" signify.

geometric. The Splice Garden, while looking like a fully natural (though very controlled) garden space, is in fact made of totally inorganic materials. The geometrically shaped bushes are in fact, all plastic, and the grass is Astroturf. Her stated intention was to "create a garden through abstraction, symbolism, and reference". The Splice Garden was created for the Whitehead Institute in Cambridge, Massachusetts, USA. The Whitehead Institute is a microbiology research centre and the garden is a "cautionary tale about the dangers inherent in gene splicing: the possibility of creating a monster" [22]. If we look at Martha Schwartz's work in the Center for Innovative Technology we see that it includes a circular garden where visitors to the centre drive their vehicles to park when they arrive for a meeting. The cars then park on flagstone "welcome mats" before the visitors entering the building. This imagery and layout is very reminiscent of the imagery in the Knowledge Garden and the history of this architectural design is pertinent to our discussion. The clients for the work objected to some aspects of the design because "It was expected the building could convey an idea or exhibit a radical form, but the landscape was to be the non-intellectual, passive realm in which the building would stand in contrast" [22]. Despite the changing way we are beginning to view nature there are still some who think that intellectual pursuit and technological excellence are to be separated from natural imagery (see Fig. 14.2).

14.3 The Knowledge Garden

We have discussed in some detail the kind of qualities we would want in our organic social navigation system. Now we want to turn our attention in more detail to the work in progress at BT Labs in Suffolk and how it is addressing some

Figure 14.2 The Splice Garden by Martha Schwartz.

of these issues. The Knowledge Garden aims to provide a unified two- and three-dimensional shared, dynamic information interface and has been built from the following underlying building blocks:

- A shared space to peripheralise information and "hang out in".
- Jasper agent technology to provide the relevant information with which to populate the space.
- ProSearch technology to seed the garden with proactively gathered information [26].
- ProSum technology to abridge the information automatically for quick reference [26].
- APIA availability software to provide users with information about who is around to talk online.
- Autogenerative landscape algorithms to create a mutable, navigable, meaningful space.

With these building blocks the Knowledge Garden provides the following features:

- An easy-to-use, engaging, unified information interface.
- Agent generated navigation tools.
- Collaborative, communal interaction and communication.
- Live, self-updating information.
- Agent-brokered information.
- Shared access to information places.

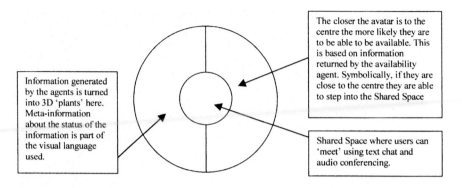

Figure 14.3 A schematic view from above the Knowledge Garden.

- Persistence.
- Communication (audio, whiteboards, etc.) for shared place.

The Knowledge Garden takes the form of a circle with an outer circle split into two. One half of the outer circle is where the APIA availability software returns its information while the other half is where the Jasper and ProSearch agents return their results. The centre is a shared space where users can meet online and swap information or discuss information (see Fig. 14.3).

The closer the avatar is to the centre the more likely they are to be able to be available. This is based on information returned by the availability agent. Symbolically, if they are close to the centre they are able to step into the Shared Space

The Knowledge Garden concept is defined, first, by the capability to navigate a shared, self-generating hierarchical information structure from which the user can browse and view HTML-based documents. This provides a constantly updating and persistent information window on the desktop, utilising a unified two- and three-dimensional interface to their best advantage. In the Knowledge Garden the user can navigate directly to specific information resource in a shared three-dimensional "place" containing representations of context/activity driven and user-defined information. The information resources visible in the space are automatically generated from information supplied by the Jasper, ProSearch and ProSum information agents [27].

The Knowledge Garden was built on previous BT projects such as the Portal[4] and the Mirror[5] and extended the learning done on these into a framework for the user interface. The Portal was an experimental VRML (virtual reality modelling language) Internet site acting as a "virtual visit" to the research projects at BT Labs. It was written in VRML version 1, but used PERL programming language to auto-generate geometry. The geometry was designed to give a sense of which projects were being visited most and who had visited. It created a kind of asynchronous shared space by leaving a "ghost" avatar on a project area that could be clicked on to see what country the person visited from.

---

[4]http://virtualbusiness.labs.bt.com/vrml/portal/home/index.html
[5]http://virtualbusiness.labs.bt.com/SharedSpaces/

Figure 14.4 Image of the portal.

The "ghost" avatar faded out over a number of days. Thus online presence in a project area could be seen over time (see Fig. 14.4).

The Mirror was a ground-breaking collaborative experiment in Inhabited television, created by BT, Sony, Illuminations and the BBC. Six online worlds were available to over 2000 viewers of the BBC2 series "The Net" in January and February 1997. Social chat and interaction were mixed with professional content and programming to create online communities. The worlds of the Mirror reflected the themes of the six broadcast television programmes in their overall settings and individual audio-visual elements. Moreover, they were also designed to experiment with specific aspects of Inhabited television content, to explore which would be most appealing to both new and experienced participants. The six virtual worlds were built around the following themes: space, power, play, identity, memory, and creation. Creation world, in particular, was interesting to this discussion. It featured ecological imagery and a changing cycle of day, dusk, night and dawn played out against a background of changing music. The space also featured an art show where users were invited to upload their own art and an opening night was held online which held the attention of the visitors for several hours (see Fig. 14.5).

The lessons learned in Portal and in the Mirror, especially the generation of automatic content and the maintenance of a sense of community, form part of the design rational of the Knowledge Garden. Another key driver for the design of the Knowledge Garden was the maturity and availability of the underlying information agents Jasper and ProSearch.

14.3.1 Jasper Shared Information Store

Given the immense amount of information available on the World Wide Web (WWW), when information of interest is found, it is preferable to avoid copying

Figure 14.5 Image of the Mirror.

entire documents from the original location to a local server. In addition, it is important to help people manage the distribution of information among their colleagues, after all, one person's vital e-mail is another's junk mail. In the Jasper information agent [27] a defined set of users can "store" pages of interest. Jasper extracts the important details from these documents, including the summary and keywords, as well as the document title, its URL, and the date and time of access. This information is maintained locally and has several purposes. The summary gives a user an idea of the content of the document without the need to retrieve the remote pages. A quick link to the full remote information is also provided. Jasper locally indexes the pages that have been stored. Over time, a rich set of pages is built up by Jasper from different users' entries.

When a user stores a page in the Jasper system, that user's agent automatically informs other users with matching profiles, by e-mail. A reference is also added to every users "What's New?" page along with a predicted interest rating. Furthermore, if the document does not match the profile of the user who stored it, then Jasper prompts that user to update their profile accordingly and suggests suitable additional keywords. In this way Jasper learns more about the user's interests as the system is used.

Of course, systems such as this are prone to the "cold start" problem. That is, no one will use it as no one is willing to put the effort into starting it. This problem can be partially addressed by the addition of the ProSearch agent which proactively searches on the Internet or intranet for information as defined by your interests in your Jasper profile. The combination of the Jasper and ProSearch agents with the Knowledge Garden concept allow for changes in the output from the agents. More detail of what these changes are is described below.

14.3.2 Information Representation within the Garden

Within the shared environment of the Garden the ability to generate WWW-based information representations is key. Jasper and ProSearch, described above, are used to provide shared information resources and the ability to generate group- or personal-based searches and representations within the Garden add another opportunity for social navigation inside the shared three-dimensional spatial interface (see Fig. 14.6).

The Knowledge Garden exploits the pages stored in the Jasper system by using it to furnish our "place to hang out in" with "incidental" information and in which users can navigate through in a logical, intuitive and accessible way. This arrangement of information is achieved by using the meta-information in the Jasper store and results from an automatic clustering process which groups together URLs (WWW pages) containing related information. The resulting cluster structures of the Jasper store are thus used to create a front end onto the Jasper system, using VRML. This clustering work is described in more detail in [28]. It is not expected that this front end will replace the existing front end to Jasper, which is of a traditional, two-dimensional web page arrangement. Rather, the Knowledge Garden will augment the access to Jasper by providing people a more serendipitous method of accessing the stored information.

The Knowledge Garden is also divided into sectors, visible in Fig. 14.7. Each sector represents a different information source. The user group who own the garden space define the sources that appear in their garden. That is, they decide on the resources that will supply their garden with information and dedicate a portion of the space to that resource. Current sources of information supported are the Jasper information agent, a news feed sector and a "Grow your own

Figure 14.6 The Knowledge Garden.

searches" area supported by the ProSearch agent. An algorithm takes the results generated by each agent and automatically turns it into a notional three-dimensional plant shape. The algorithm also generates a path through the information to help the user with navigation. This takes the form of a list of views from which the user can choose and which cause the viewpoint on to the garden to be moved to a good view of the information resources.

In Fig. 14.7 we can see clustered groups of documents (Data Mining, Education, Electronic Commerce, and so on). Each document is represented by a stalk with a coloured square at its end. The colour coding of the squares on the stalks represents the status of the particular document: red squares indicate that a document has been updated since it was stored in Jasper, black that the link is "dead" (that is, the URL associated with this document is no longer valid), while a blue square indicates that neither of these conditions apply. The smaller squares halfway up the document stalk represent a locally held summary of the information, as generated by Jasper. From the opening view of the Knowledge Garden these plants and their stalks can be seen and an overall "feel" for the status of your information can be gathered easily. The more red we have in our garden the more "new" information we might want to look at. If there is a lot of black in the garden then our information resources are both out of date and diminishing and should be pruned and updated. The idea is to keep the garden generally blue in colour indicating that we are up to date with our information resources. All this can be seen at a glance from the opening view on to the space.

As the user navigates towards a particular sector, more detail becomes visible. This feature makes use of the ability of the VRML specification to swap more detailed information into the space as the user gets closer to any particular place. As in the real (and analogue) world the closer we get to things the more detail we

Figure 14.7 Knowledge Garden document cluster.

see. This feature in the garden has benefits for the user in that from a distance the most important information can be seen. As the user gets closer to the area of interest the computer displays more and more detail, showing less "important" or less headline information to the user without them specifically asking to see more. This has also been done to maintain a high frame rate on the users computer. As VRML uses real-time rendering we need to ensure that there is only enough information on the screen as is necessary for the user. If too much information is on screen then the world can slow down to an unusable speed. The plants are geometric in shape for the same reason, to keep the complexity of the space to an efficient level.

When a plant "bud" is selected, the associated URL is loaded into a separate two-dimensional browser window. The selected stem then starts to wave, analogous to touching stems on a real plant as one walks through a real garden. The wave period decreases over time. This visual representation allows, for instance, a "busy" or "hot" plant to be immediately visible to other users by virtue of the number of moving stems.

The two-dimensional browser is separated out from the garden because computer-based three dimensions is of little use for reading text in anything other than small quantities. At some point, once the users attention has been attracted in the three-dimensional world, it makes sense to go to a two-dimensional view of the information. Should the user click on the small "bud" half way up the stalk of the plant they will see the summary information rather than the original document. This is done because sometimes it can take a few minutes to download a document from a remote site and it may be quicker to look at a locally stored summary before deciding if the full document needs to be accessed and read.

To instigate a new and *persistent* search the user clicks on the ProSearch link in the garden and from there can create a new search based on keywords. So, for example, if the document was about Teleworking the query extracted might be:

{ teleworking, cscw, mobility, flexible working}.

This query is then passed to a search engine that searches WWW for documents that match the query. After some post-search filtering, the ten most relevant documents are then represented as stalks on a plant that grows in the three-dimensional space. As the ProSearch agent starts to return documents over time the user can "cut" results which are irrelevant. Such unwanted results are a common occurrence in online searching scenarios. When the user cuts such an unwanted "bud" ProSearch automatically looks at the offending page for possible keywords which the user might not have wanted. This information is presented to the user where they can decide to add "exclude" keywords to the underlying search string. This is an interesting point to note. Typical users find it difficult to create refined search strings and yet in the garden the simple act of cutting and pruning helps the user to train their own searches. Taking the gardening metaphor further, users can also take "cuttings" from information plants and grow their own copy within their personal environment. Another feature of online searching scenarios is finding information we did not know we were interested in until we happen across it. In such a scenario the user typically wants to keep the search focused on the original search task. If the user takes a cutting of the interesting "bud" then the document represented by that "bud" is analysed and a query in the form of a set of key words is extracted from it. ProSearch identifies any keywords in the page *additional* to the original search and the user is invited to refine their search as another plant. The usefulness of the organic metaphor is clear. The users intuitively grasp the idea of training a plant and apply it to their "searching" problem to refine a persistent information search.

14.3.3 People Representation within the Garden

Territorial well-being requires a sense of shared as well as of private space, a feeling of community or neighbourhood in a space that is neither "mine" nor "yours" but "ours" [29].

In Fig. 14.7 we can see representations of people at the far side of the space. Increasingly people are becoming familiar with the notion of the online embodiment, the "avatar". The avatar is driven by the user in that they move it around in three-dimensional space through the direct manipulation of their computer mouse. In the Knowledge Garden the representations of users are different from this notion of avatars. The Garden makes use of two types of avatar one is the typical avatar model, direct manipulation of a representation of your online presence. This is used in the centre of the Garden, in the shared meeting space. Another use of embodiment in the garden is based on the realisation that "In cyberspace we are where our attention is focused, but we have no presence until we are visible in public" [30]. This second type of avatar is used to give other users of the space a sense of how present you are online. We acknowledge that presence online is not a simple "are you or aren't you" question but is more usefully thought of as *degrees* of presence online. If I know (through visual cues) that you are "80% present" in the space then maybe I can entice you into full presence through trying to contact you. Similarly, if I can see that you are only "10% present" I can sensibly leave a message for you instead. The rate of change of presence online can also be useful to us. We can *see* a person who is busy or a person who is active through the rate-of-change of their availability. The use of this type of avatar allows the user to "catch sight" of people online as they move between tasks, rather as they would in the office environment, in the corridor or the coffee area. In the real world "Proximity is negotiated as the need arises ... such discussion tends to happen particularly when matters appear to become sensitive, complex, or uncertain" [10]. In the Knowledge Garden proximity can be gauged and acted on simply and easily through the information afforded by the availability display.

This part of the garden displays the availability of people, it shows the likelihood of they being able to enter into a interaction with you or with others. How does it do this? As the user, at his or her terminal, interacts with applications an agent sends "availability" cues to the Knowledge Garden. This agent is called the *availability of people information agent* or APIA. The cues are based on a collection of information sources none of which on its own tells us much but taken together can give useful information. The APIA agent returns information on desktop interaction like mouse movement, opening and closing windows, it can access the diary of the user to see if we are in a meeting or look and see if we have many phone messages waiting. The information returned drives the position of the avatar in the space. The more available or "there" the agent thinks the person is the closer it puts the avatar to the centre of the Garden. Symbolically the avatar or person is therefore able to "step into" the centre where a synchronous meeting can be held. Of course online, just like in the real world, our availability can be misunderstood. I can think you look busy when you are not really too busy to talk to me, and likewise I can think you available to talk and be wrong. When there are mismatches in perceived availability we negotiate our communication between our desire to talk and the others desire not to be drawn into communication. In the APIA system we can always over-ride the representation if we desire to be uninterrupted.

14.3.4 Information Sharing within the Garden

Once the users have "stepped" into the centre of the garden they are able to enter into a synchronous shared collaborative working space. By making the Knowledge Garden multi-user, users can not only access the rich variety of information resources within the Garden, but collaborate with other users across the Internet. The avatar representations of users are not just visual representations, but contain communications and profiling information about them. There is an interesting parallel with Paul Rankin's "Star Cursor" work at Philips Research [31].

Thus, when avatars meet in the shared space in the centre of the garden, they can communicate via a rich variety of media. The communications channels are set up automatically within the Garden. These include:

- text
- speech (both Internet Telephone and PSTN telephone)
- audio
- white board
- application sharing.

Users endeavouring to research information on a particular topic not only have access to the wealth of links stored within Jasper and any other information sources represented in the garden, but can be automatically put in touch with (perhaps unknown) colleagues. Once these colleagues have been put in touch they can then go on to meet in the shared environment.

14.4 (By Way of Some) Conclusions

The Knowledge Garden is an example of using an ecological approach to social navigation. The work is experimental and without user analysis data it is hard to assess at present. However, we believe that such Arcadian information environments encourage a fresh approach to social navigation. At BT Labs elements of the ideas in the Knowledge Garden are continuing in new CSCW applications with very positive results beginning to appear. The Knowledge Garden concept, to an extent, relies on a type of computing power and a ubiquity of network access that will not be possible for some time yet. This however, means that there now is the time to explore and refine such ecological approaches to shared information spaces.

It is important to reiterate that we do not envisage an informational world devoid of other people. Rather, like our informational searches in the real world, we acknowledge that seeing other people interacting with information artefacts, meeting them in a place where we tend our informational resources, and being able to strike up casual conversations with people are not a side issue which may make the interface to an "informational tool" more friendly. Rather, they are a crucial and central aspect to searching for, understanding and using information. Our information worlds are social worlds.

These developments are in an early stage, so this chapter is necessarily a "work in progress". However, we feel that this "ecological" approach shows promise and may be heuristic further than the present domain. There are many issues with which it and systems like it will have to cope. For example, taking Harper's work in the International Monetary Fund [5], how well will these more autonomous systems fit in with the work practices of an information professional like the desk officer? We cannot say truthfully ourselves that we have dealt with more than a small proportion of the issues that have been raised by social navigation and this book. The Knowledge Garden and its accompanying technology is but a start.

There will be many many challenges to such systems as they develop, especially if we think beyond the keyboard and screen technology used at present. Will these types of systems have new ways to deal with degradation of performance over low-bandwidth networks? How will we cope with scalability questions? Just what is an optimal size of "community" to work with when one is information seeking with others? How exactly will it work across the divide of virtual and "real" worlds; that is, from the Knowledge Garden to information artefacts and settings which we currently use and which we inhabit?

The authors wish to open the concept of Arcadian information spaces for discussion and debate to the wider community.

Acknowledgements

The authors gratefully acknowledge Chin Weah Chia, Martin Crossley, Rob Taylor-Hendry, Amanda Oldroyd, Matt Shipley, Phil Clarke, Andrew Hockley, Luis Collins, Chris Hand, Elaine Raybourn, Richard Harper, John Davies and Mirko Ross.

References

1. Dourish, P. and Chalmers, M. Running out of space: models of information navigation. *HCI'94*, Glasgow, 1994.
2. Benyon, D. and Höök, K. Navigation in information spaces: supporting the individual. *INTERACT'97*, Sydney. London: Chapman and Hall, 1997.
3. Naughton, J. Internet column. *The Observer*, 17 January, 1999.
4. Suchman, L. Making work visible. *Communications of the ACM* **38**(9): 56–65, 1995.
5. Harper, R.H.R. *Inside The IMF: An Ethnography of Documents, Technology and Organizational Action*, Computers and People Series. London: Academic Press, 1998.
6. Moran, T.P. and Anderson, R.J. The workaday world as a paradigm for CSCW design. *CSCW '90*. New York: ACM, 1990.
7. Button, G. and Harper, R. The relevance of "work-practice" for design. *Computer Supported Cooperative Work (CSCW)* **4**: 263–280, 1996.
8. Star, S.L. and Ruhleder, K. Steps towards an ecology of infrastructure: complex problems in design and access for large-scale collaborative systems. *CSCW'94*, Chapel Hill, NC. New York: ACM, 1994.
9. Hinrichs, E. and Robinson, M. Study on the supporting telecommunications services and applications for networks of local employment initiatives (TeleLEI project). GMD German National Research Centre for Information Technology Institute for Information Technology, Sankt Augustin, 1997.
10. Boden, D. and Molotch, H.L. The compulsion of proximity. In *NowHere: Space, Time and Modernity*, D. Boden and R. Friedland, editors. Berkeley, CA: University of California Press, 1994.
11. Star, S.L., Bowker, G.C. and Neumann, L.J. *Transparency at Different Levels of Scale: Convergence Between Information Artifacts and Social Worlds*. Urbana-Champaign, IL: Library and Information Science, University of Illinois, 1997.
12. Latour, B. *Science in Action*. Cambridge, MA: Harvard University Press, 1987.
13. McClosky, D.N. *The Rhetoric of Economics*. Madison, WI: University of Wisconsin Press, 1985.
14. Gilbert, N. and Mulkay, N. *Opening Pandora's Box*. Cambridge: Cambridge University Press, 1984.
15. Woolgar, S. *Science: The Very Idea*. London: Routledge, 1988.
16. Silverman, D. *Harvey Sacks: Social Science and Conversation Analysis*. Oxford: Polity, 1998.
17. Grudin, J. Why groupware applications fail: problems in design and evaluation. *Office: Technology and People* **4**: 245–264, 1989.
18. Heath, C.C. and Luff, P. Collaboration and control: crisis management and multimedia technology in London Underground line control rooms. *Computer Supported Cooperative Work* **1**: 69–94, 1992.
19. Heath, C.C. *et al.* Unpacking collaboration: the interactional organisation of trading in a City dealing room. *ECSCW '93*. Milan, 1993.
20. Hunt, J.D. The garden as virtual reality. *Die Gartenkunst Magazine* 2, 1997.
21. Ackerman, M.S. Answer Garden 2: merging organizational memory with collaborative help. *Computer Supported Cooperative Work CSCW'96*, Boston, MA, 1996.
22. Schwartz, M. *Transfiguration of the Commonplace*. Washington, DC: Spacemakers Books, 1997.
23. Anderson, B. and McGrath, A. Strategies for mutability in virtual environments. *BCS Conference: Virtual Environments on the Internet, WWW, and Networks*, Bradford, 1997.
24. Jackson, J.B. *The World Itself: Discovering the Vernacular Landscape*. New Haven, CT: Yale University Press, 1984.
25. Cubbit, S. *Digital Aesthetics*. London: Sage, 1998.

26. "Information agents for WWW" John Davies, Scott Stewart and Richard Weeks in *Software Agents and Soft Computing* edited by Nwana, H and Azarmu, N. Springer, London, 1997.
27. Davies, N.J., Weeks, R. and Revett, M.C. An information agent for WWW. *Fourth International Conference on World Wide Web*, Boston, MA, 1995.
28. Davies, N.J. et al. Using clustering in a WWW information agent. *Eighteenth BCS Information Reterieval Colloquium*. British Computer Society, 1996.
29. Solomon, J. *The Architectural Sign: Semiotics and the Human Landscape*. New York: Harper and Row, 1998.
30. Waterworth, J. Experiential design of shared information spaces. *Workshop on Personalised and Social Navigation of Information Space*. Stockholm: Swedish Institute of Computer Science, 1998.
31. Rankin, P.J. et al. StarCursors in ContentSpace: abstractions of people and places. *Workshop on Personalised and Social Navigation of Information Space*. Stockholm: Swedish Institute of Computer Science, 1998.

Index

Education Policy

Education policy is high on the agenda of governments across the world as global pressures focus increasing attention on the outcomes of education policy and on their implications for economic prosperity and social citizenship. However, there is often an underdeveloped understanding of how education policy is formed, what drives it and how it impacts on schools and colleges. *Education Policy: Process, Themes and Impact* makes these connections and links these to the wider challenges of educational leadership in a contemporary context.

The book is divided into three sections which explore three key linked aspects of policy:

- 'Policy and Education' focuses on the development of policy at the level of both the nation state and the individual institution.
- 'Themes in Educational Policy' explores the forces that shape policy with a particular emphasis on the themes of human capital theory, citizenship and social justice and accountability.
- 'The Impact of Educational Policy' illustrates how policy develops in practice through three research-based case studies, which highlight the application of policy in a range of situations.

The authors develop a powerful framework for policy analysis and seek to apply this to the formulation and implementation of policy in a range of international settings. In so doing they make an important connection between theoretical frameworks of policy analysis and the need to anchor these within an evidence base that is grounded in empirical research.

Education Policy: Process, Themes and Impact is part of the **Leadership for Learning** series that addresses major, contemporary themes within educational leadership and management, including: policy, leadership, human resource management, external relations and marketing, learning and teaching, and accountability and quality. The series provides a valuable resource for students, practitioners, middle managers and educational leaders in all sectors, both in the UK and internationally, who are engaged on masters and doctoral degrees, or undertaking leadership training and preparation programmes.

Les Bell is Professor of Educational Management and Director of the Doctorate of Education programme. **Howard Stevenson** is a lecturer in Educational Leadership and Management. Both are based at the Centre for Educational Leadership and Management, University of Leicester.

Related titles

Change Forces with a Vengeance
Michael Fullan

Developing the ICT Capable School
Steve Kennewell, John Parkinson and Howard Tanner

Diary of a Deputy
Susan M. Tranter

Effective Leadership for School Improvement
Christopher Chapman, Christopher Day, Mark Hadfield, Andy Hargreaves, Alma Harris and David Hopkins

Improving Induction: Research-based Best Practice for Schools
Maxine Bailey, Sara Bubb, Ruth Heilbronn, Cath Jones and Michael Totterdell

Leadership for Change and Reform
Robin Brooke-Smith and Michael Fullan

Managing Teacher Appraisal and Performance
Carol Cardno and David Middlewood

School Leadership in the 21st Century
Brent Davies and Linda Ellison

Self-evaluation in the Global Classroom
John MacBeath and Hidenori Sugimine

Leadership for Learning
Series Editors: Les Bell, Mark Brundrett and Clive Dimmock

Education Policy
Process, Themes and Impact

Les Bell and Howard Stevenson

Routledge
Taylor & Francis Group

LONDON AND NEW YORK

First published 2006
by Routledge
2 Park Square, Milton Park, Abingdon, Oxon OX14 4RN

Simultaneously published in the USA and Canada
by Routledge
270 Madison Ave, New York, NY 10016

Routledge is an imprint of the Taylor & Francis Group, an informa business

© 2006 Les Bell and Howard Stevenson

Typeset in Sabon by
HWA Text and Data Management, Tunbridge Wells
Printed and bound in Great Britain by
TJ International Ltd, Padstow, Cornwall

British Library Cataloguing in Publication Data
A catalogue record for this book is available from the British Library

Library of Congress Cataloging-in-Publication Data
A catalog record for this book has been requested

ISBN10: 0–415–37771–4 (hbk)
ISBN10: 0–415–37772–2 (pbk)
ISBN10: 0–203–08857–3 (ebook)

ISBN13: 9–78–0–415–37771–3 (hbk)
ISBN13: 9–78–0–415–37772–0 (pbk)
ISBN13: 9–78–0–203–08857–8 (ebook)

This book is dedicated to Sue and Kate in appreciation of their tolerance and support and to Steven, Georgina and Tom with best wishes for their future

Contents

Illustrations

Figures

Tables

Series editors' foreword

Leadership and learning are proving enduringly crucial concepts in contemporary debates on policy and practice regarding improving performance and achievement in education. In this series, we marry the two. Leadership in education is, after all, about steering educational organizations in ways that improve student learning. In putting this series 'Leadership for Learning' together, the Editors had a number of aims. First, we saw the need for a collection of books that addressed contemporary and major themes within Educational Leadership and Management. Second, the approach to these themes needed to be both scholarly and up to date in engaging contemporary academic debates and practitioner problems and issues. Third, promotion of a scholarly approach in turn meant that we attached considerable importance to coherence of argument in each volume. This was more achievable, we decided, by having a series of authored and co-authored rather than edited books. Fourth, the series should be research-based, with discussion where possible grounded in empirical evidence and contemporary research. Fifth, the volumes needed to capture and engage contemporary practical problems and issues experienced by practitioners in all sectors of education – primary, secondary, post-compulsory and even higher education. Since many of these problems and issues cross international boundaries, the series should have an international appeal. Finally, the practical problems and issues were best engaged, we believed, through a rich variety of lenses, including competing and complementary theories and concepts, some of which might be contested. We wanted to capture Educational Leadership and Management as a field riveted with a rich diversity of theory, research evidence, views and interpretations – but above all, a field of great importance to improving the quality of educational organizations and the performance and achievements of students and professionals who work within them.

In achieving the above aims, we have identified six themes, each of which provides the basis for a volume. In part, the conceptualization of the texts and their themes addresses the emerging international agenda for leadership development – both in academic institutions and by national accreditation bodies such as those in the UK, USA and Australasia. The six are: policy, leadership, human resource management, external relations and marketing, learning and teaching, and accountability and quality. To write the volumes, we have assembled an impressive list of authors with the proven experience, expertise and ability.

The intention underpinning the series is to provide a valuable learning resource for a wide and diverse set of people. The volumes are directly relevant to students and educational leaders, both in the UK and internationally, many of whom are engaged on masters' and doctoral degrees such as those organized by the University of Leicester's Centre for Educational Leadership and Management (CELM). More widely, they will appeal to academics and researchers in education and to a large practitioner body of teachers, middle and senior managers, including headteachers and principals in primary, secondary, post-compulsory and higher education in many countries. Large numbers of aspiring and experienced leaders undertaking leadership training and preparation programmes, such as those of the National College for School Leadership in the UK, will also find the series invaluable. We dedicate this series to all leaders and learners and those willing to lead by learning.

The focus of this the first text in the series is on policy in education. This is a substantial and scholarly analysis of a topic that has been comparatively neglected in recent years. The authors, Les Bell and Howard Stevenson, both of the Centre for Educational Leadership and Management at the University of Leicester, create an impressive conceptual analysis of the topic which, in turn, informs and illuminates an exploration of their own extensive research. The writers argue that it is vitally important to recognize that educational leadership is shaped decisively by the wider social and political environment, and by the power relations within an organization. A key purpose of the volume is, therefore, to explore the relationships between policy development at institutional level, the impact of local context and the influence on these of the macro-policy environment. Those in leadership positions face a particular challenge as they often represent the interface between the organization and the external policy environment. Thus the importance of policy on leadership and the role of leadership in mediating policy is manifest in the whole of this wide-ranging text.

The series editors are delighted that the first text in the series sets out a challenging and articulate analysis of policy issues both nationally and internationally. We are confident that this text will provide a theoretical framework and benchmark for the subsequent texts in the series.

Clive Dimmock, Mark Brundrett and Les Bell

Acknowledgements

We would like to acknowledge the help given to us in writing this book by all our colleagues at the Centre for Educational Leadership and Management, School of Education, University of Leicester. We are very grateful to Bernard Barker and Ann Briggs for their perceptive comments on earlier drafts. Thanks are also due to our students for their helpful suggestions. We are especially grateful to Kusi Hinneh, Peter Makawa, Joshu Mose, Sylvester Munyenyembe, Beatrice Uchenna Amadi-Ihunwo, John Rutayisire, Beverley Topaz, Christina Xie and our many students in Hong Kong, Malaysia and Singapore.

We are especially grateful to our friend and colleague in Hong Kong, Daniel Chan, who gave us access to his research data, gave us permission to publish his work and contributed to the writing of Chapter 6. Penny Brown has provided us with detailed critical yet supportive comments on various drafts. We have benefited from her considerable proofreading skills and her technical support in helping us to prepare the final manuscript. We are very grateful for all her hard work on our behalf. We would also like to thank Sandra di Paulo for helping us to copy and print the various drafts of this book and the series editors Mark Brundrett and Clive Dimmock for encouraging us to write this book in the first place.

Much of the material in Chapter 3 was presented at the Athens Institute for Educational Research Conference and was subsequently published as Bell, L. (2004a) 'Throw Physic to the dogs. I'll none of it! Human Capital and Education Policy – An Analysis' in Lazaridou, A. (ed.) *Contemporary Issues on Educational Administration and Policy* Athens Institute for Education and Research, 2004: 187–208. We are grateful for permission to use that material in this volume. An earlier version of Chapter 6 was presented in a paper, 'The role of principals in strategic planning in primary schools in Hong Kong and England: a comparison' at the Commonwealth Council for Educational Administration and Management Regional Conference, *Educational Leadership in Pluralistic Societies* Hong Kong and Shanghai, 20–26 October 2004 with Daniel W. K. Chan and was published in *International Studies in Educational Administration* in September 2005. The research on which Chapter 6 is based was partly funded by the National College for School Leadership (NCSL) International Fellowship Scheme. We are grateful for this support. We are also grateful for the permission of the journal's editors to use the material in this volume. Chapters 7 and 8 both draw on funded research initiatives and we are grateful to the EAZ that is the focus of Chapter 7, and to the NCSL in Chapter 8, for their support and co-operation with these projects. Particular thanks go to our colleagues in both these ventures – Phil Hingley (EAZ) and Clive Dimmock, Brenda Bignold, Saeeda Shah and David Middlewood (NCSL).

Introduction

Education policy, themes and impact

This book is about educational policy but it is not an educational policy book in the traditional sense. It is recognized that:

> There was a time when educational policy as policy was taken for granted ... Clearly that is no longer the case. Today, educational policies are the focus of considerable controversy and public contestation ... Educational policy-making has become highly politicised.
>
> (Olssen et al. 2004: 2–3)

Nevertheless, this book does not seek to explore in detail the extensive philosophical and ideological underpinnings that have shaped educational policy over time although it does consider briefly liberalism, neoliberalism and the emergence of the new right. Nor does this book examine in detail governmental policy-making processes. It does not explore the minutiae of the legislative procedures that are used to formulate and implement policy. It does not examine extensively the complex relationships between the state, the local administrative bodies and educational institutions. These matters are not ignored. Indeed, they provide a coherent framework for considering policy at a range of different levels and developing an international perspective, albeit a limited one, on educational policy, its themes and its impact.

This book sets out to examine an important but limited number of interconnected themes that can be identified within educational policy making over the last two decades. Two main themes, human capitalism and citizenship and social justice are linked with a third set of themes, markets, choice and accountability, to provide an analysis of the dominant discourses that have shaped educational policy in many countries across the world. Within this context, aspects of school leadership and management will be considered in order to establish the extent to which the work of school leaders is shaped by such themes and how far school leaders can interpret, modify or create policy at an institutional level.

Policy studies in education have tended to take one of three forms:

1   The development of broad analytical models through which the policy process can be understood and interpreted.
2   Analyses of a range of policy issues.
3   Critiques of specific policies.

These relatively fragmented approaches often fail to provide a cogent account of the policy process within a clearly articulated framework for analysis. It is often difficult, therefore, for those studying policy and for those working in schools that are subject to educational policies to make sense of the policy contexts within which they have to operate. Nevertheless, it is important to recognize that those working in schools are not merely passive receivers and implementers of policy decisions made elsewhere. In many cases, they are able to shape the policy process, especially at institutional level.

The main purpose of this book, therefore, is to analyse such policy issues and policy implementation. It is based on the assumption that the policy process may pass through a variety of stages and can take place at a number of different levels. Policy development therefore is not a simple case of understanding the priorities, or indeed the whims, of governments or individual school leaders. Policy must be seen as a dialectic process in which all those affected by the policy will be involved in shaping its development. Policy development is therefore both a continuous and a contested process in which those with competing values and differential access to power seek to form and shape policy in their own interests. To this end, a model for analysing policy formulation and implementation is established which informs the analysis throughout this book.

This book will also seek to bring a limited but important international dimension to this analysis within the framework of the model. Bottery (1998 and 2000) has sought to define policy in terms of global trends and to explore the impact of those trends on the professional values of educators. In so doing, he is one of the few to try to develop an international perspective on educational policy. Although the importance of globalization is recognized in this book, the intention here is to provide a more detailed analysis of specific trends that appear to have a part to play in shaping education policy in a number of different international contexts. It is not intended to undertake a comprehensive international analysis here. Rather, where relevant international examples are available, they are considered as part of the analysis. This inevitably means that choices have had to be made and that, in some chapters, the international dimension is much more evident than it is in others.

The book is divided into three parts. Part I, *Policy and Education*, explores the nature of policy and begins to identify some macro-level issues related to policy formulation and implementation. The first chapter in this section, Chapter 1, poses the question, 'What is educational policy?' and introduces the central argument of the book, that policy is derived from values that inform the dominant discourses in the *socio-political environment* and the values that are derived from that discourse. Policy trends emerge based on these discourses that establish the *strategic direction* for policy and translate this into broad policy that is then applied to different domains such as health, economy and education. The parameters for policy in any one of these spheres of activity is defined by the *organizational principles* and the *operational practices and procedures* which are the detailed organizational arrangements that are necessary to implement the policy at the regional or even institutional level. The second chapter in the first part, Chapter 2, examines the concept of the state and the relationship between the state and its institutions. It is argued that the nature of educational policy is, to some extent, derived from assumptions about political processes. Policies shaped by pluralism may be significantly different from those determined from a structuralist perspective. The relationship between the educative process and the state and assumptions about the purposes of education all shape the nature of policy.

Part II, *Themes in Educational Policy*, considers some of the main themes that appear to drive educational policy making in many countries. The opening chapter, Chapter 3, explores one of the most pervasive of these themes, the relationship between education and human capital. It argues that as the emphasis on economic utilitarianism as a rationale for educational policies increases in significance, then equity issues become subservient to the economic imperative. Closely linked to, but different from, arguments about economic utility are concerns for citizenship and social justice. Chapter 4 shows how the contested notions of citizenship and national identity are often informed by both globalization and economic utilitarianism to provide a rationale for a range of educational policies, and considers how this might influence perceptions of equity and social justice. In Chapter 5, a further theme is identified as the trends towards greater accountability, increased choice opportunities and developed autonomy are explored. Accountability, autonomy and choice emerge as themes in educational policy in a number of different forms. Whilst policies to promote accountability have developed differently in different countries, accountability now assumes a dominant position in the global educational agenda. This chapter explores the variety of forms in which accountability is manifested in specific policy contexts, focusing on accountability through the operation of market forces, choice, school-based management and performance appraisal.

Part III, *The Impact of Educational Policy*, looks at the implementation of specific policies in particular contexts at the local and institutional levels. It is important to recognize, however, that policy responses are shaped by cultural context. Therefore, the themes and specific policies chosen reflect issues that are of international concern and they are illustrated with reference to research drawn from a range of different countries and contexts. The first chapter in this section, Chapter 6, draws on research on strategic leadership and management in primary schools in the UK and Asia. This chapter explores the deployment of forms of strategic planning as part of the policy process in Hong Kong and England and shows how the strategic direction of policy can be modified or even challenged through organizational procedures and operational practices adopted at school level. In Chapter 7, the impact of a specific policy introduced in England, Education Action Zones, is explored, showing how a national policy initiative is often largely dependent for its implementation on local decisions and locally determined procedures. Chapter 8 draws on empirical evidence to assess how school leaders develop micro-level policies in response to social and cultural diversity in many communities. The final chapter re-considers the nature of educational policy in the light of the themes and case studies explored above.

The entire book is based on the assumption that education policy making is a dynamic process in which the nation state exerts power and deploys resources in conjunction with regional, local and even institutional agencies. The nature of the relationship between these participating, perhaps competing, agencies may change over time according to the dominant discourses in the socio-political environment and the resultant policy decisions. For example, in England the policy process has passed through at least four stages over the last 40 years (Bell 1999a). The Social Democratic Phase, 1960–73, was typified by a partnership between national and local agencies and relative autonomy over the curriculum at institutional level. The Resource Constrained Phase, 1973–88, saw a breakdown of the relationship between national and local agencies with more control shifting to the centre and more direct relations being established between the state and institutions. The Market Phase, 1988–97, was exemplified by the diminution of the powers and functions of local agencies, devolution of financial autonomy to institutions within a centrally controlled curriculum and rigorous accountability mechanisms. In the Excellence Phase, 1997 to date, the central control of the state and the tightly monitored accountability mechanisms remain but have been extended to include pedagogy and pupil performance. In other countries similar shifts can be found. The pattern in New Zealand and some states in Australia has been similar to that in England (Grace 1997). In Hong Kong, the establishment of a Special Administrative Region has increased devolution to the institutional level and loosened some aspects of central control (see Chapter 6). In France, the state is devolving greater autonomy over pedagogy to the institutions, while in Singapore this is also happening, but within a much more centralized system. Thus, the policy-making process must not be treated as a set of immutable relationships between its constituent parts. Rather, it is an ever-evolving pattern of relationships and it is to closer consideration of these relationships that we now turn in Chapter 1.

Part I

Policy and education

The first part of this book considers the nature of policy and the particular nature of education policy within its wider social, political and economic contexts. It considers theories of the state, the levels at which policy is developed and implemented, issues related to power and influence in policy formulation and the importance of values in shaping and implementing policy. An interest in the macro-policy environment inevitably focuses attention on the role of the state. Although there is significant variation in state formations between nations, the state almost universally has a key role in the provision and/or regulation of education services. 'State policy', whether national or local (or increasingly supra-national), therefore has a considerable impact on shaping what happens on a daily basis in schools and colleges, and the lived experiences of those who study and work in those establishments. The role and influence of the state in educational policy is explored further in Chapter 2, as is the relationship between the state and individual educational institutions. To what extent do those who work in educational institutions enjoy the latitude to generate and develop institutional policies that may be at odds with state agendas? In this first chapter, it is argued that policy development is not a self-contained, linear or rational process – rather it is likely to occur at a range of levels almost simultaneously. This has implications for the organization of educational institutions and for their leadership and management.

1   What is education policy?

Introduction

In recent years interest in 'leadership' has burgeoned and consequently studies
of educational leadership have proliferated. Research around the world
is contributing to an increasingly rich understanding of how educational
institutions are led and managed. However, it is important to recognize
that educational leadership does not exist in a vacuum – it is exercised in a
policy context, shaped decisively by its historical and cultural location. It is
important, therefore, that studies of leadership adequately reflect this wider
policy environment:

> it is essential to place the study and analysis of school leadership in its
> socio-historical context and in the context of the moral and political
> economy of schooling. We need to have studies of school leadership
> which are historically located and which are brought into a relationship
> with wider political, cultural, economic and ideological movements in
> society.
>
> (Grace 1995: 5)

Grace (1995) argues against a reductionist approach to the study of
educational leadership, in which quasi-scientific management solutions
are developed with little regard for contextual specificity. There is also a
tendency to detach studies of leadership from studies of power (Hatcher
2005). Rather, it is important to recognize that educational leadership is
shaped decisively by its wider environment, and by the power relations
therein. The nature of that environment will be formed by a multiplicity
of factors unique to each institution – these may range from local 'market'
conditions to the impact of global economic pressures. What is certain is that
within education, across phases and across continents, the policy context
impacts decisively on shaping the institutional environment. A key purpose
of this volume is to explore the relationships between policy development
at an institutional level, the impact of local context and the influence on
these of the macro-policy environment. An interest in the macro-policy

environment inevitably focuses attention on the role of the state. Although there is significant variation between nation states in virtually all countries the state has a key role in the provision and/or regulation of education services. 'State policy', whether national or local (or increasingly supranational), therefore has a considerable impact on shaping what happens on a daily basis in schools and colleges, and the lived experiences of those who study and work in those establishments. The role and influence of the state in educational policy is explored further in Chapter 2.

All those working in schools and colleges must make sense of their policy context. Policy agendas require a response as those in the institution are faced with the task of implementing policy directives. Those in senior leadership positions face a particular challenge as they often represent the interface between the organization and the external policy environment. Key decisions must be made relating to the interpretation and implementation of external policy agendas – those decisions will in turn reflect a complex mix of factors including personal values, available resources and stakeholder power and perceptions. Understanding and anticipating policy therefore becomes a key feature of 'leadership' (Day *et al.* 2000) – understanding where policies come from, what they seek to achieve, how they impact on the learning experience and the consequences of implementation are all essential features of educational leadership. To some extent it may be argued that in recent years studies of 'leadership' have supplanted studies of policy. This in part reflects the emergence of a managerialist agenda in which institutional leadership and management is often reduced to a technical study of the 'one best way' to deliver education policy objectives determined elsewhere within the socio-political environment and legitimated by a dominant discourse which may be located outside the immediate sphere of education (Thrupp and Willmott 2003). Policy is treated uncritically and denuded of its values, neglecting to assess how policy impacts differentially on different social groups. The importance of policy, as distinct from leadership, is recognized in this volume, but a simple dichotomy between leadership *or* policy is avoided – the key issue is to explore the relationship between the interdependent themes of leadership, policy and power. This volume acknowledges the importance of leadership, but seeks to make the case that leadership must be located within a policy context. A failure to fully understand the complex ways in which policy shapes, and is shaped by, leadership fails adequately to explain the actions and practices of leaders at both the organizational and operational levels.

Key practitioners in schools and colleges, rather than being passive implementers of policies determined and decided elsewhere, are able to shape national policy at an early stage, perhaps through their involvement in interest groups, professional associations or their favoured position in government policy forums and think-tanks. In other cases, influence may be exerted at an institutional level as the organizational principles and operational practices through which policy is implemented are formed and re-formed. Leaders in educational institutions, therefore, are both policy implementers and policy generators.

For these reasons it can be more accurate to describe a process of policy development, rather than use the more traditional, but less helpful, term of policy making. Sharp distinctions between policy generation and implementation can be unhelpful as they fail to account for the way in which policy is formed and re-formed as it is being 'implemented'. The term policy development also more accurately conveys the organic way in which policy emerges. This is not to argue that policies develop in entirely serendipitous ways. On the contrary, an important theme of this book is to argue that policy is decisively shaped by powerful structural forces of an economic, ideological and cultural nature. Nevertheless the crucial role of human agency in the development of policy must be recognized. Furthermore, if institutional leaders do not mechanically implement policy from the state, nor do those studying and working in educational institutions mechanically implement the policies of their institutional leaders. Policy is political: it is about the power to determine what is done. It shapes who benefits, for what purpose and who pays. It goes to the very heart of educational philosophy – what is education for? For whom? Who decides? The point is well made by Apple:

> Formal schooling by and large is organized and controlled by the government. This means that by its very nature the entire schooling process – how it is paid for, what goals it seeks to attain and how these goals will be measured, who has power over it, what textbooks are approved, who does well in schools and who does not, who has the right to ask and answer these questions, and so on – is by definition political. Thus, as inherently part of a set of political institutions, the educational system will constantly be in the middle of crucial struggles over the meaning of democracy, over definitions of legitimate authority and culture, and over who should benefit the most from government policies and practices.
>
> (Apple 2003: 1)

The questions posed by Apple are steeped in values – the values of individuals and the values embedded in wider societal institutions and structures. It is through these values that policy develops. This conception of policy seeks to reflect the complexity of the policy-development process. The argument here is that it is not possible to understand what is happening in our educational institutions without developing an understanding of policy that reflects both its multi-stage and multi-tier character. This process may be considered to have neither a beginning nor an end. Schools and colleges are constantly engaged in developing their own policies as they seek to both pursue their own internal objectives and respond to the external policy environment. Policy making as a process is therefore not something that happens exclusively 'up there', but is something that happens 'down here' too. Those working in schools and colleges are simultaneously engaged in making sense of the policies of others, and forming policies of their own – two processes that in reality are more than interdependent – they are separate elements of a single process. Developing a conceptual understanding of these processes is a pre-requisite for developing a better informed theoretical and empirical understanding of what is happening in our schools and colleges. This provides the basis for the study of policy – policy analysis.

Policy analysis as the study of policy

The central concern of this volume, policy analysis, can take a number of forms, for example the development of broad analytical models through which the policy process can be understood and interpreted, analyses of a range of policy issues or critiques of specific policies. Gordon *et al.* (1997) identify several types of policy research, each of which falls within a continuum which they characterize as either analysis *for* policy, or analysis *of* policy. This is represented in Table 1.1.

Policy Advocacy – refers to research which aims to promote and advance either a single specific policy, or a set of related policies.

> In some cases policy advocates argue from their findings toward a particular conclusion, which is offered as a recommendation. In other cases, where a very strong commitment to a particular course of action predates the research, whatever analysis was conducted may have been designed, consciously or unconsciously, to support the case to be argued.
>
> (Gordon *et al.* 1997: 5)

Information for policy – this type of research aims to provide policy makers with information and advice. It is premised on the need for action (tackling a commonly perceived problem, for example) and may suggest the introduction of a new policy or the modification of an existing one.

Policy monitoring and evaluation – this is a common form of policy research, particularly in the current climate of high level accountability and the need to justify actions undertaken. Gordon *et al.* (1997) point out that public agencies frequently perform monitoring and evaluation functions in

Table 1.1   Analysis *for* policy and analysis *of* policy

Analysis *for* policy		Analysis *of* policy		
Policy advocacy	Information for policy	Policy monitoring and evaluation	Analysis of policy determination	Analysis of policy content

Source: Gordon *et al.* 1997: 5

respect of their own activities, although some may be facile and uncritical. Most monitoring and evaluation research is concerned with assessing impact, although it can go beyond this with a deliberate aim of influencing the development of future policy. Evaluative research will often make claims to objectivity, but it is important to recognize that 'evaluation is a motivated behaviour' (House 1973: 6) and the highly politicized environment within which policy evaluation research takes place can present very distinct methodological challenges for researchers of educational policy (Walford 1994 and 2001).

Analysis of policy determination – here the emphasis is very much on the policy process – not on the impact of policy, but on how policy developed in the precise way that it did. Such research can give a vital insight into explaining how and why specific policies emerged in the final form they adopted.

Analysis of policy content – Gordon *et al.* (1997) argue that this research is conducted more for academic interest rather than public impact and here the emphasis is on understanding the origin, intentions and operation of specific policies. The common approach to this type of research is to utilize a case-study format and this raises important questions about the appropriateness of methods in policy research (Halpin and Troyna 1994).

These distinctions can be helpful in identifying approaches to policy research, but they do not, on their own, shed light on the complexity of policy development processes. Policy analysis within education must be capable of recognizing the many different levels at which policy development takes place, the myriad range of educational institutions involved and the importance of specific cultural contexts. For example, legislation passed by a central state is clearly an example of 'government policy'. A policy developed at an individual state school may fall within a broader heading of public policy, but what of policy developed in a private school, independent from, but regulated by, the central state? A model of policy analysis must be capable of illuminating policy development in all these diverse and various contexts. Taylor *et al.* (1997) suggest that a simple summary of policy analysis is the study of what governments do, why and with what effects. This can be a helpful starting point as long as it is recognized that such analyses must embrace institutions at all levels of the education system and must be capable of including institutions that are effectively part of a public system, even if they are not formally in the public sector. Taylor *et al.* (1997: 37) go on to identify a number of questions that can form the basis of policy analysis, and which are capable of handling the diversity of contexts identified. These are:

- What is the approach to education? What are the values relating to the curriculum, assessment and pedagogy?
- How are the proposals organized? How do they affect resourcing and organizational structures?

- Why was this policy adopted?
- On whose terms was the policy adopted? Why?
- On what grounds have these selections been justified? Why?
- In whose interests? How have competing interests been negotiated?
- Why now? Why has the policy emerged at this time?
- What are the consequences? In particular, what are the consequences for both processes (professional practice) and outcomes?

Using these questions as a starting point, Taylor *et al.* (1997) develop a framework for policy analysis. This focuses on three aspects of policy: context, text and consequences.

Context – refers to the antecedents and pressures leading to the development of a specific policy. This requires an analysis of the economic, social and political factors that give rise to an issue emerging on the policy agenda. However, it goes beyond this and includes a study of the role played by pressure groups and social movements that may have forced policy makers to respond to the issue in the first place. At this point it is important to understand how the policy may relate to previous policy experience – to what extent does it build on, or break with, previous policy? Clearly, an analysis of context can take place at any level. Policies at the state or institutional level (or indeed anywhere in between), will have their own context and including this within the analysis is vital if the aim is to build up as full a picture as possible of the policy process.

Text – broadly refers to the content of the policy itself. How is the policy articulated and framed? What does the policy aim to do? What are the values contained within the policy? Are these explicit, or implicit? Does the policy require action, if so what and by whom? It may be worth highlighting that analysis of the policy text is not a simple and straightforward activity. There is considerable scope for interpretation, even in the most explicit of policies, and it is as important to identify the 'silences' (what is not stated) as well as what is clearly and openly articulated.

Consequences – if policy texts are open to differing interpretation by practitioners then this is also likely to result in differences in implementation. Such differences will then be magnified, as the unique conditions prevailing in each institution further shape the implementation of the policy. Distortions and gaps appear in the implementation process, resulting in what is best described as 'policy refraction'.

Taylor *et al.*'s (1997) analytical framework focusing on the context, text and consequences of policy offers a model for policy analysis that will be used throughout this volume. However, in order to understand more fully how educational policy shapes and is shaped by the actions of those who have the responsibility for implementing it, further dimensions need to be added to this analytical framework. These take account of how the content of policy emerges from the economic, social and political factors that give rise to an issue, explore more fully the consequences of policy and focus in more on the processes of moving from policy formulation to policy in practice.

The proposed addition to this framework has four levels: the *socio-political environment* from which policy, based on the dominant discourse, is derived and within which its overarching guiding principles are formulated; the *strategic direction* which emanates from the socio-political environment and which broadly defines policy and establishes its success criteria as they apply to spheres of activity such as education; *organizational principles* which indicate the parameters within which policy is to be implemented in those spheres of activity; and *operational practices*, based on the organizational principles, which are the detailed organizational arrangements that are necessary to implement the policy at the institutional level and to translate such policy implementation into institutional procedures and specific programmes of action. Thus, in terms of translating policy into practice, the four levels are in a hierarchical relationship, the first two being concerned with policy formulation and the second two with policy implementation. The four levels are nested (Barr and Dreeben 1983) in the sense that educational policy, derived from the wider socio-political discourse, is mediated through the formulation of a strategic direction in the national and regional context which, in turn, generate organizational processes within which schools are located and curriculum content, pedagogy and assessment determined. In this way, policy legitimized and derived from, for example, human capital theory, is translated into activities in the school and classroom.

It can be seen, therefore, that an analysis of the debates within the socio-political environment that give rise to educational policy can facilitate a

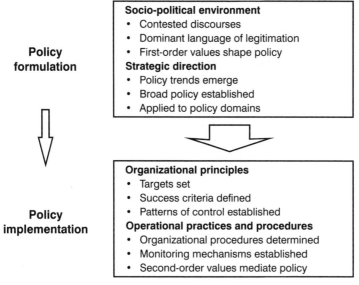

Figure 1.1 Policy into practice: a model

more detailed understanding of the context element of the Taylor *et al.* (1997) framework. The strategic direction and organizational principles provide further insight into the text of policy, its aims and purposes, while an examination of operational practices will focus attention on the consequences of policy, its interpretation and implementation. Subsequent examples will illustrate these processes at both a macro and micro level and highlight the need to see the boundaries between these analytical descriptors as fluid and porous. The conception of policy developed here is one that rarely lends itself to neat and simple models. However, before engaging in this analysis it is important to clarify more precisely what it is that is being discussed – the question that now needs to be addressed – what is 'policy'?

Perspectives on education policy

It may be attractive and convenient to be able to offer short and succinct definitions of the concepts being analysed but this is seldom possible or helpful, and a discussion of policy is no exception. The range of conceptual issues embraced by the term policy are too broad to be confined to a single, pithy definition – rather it is necessary to develop an understanding of policy that reflects the breadth and complexity that the reality of policy analysis entails. One common approach is to conceptualize policy as a programme of action, or a set of guidelines that determine how one should proceed given a particular set of circumstances. Blakemore (2003: 10), for example, presents a definition of policies as '. . . aims or goals, or statements of what ought to happen'. This distinction between objectives and 'statements of what ought to happen' echoes a similar distinction identified by Harman (1984) between policies as statements of intent, and those that represent plans or programmes of work. Hence, Harman argues policy is:

> . . . the implicit or explicit specification of courses of purposive action being followed, or to be followed in dealing with a recognized problem or matter of concern, and directed towards the accomplishment of some intended or desired set of goals. Policy can also be thought of as a position or stance developed in response to a problem or issue of conflict, and directed towards a particular objective.
>
> (Harman 1984: 13)

For Harman, therefore, it is important to recognize that policy is systematic rather than random. It is goal-oriented and it is complex – it is the co-ordination of several courses of action, and not one discrete activity. However, in both Blakemore's and Harman's argument the emphasis is on policy as a product – as an outcome. The limitation of seeing policy only as a product is that it disconnects it from policy as a process – there is a failure to see policy as *both* product *and* process (Taylor *et al.* 1997). Furthermore, this conceptualization of policy is de-coupled from the context from which it is taken – what is the purpose of the policy? What is it trying to achieve? The notion of policy as the pursuit of fundamentally political objectives is recognized in Kogan's study of educational policy making in which he refers to policies as the 'operational statements of values' – the 'authoritative allocation of values' (Kogan 1975: 55) and this helpfully locates policy within a context of wider fundamental questions that have already been identified – what is education for? Who is education for? Who decides?

The service that Kogan provides is to place values at the centre of understanding policy. Kogan identified four key values that underpin and inform educational policy – educational, social, economic and institutional values. In his study he distinguishes between basic and secondary values with educational, social and economic values being considered as instrumental, or basic, and institutional values being considered as consequential or secondary. Kogan asserted that a basic value is one that 'requires no further defence than that it is held to be right by those who believe it' (1975: 53). Secondary, or instrumental, values are justified by the extent to which they support, or further, the advancement of basic values. 'The basic values are self-justificatory "oughts". The secondary values are concepts that carry the argument into the zone of consequences and instruments and institutions' (Kogan 1975: 54).

The suggestion of a hierarchy of values is helpful in so far as it can shed light on the relative importance of different factors that drive policy – some of these policy drivers are discussed in more detail in Part 2 of this volume. Kogan's highly authoritative study of the policy-making process in England and Wales presents a thoroughly researched study of the development of education policy in a period characterized largely by cross-party consensus and a commitment to the expansion of educational provision (Simon 1991). This was a period of social and political consensus in which an accommodation between capital and labour, the emergence of Keynesian economic management and the development of a substantial welfare sector provided the basis for a social democratic settlement. This phase of broad political consensus was therefore effectively a period of values consensus in which it was argued that the development of welfarism had presaged a new era of citizenship based on the developments of new social rights (Marshall 1950). As a result of this consensus, studies of policy such as Kogan's, viewed the values underpinning policy as largely unproblematic. Policies were presented in terms of achieving objectives, or solving problems – negotiation and compromise through the policy process would result in a coalescing of views and values. The process was both technical and rational. The corollary of this analysis was a largely linear view of policy development whereby problems were identified, solutions developed and strategies and interventions then implemented. Such an approach to policy making is located within the pluralist tradition that sees the role of political institutions as reconciling the competing demands and expectations of different interest groups. Conflict is not denied, but it is not seen as inevitable, and it is certainly

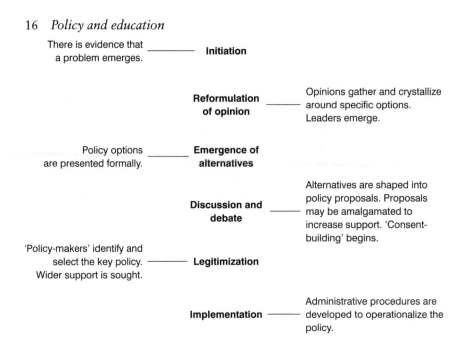

Figure 1.2   A linear model of policy development

seen as manageable. The consequence of this analysis is a logical, sequential approach to policy making typified by Jennings's (1977) model in Figure 1.2.

This view of policy, and the policy process, has several strengths. Its emphasis on the internal workings of policy-making bureaucracies, especially at a governmental level, can provide an important spotlight on the internal workings of public administration. It can also reflect the key influence of important actors in the policy-making process. However, in several respects it provides an inadequate model of what constitutes policy, and how policy is both shaped and experienced by those involved at all stages in the policy process. Policy emerges from political pressures and is contained within a political system whose purpose is to transform 'group conflict over public resources and values into authorized courses of action concerning their allocation' (Harman 1984: 16). Conflict is recognized, but exists within tightly defined parameters. Power is acknowledged, but it is rarely problematized. Sources of power are rarely discussed and little attention is paid to the (unequal) distribution of power. Similarly, the pluralist emphasis on institutional policy processes tends to privilege the generation of policy, but has less to say about implementation.

It may be accurate to characterize the period about which Kogan (1975) was writing as a period of consensus but there can be no such claim made today. Traditional assumptions are challenged by rapid economic and social change. Change brings winners and losers. If policy is to be conceived in terms of the operational statements of values there must be a recognition that those values are continually being contested, with ensuing conflicts ebbing and flowing. It is important to recognize that policy must be viewed as both product and process (Taylor *et al.* 1997), and that conflicts over values are played out as much, indeed more so, in the processes of policy development as in the policy text itself. Bowe *et al.* (1992) argue that within the traditional pluralist framework the artificial separation of generation from implementation, and the privileging of the former at the expense of the latter, results in an over-simplified model of the policy process that fails to reflect the complexity and 'messiness' of policy formulation and implementation. They argue that linear approaches to policy making, such as those presented by Jennings, 'portray policy generation as remote and detached from implementation. Policy then "gets done" to people by a chain of implementers' (Bowe *et al.* 1992: 7).

In contrast Bowe *et al.* (1992) argue that policy as both product and process is continuous and that policy is still being made, and re-made, as it is being implemented.

> In a very real sense generation and implementation are continuous features of the policy process, with generation of policy . . . still taking place after the legislation has been effected; both within the central state and within the LEAs and the schools.
>
> (Bowe *et al.* 1992: 14)

Bowe *et al.* (1992) assert that as policy is 'made' it is constantly being recontextualized and therefore rather than policy development as a linear process it should be seen as a cycle, made up of 'policy contexts'. This critique points to a wider conceptualization of policy that takes as its starting point the notion of policy as the operationalization of values, but recognizes that there is no automatic consensus around what those values might be. Ball (1994) seeks to conceptualize policy as both text and discourse. Policy as text emphasizes the manner in which policies are presented and interpreted – in literary terms, how the policy is written and read. The literary analogy can be helpful in illustrating how policies may have both multiple authors and multiple readers. Authorship of the text involves encoding policy in complex ways – 'via struggles, compromises, authoritative public interpretations and reinterpretations' (Ball 1994: 16). However, the de-coding of policy texts by multiple readers ensures a multiplicity of interpretations. Readers have their own contexts – their own histories and values. All of these factors shape how policies may be interpreted by readers:

> The physical text that pops through the school letterbox, or wherever, does not arrive 'out of the blue' – it has an interpretational and representational history – and neither does it enter a social and institutional vacuum.
>
> (Ball 1994: 17)

Ball's notion of policy as text emphasizes the capacity of those writing and reading the policy to shape its form at the strategic, organizational and operational levels. This highlights the scope for actors in the policy process to exert some element of agency over the development of policy. In determining actors' responses to policy there is 'creative social action, not robotic reactivity' (Ball 1994: 19). However, whilst acknowledging the scope for individual and collective agency, there is also the need to recognize that policy responses are also shaped by wider structural factors and these powerfully circumscribe the capacity of individual actors to shape policy. This introduces Ball's notion of policy as discourse in which he argues that the way in which policies are framed and the discourses that develop around policies, shape and constrain the scope for individual agency. Ball draws on the work of Foucault and argues that discourses provide a parameter within which notions of truth and knowledge are formed. The actions of actors take place within such parameters. However, the factors that shape such discourses are not value neutral, but reflect the structural balance of power in society: 'Discourses are about what can be said, and thought, but also about who can speak, when, where and with what authority' (Ball 1994: 21). Such an approach to policy analysis recognizes the importance of human agency – the capacity of individuals to fashion their own future – but is arguably better placed to reflect powerful structural pressures, such as the economic imperative to develop human capital, that have a decisive influence on driving policy. The chapters in Part 2 of this volume illustrate the importance of these discourses in shaping the socio-political environment within which policy develops.

Ball's two-dimensional approach to policy reinforces the need to see policy as both product and process. Policy can now be seen as not only the statements of strategic, organizational and operational values (product) but also the *capacity* to operationalize values (process). Conceptualizing policy in these twin terms emphasizes the intensely political character of policy. Policy is about *both* the identification of political objectives, *and* the power to transform values into practice through organizational principles and operational practices. This emphasis on policy as process recognizes that values by definition are not neutral. They are contested and often the subject of negotiation, compromise and conflict. In this conception of policy, power, conceived of as largely unproblematic in pluralist analyses, moves centre stage. For these reasons it becomes important both to articulate a conception of power and to offer some insights into the exercise of power.

Policy as the operationalization of values – understanding the nature of power

Understanding the link between educational leaders and the development of policy is a central concern of this volume. It has already been argued that policy development is not a neat process in which educational leaders simply digest policy from above and translate it into practice in the institution. Rather, policy development is fuzzy, messy and complex. It is the product of compromise, negotiation, dispute and struggle as those with competing, sometimes conflicting, values seek to secure specific objectives. Educational leaders are not simply faced with making sense of policy 'from above', but also the demands and aspirations from those below. Individuals and collectivities within organizations will naturally seek to shape policy and these pressures create a pincer movement in which educational leaders must seek to reconcile both external and internal pressures for, or in opposition to, change. In such circumstances the capacity of organizational leaders to secure policy changes, or resist them, will reflect a complex balance of power between the leader and those from within and outside the organization. It is important therefore to set out a broader conceptualization of power that can be helpful in explaining policy development processes at micro and macro levels.

Pluralist conceptions of power have tended to conceive of power as the ability of an individual to assert their will over the resistance of another. This draws on the Weberian notion that 'power is the probability that one actor within a social relationship will be in a position to carry out his own will despite resistance, regardless of the basis on which this probability rests' (Weber (1922) quoted in Dahl 2002: 10).

The political theorist Robert Dahl (1957) has argued therefore that C (the 'controlling unit') can be considered to have power over R (the 'responsive unit') in so far as C can compel R to do something that R would not otherwise do. This conception of power emphasizes the importance of decision-making processes, and accords power to those whose will prevails in these decision-making processes. The corollary of this is the need to establish decision-making structures that accord equal access to policy-making procedures. Such a framework provides a useful entry point to discussions about power, but critics have argued that it provides an over-simplistic analysis of decision-making structures. Bachrach and Baratz (1962) for example argue that the focus on formal decision-making structures neglects the extent to which power is reflected in the selection, or perhaps more importantly the non-selection, of issues about which decisions are being made:

> Of course power is exercised when A participates in the making of decisions that affect B. But power is also exercised when A devotes his energies to creating or reinforcing social and political values and institutional practices that limit the scope of the political process to public consideration of only those issues which are comparatively innocuous to A. To the extent that A succeeds in doing this, B is prevented, for all practical purposes, from bringing to the fore any issues that might in their resolution be seriously detrimental to A's preferences.
>
> (Bachrach and Baratz 1962 in Haugaard 2002: 30–1)

Bachrach and Baratz's A corresponds to Dahl's C and their B to Dahl's R. Bachrach and Baratz (1962) therefore place considerable emphasis on the capacity of those with power to control the policy agenda by determining precisely what issues are, or are not, opened up for discussion and debate, and ultimately for possible decision making. In such circumstances there may be pressures for change (from within, or without, the organization) that those with power are able to effectively exclude from the agenda. This provides an important example of how those with power are able to begin to shape the socio-political discourse within which policy debates are framed. Such an approach clearly develops the more limited pluralist perspective of power by emphasizing 'two faces' of power (Bachrach and Baratz 1962). However, Lukes (1974) is critical of both the pluralists and Bachrach and Baratz for focusing excessively on the observable behaviour of individuals. According to Lukes, power is often more subtle and more elusive than the pluralists and their early critics suggested. First, power is often exercised through collectivities of individuals and not individuals acting independently. Indeed, individuals in such collectivities may not appreciate they are in a position of power at all. This emphasis on collective power can create organizational bias that can shape decision-making spaces, but often in opaque ways. Secondly, Lukes questions whether power is only evident when there is apparent conflict, that is when A is compelling B to do something that B is opposed to doing. Might power be exercised in situations where both A and B appear to pursue the same objectives? These concerns led Lukes to add a third dimension to the developing conception of power in which he argues that A has power over B if A is able to influence B's thinking in such a way that B wants what A wants. This is not a case of crude brainwashing but rather the subtle and largely systemic way in which individuals' views about their interests, and indeed their values, are shaped. It highlights the importance of the socio-political environment within which notions of 'common sense' (Gramsci 1971) are formed and in which policy is subsequently developed. It particularly focuses on the way in which the state, and agencies such as the media, are able to construct the parameters within which policy debate takes place.

Lukes's three dimensions of power provide a useful framework for analysing power at both a state and institutional level and can usefully be applied to the model of policy development presented in this chapter in Figure 1.2. It highlights both the multi-faceted nature of power, and the manner in which power may be exercised collectively and systemically. The emphasis on power as not only an ability to shape the policy agenda, but as the capacity to shape how that agenda is perceived, highlights the centrality of understanding how policy 'problems' are presented and defined. Policy 'solutions' are then shaped decisively by those who are able to define the problem, and set the parameters within which solutions might be considered possible. This provides a broad conceptualization of power that is utilized to support policy analyses throughout this volume. However, such an approach still suggests a hierarchical down flow of power, and this fails to adequately reflect the way in which policy may be formed and re-formed by challenge from below, as well as by imposition from above. The four-stage model of policy development presented in this chapter points to a policy trajectory that flows downwards from the central state. This reflects the pivotal importance of the state and state power in shaping education policy (discussed more fully in Chapter 2). However, such a policy trajectory is never clear cut. Policy may be contested and challenged as it is developed and will be shaped and re-shaped by pressures from below as well as pressures from above. It is helpful, therefore, further to develop an analysis of power that recognizes that flows of power can be multi-directional, rather than simply and mechanistically flowing from the top down. Bacharach and Lawler's (1980) differentiation between authority and influence as two distinct forms of power can provide a useful contribution to the analysis at this point. Authority is bounded by bureaucratic rule-making processes with a clear expectation that subordinates will implement the decisions of superiors – willingly or unwillingly. Its power source is invested in the role an individual holds and their location in the hierarchy. A headteacher clearly has a significant level of authority based purely on their role and the commonly shared views about the legitimacy of this role. These perceptions are not static – they are in turn contested – but they exist in some form at some point in time. Authority represents a downward flow of power. In contrast, influence may be considered multi-directional and therefore includes the possibility of those subordinate in the policy-development process being able to shape the decisions of those more senior in the hierarchy. Sources of influence are more diverse and more fluid than sources of authority. Bacharach and Lawler (1980) suggest they may include personal characteristics, expertise (the possession of specialized information) and opportunity (derived from a strategically important location in the structure or organization). Recognizing the importance of influence as a form of power allows for a more complete picture of policy making at several levels – one in which decisions are seen as the outcome of continuous interaction between individuals and collectivities:

> While authority may be a prime source of social control, influence is the dynamic aspect of power and may be the ultimate source of change. Those in authority typically want to restrict the influence of subordinates; subordinates typically want to use influence to restrict the exercise of authority by superiors … while authority is inherently an aspect of hierarchy, influence is not. The context of influence need not be superior-subordinate relations; in fact, influence is the mechanism through which divergent subgroups without authority over one another may compete for power within an organization.
>
> (Bacharach and Lawler 1980: 30)

Bacharach and Lawler's study is concerned with identifying the conditions in which work groups and interest groups in organizations combine together, and from which coalitions emerge. Such groups may be informal and *ad hoc*, or may be more formalized, allowing for example for the possibility of acknowledging trade union influence on policy development at both state (Barber 1992) and institutional levels (Stevenson 2005). The nature of such coalitions will inevitably be bound by context. Workplace trade union organization for instance may be accepted and relatively commonplace in some contexts (Stevenson 2003a), it may be virtually absent in others. All of these differences will influence the extent to which policy may be re-shaped at an institutional level in differing contexts.

Bacharach and Lawler (1980) highlight the dynamic and shifting nature of coalition building. Coalitions emerge, develop and potentially fade in response to shifts in the local context. Principally, coalitions are formed to generate influence either in pursuit of, or opposition to, change. Change is the impulse that drives this dynamic. Change, by definition, undermines the *status quo*. Existing practices are often questioned, traditional assumptions can be threatened and values may be challenged. Change is seldom neutral – there are winners and losers, those who benefit from proposed policy changes and those who pay. It is therefore a process that requires action and will generate reaction. The tensions and conflicts that flow from these responses therefore need to be seen as inevitable and not irrational, as Ball (1987) argues:

> ... it is not surprising that innovation processes in schools frequently take the form of political conflict between advocacy and opposition groups. Either in public debate or through 'behind the scenes' manoeuvres and lobbying, factional groups will seek to advance or defend their interests, being for or against the change. Negotiations and compromises may produce amendments to initial proposals, certain groups or individuals may be exempted, trade-offs arranged, bargains arrived at.
>
> (Ball 1987: 32)

Policy can therefore be presented in part as the analysis of change and the way in which change is managed. Change may be inevitable – but there is no inevitability about how change is experienced. Those with power are often able to shape the way the 'real world' is perceived – to define the problem, to set the limits within which solutions might be acceptable and even to select and impose specific solutions. The value of the analyses presented above is to link policy and change, but to recognize that change is not neutral. Policy, as one of the ways in which people experience change, will inevitably be contested, and its outcomes shaped by the consequences of macro and micro-political processes in which competing groups seek to shape and influence policy. Empirical studies in Part 3 of this volume provide examples of how policy is shaped by these micro-political manoeuvrings.

Conclusion

This chapter has highlighted the complex nature of policy, and pointed to the inadequacy of overly simple definitions of policy and descriptions of the policy process. What is understood by 'policy', how it is conceptualized, requires a broad understanding of a range of inter-related processes. What is often presented as policy is frequently no more than a statement of intent, a plan of action or a set of guidelines. At one level the purpose of such policies may appear clear, but it is important to locate policy within a wider context. Policy is about the power to determine what gets done, or not done. These are profoundly political issues. Those presenting policy will interpret its content differently, and those receiving policy will do similarly – a single 'policy' may be better understood therefore as a plurality of policies that emerge and develop as the policy process moves from formulation to implementation. The model presented in this chapter, and applied throughout this volume, identifies a four-stage process that begins by recognizing the importance of the wider socio-political environment in shaping the discourse within which policy debate is conducted. From within this discourse, a strategic direction develops in which specific educational policies become more clearly defined, and success criteria are established. As policy texts emerge with greater clarity this in turn shapes the organizational principles, and ultimately the operational practices, that shape the experience of policy at an institutional level.

In reality this is not a tidy linear process in which policy progresses obediently from one stage to the next. Differences in emphasis, differences in interpretation and differences in attitudes to policy are ever present and will in large part reflect the differences in values that underpin policies – policy being seen, in Kogan's (1975) terms, as the authoritative allocation of values. Conflicts over policy represent struggles between opposing values sets. This chapter has highlighted the centrality of values to an understanding of policy and the need to see policy as both product and process. Policy therefore can be seen as *both* operational statements of values, *and* as the capacity to operationalize values through the ability to exert influence at key points in the four stages of policy development. Those values are clearly the values of individuals – values are, after all, those beliefs and principles that individuals hold most dear. They provide both a lens through which the world is viewed, and they provide a moral compass that shapes actions and responses to the environment (Begley 2004). However, values do not float free of the environment in which they are enacted. Values are constantly being shaped, formed and re-formed. Pluralist approaches to policy development emphasize the extent to which values are able to shape policy, but it is also important to recognize how policy can shape values (Bottery 2004a). This is an iterative process (Giddens 1984) that raises important questions about where power lies. To what extent are individuals free to shape policy, or to what extent might the influence of individuals be shaped by more powerful structural factors? This chapter has identified a number of concepts relating to the nature of power that help develop an understanding of policy as both a product and a process in which access to resources of power can decisively shape the development of policy. The centrality of power in the policy process highlights the need to explore further the sites of policy development – principally the state and the institution. What is the relationship between the two and how does the relative balance of power impact on the development of policy at both state and institutional level? These issues are explored in the following chapter.

2 Investigating the sites of policy development

Introduction

In Chapter 1 it was argued that policy-making processes do not lend themselves to a simple dichotomy between formulation and implementation. It can be helpful to distinguish between the two but is more useful to see both elements as part of a seamless process in which implementation is as important as formulation. Policy development provides a more useful term to describe policy as not only product, but also as a process that rarely has an identifiable beginning or end. However, it is important to understand the context in which policy development takes place – how policies emerge, how they form and take shape, and how they become lived through the actions of those engaged in the policy-development process. This chapter focuses on two key sites of policy development – the state, and the individual institution, and seeks to make connections between the two. It introduces the question, addressed in different ways throughout this volume, about the extent to which those working in educational institutions may be considered to enjoy any meaningful autonomy to develop organizational principles and operational practices independent from the state. The state is often represented as the source of educational policy. It is indeed the case that much policy experienced by educational institutions located in both the public and private sectors derives directly from state legislation and directives – a point emphasized by Dye's (1992) description of policy analysis as the study of what governments do, and why. Furthermore, the tendency towards policy centralization (Simon 1988 and 1991), evident in many countries (Smyth 1993), has emphasized the need to reflect the pivotal role of central government in shaping policy. However, the link between state policy and institutional practice remains complex and it is important, therefore, to develop a broader understanding of the state that acknowledges the myriad functions and purposes of state activity.

However, it is also important to focus on the individual educational institution because this is the point that represents the interface between the wider policy environment and the individual learner. Those working in institutions must both make sense of policy from outside, and generate policy within. These are of course not disconnected processes. Policy within is shaped decisively by policy from without. But what is the nature of the link between the two? To what extent are those working in institutions able to shape their own policy agendas, or to what extent can institutional policy be considered to be driven by state policy? How are values manifest in state policy and how far are 'state values' reflected in the organizational principles and practices of individual institutions? These questions raise fundamental issues about the nature of power at both state and institutional level. This chapter explores the respective roles of the state and institutions, and the inter-relationships between leadership and power, in the policy-development process.

The state and policy development

The concept of a pluralist approach to policy making introduced in Chapter 1 can be located in a broader theoretical approach to analysing the role and purpose of the state. Pluralist approaches to the state and policy development have tended to represent the dominant discourse within much policy analysis (McNay and Ozga 1985); however, the relevance of such analyses are limited by a number of factors, not least the almost exclusive focus on western liberal democratic systems. Pluralist conceptions of the modern state emphasize the role of state institutions in representing and reconciling the competing and sometimes conflicting interests in society. In modern societies where mass participation in democratic institutions is not practical, institutions need to be developed that are able to give voice to diverse interests, and to provide mechanisms for resolving tensions between interest groups. The pluralist model presents the role of government as using democratic processes to ensure that state policies reflect majority views within society. In this sense a key role of the state is to reconcile competing values positions, and to cohere these in to a consensual articulation of communal or societal values. The pluralist perspective therefore places a premium on the capacity of people participating in political processes to shape policy as 'operational statements of values' (Kogan 1975). State decisions derive their legitimacy from the robustness of the democratic processes involved. In this system political parties and pressure groups are crucial to the democratic process. These organizations articulate the collective aspirations of different interest groups and represent these views in governmental institutions. Those who are more effective at securing their objectives may be considered to be more powerful. Power is conceived in relatively limited terms as the capacity for one individual, or group, to compel another individual or group to take action that they otherwise would not have done (Dahl 1957). This has led some pluralists to argue that a study of policy-making processes, and more specifically the outcomes of these processes, can allow researchers to make judgements about where power lies (Polsby 1963).

Pluralists do not argue that power is equally distributed. For example, Dahl's (1982) later work develops and presents a more sophisticated understanding of the unequal balance of power between different interest groups. However, pluralism emphasizes the importance of having democratic structures that provide access for all social groupings to decision-making bodies and the policy process. These structures allow social groupings to compete for influence in their bid to shape state policy. Political decisions that flow from this competition for influence are then the result of complex bargains and compromises that have been struck in order to gain sufficient support for the policies to be advanced further. This analysis casts the state in the role of rational arbitrator, seeking to accommodate the diverse and competing interests that are articulated by different social groupings. The state is not the representative of any particular interest group, but rather acts to balance interests between groups. Tensions between social groupings are not denied within the pluralist model on the contrary, the state is seen as having a key role in reconciling conflicting interests. However, conflict is expected to be exercised within the 'rules of the game'. Such rules, effectively the 'rule of law', are considered neutral and not to favour any specific interest group. Before questioning the adequacy of this approach it is important to set out more precisely what is considered as 'the state'.

Thus far the terms 'state' and 'government' have been used interchangeably, but it is important to distinguish between the two and recognize that the activities of the former are far broader, and more significant, than the latter. Dale (1989: 54) refers to 'state apparatuses' and defines these as 'specifiable publicly funded institutions'. This definition of the state generates a list of state institutions that includes government ministries, but goes beyond this to include the military, the police and the judiciary. In the case of some of these institutions, there may appear to be considerable autonomy between the state and government. Governments, largely organized through political parties and coalitions of parties, represent public interests outside the state and may be considered 'to mediate the State and its subjects together' (Dale 1989: 53).

It is also important to recognize that the state can be considered to operate at a number of different levels. Again, traditionally, the 'state' refers to the nation state – those state institutions that function at the level of the whole nation, and in governmental terms are associated with national legislative bodies or parliaments. However, any conceptualization of the state must also embrace those regional and local institutions that are also publicly funded institutions and therefore part of the state apparatus. In many countries this distinction is often at its clearest in the separation between the institutions of central government, and those of regional or local assemblies with the latter often playing a significant role in the development of educational policy in particular.

More recently, it has become clear that the traditional notion of the state that begins by identifying publicly funded national institutions looking downwards to the regional and the local is inadequate (Bottery 2000). Global economic and political developments have brought forth supranational institutions that perform many state functions in terms of policy development, but which function across nation states, rather than within them. In some cases, such as the European Union (EU) these institutions have been established for some time but are increasingly beginning to resemble traditional governmental institutions with developing constitutional arrangements. However, in other cases, institutions with inputs from nation states appear to be more disconnected from traditional state apparatuses, but can have a similar, indeed greater, impact on the development of policy at a national and local level. Perhaps the most significant example of such a body is the World Trade Organization (WTO). Many of these institutions now have the capacity to exert significant influence on education policy in individual nation states (in the case of the WTO this is illustrated by the increasing emphasis on liberalizing trade in services as well as goods). In developing countries the influence of the World Bank on educational policy is equally significant. For the purpose of policy analysis the implications of this more complex conceptualization of the state is important because it implicitly acknowledges the greater potential for tension between, as well as within, different elements of the state.

A focus on state purposes and institutions is clearly important but can provide only a partial picture of what is being studied. It is also important to focus in practical terms on what the state does. How does state policy manifest itself? The tools of policy are of course not value neutral, and the way in which particular policies are enacted in particular contexts are intensely political issues. Policies cannot be disconnected from the socio-political environment within which they are framed. However, before exploring the sharply differing ways in which ideological influences shape the application of policies in specific contexts it is possible to identify a range of state activities that at a basic level are common features of state activity. These can be considered to include the core activities of direct provision, taxation, subsidy and regulation. Although such activities are virtually a universal feature of state activity in any context, their application is not about purely mechanistic means of managing state resources. For example, the direct provision of education services within a system of public ownership has significant implications relating to matters of governance and control, whilst the use of taxation and subsidies not just determines what is provided, but crucially who pays and who benefits.

In an educational context the role of regulation is important because this determines the extent to which public service priorities may be exercised over those parts of the education system that are not formally within the public sector. In countries where education services are predominantly provided by non-state bodies, such as trusts, commercial organizations or religious bodies, the role of regulation becomes correspondingly more important and, although private ownership may give an appearance of greater institutional autonomy, the use and application of regulatory frameworks can ensure an extremely tight coupling between the public state and private institutions. Precisely how these policy approaches manifest themselves in practice in differing contexts highlights the need to develop models of policy analysis capable of reflecting cultural and historical contexts.

Studies of the state's role in shaping policy development must be capable of reflecting co-existing, but opposing pressures. In the first instance it is important to recognize the crucial role played by societal culture in shaping state policy. These are the pressures that account for important policy differences between nation states as factors specific to local contexts exert a decisive influence on policy. However, whilst cultural influences will shape policy in distinct and unique ways there are simultaneous pressures towards policy uniformity as global economic pressures in particular appear to drive common policies in differing cultural contexts. It is important, therefore, to explore in more detail what at first sight appear to be contradictory tendencies towards policy diversity and uniformity.

Recognizing the importance of cultural difference

Research and literature on the state is dominated by the influence of western scholars and as a consequence models capable of reflecting, for example, the experience of African and Asian contexts are limited (Apple 2003). At its worst, the conclusions of Anglo-US studies are simply extrapolated across diverse cultural contexts and their conclusions generalized with little qualification:

> Anglo-American scholars continue to exert a disproportionate influence on theory, policy and practice. Thus a relatively small number of scholars and policy makers representing less than 8% of the world's population purport to speak for the rest.
>
> (Walker and Dimmock 2002: 15)

Such a situation provides a wholly inadequate basis for analysis. Rather, what is required are analyses of the state, and models of policy development, that recognize difference and are capable of reflecting cultural context. Such analyses also need to take account of the dynamic nature of state formations and the manner in which these shift and change over time – at times variously converging and diverging with state formations in other contexts. Walker and Dimmock (2002) distinguish between a range of societal cultures. These distinctions can form the basis of an analysis of differing state formations. For example, one distinction is between 'power distributed' and 'power concentrated' societies. In the latter, state formations are likely to be more centralized, with an expectation that policy at institutional level will very closely represent the expectations of policy makers at the centre. One indication of the extent to which power is distributed is the extent to which local and regional government might be involved in the development of policy. Generally, the more 'loosely-coupled' state structures are, the more opportunity there is for policy variation at a local level. Walker and Dimmock (2002) make a case for power distributed societal cultures to tend towards greater egalitarianism, with an often correspondingly stronger commitment to redistributive state policies. This provides a link to another distinction that can be helpful in analysing state structures and their associated policy priorities – that between 'group oriented' and 'self-oriented' societal cultures. Group-oriented cultures are more collectivist in nature – 'ties between people are tight, relationships are firmly structured, and individual needs are subservient to collective needs' (Walker and Dimmock 2002: 25). It can be tempting to see the state as the obvious manifestation of a more group-oriented culture and therefore group-oriented cultures being more likely to see a significant role for the state; however, there is not necessarily a neat correspondence. For example, Nordic countries arguably tend to a group-oriented culture, and these countries have traditionally sought to provide comprehensive welfare services through the state (Rasmussen 2002, Welle-Strand and Tjeldvoll 2002). However, in contrast, there are examples of Eastern cultures, such as in Japan, that may be described as group-oriented, but where there is little tradition of state provision of welfare. Collective provision of welfare emerges in familial and occupational forms, rather than through the state.

This brief discussion of state formations in different cultural contexts does not seek to provide a comprehensive typology of state formations across a range of different contexts. Rather it highlights the need to eschew simple, one-size-fits-all approaches to analysing the state and models of policy development. There is considerable variation between cultural contexts and models of policy development must be able to take account of this diversity and complexity. However, whilst it is essential to recognize cultural difference and the way in which policy in individual nation states is mediated by cultural context, it is also important to recognize where there is similarity, and, over time, convergence. Analysing the elements of this convergence, and the global factors driving it, is as important as identifying sources of difference.

Globalization and the pressures for global uniformity

Although literature on the state and policy development is dominated by Western sources, it is arguably the economic challenge from the East that accounts for the key shifts in state formation, particularly in Western societies, in recent years and the tendency for common state policies to emerge. Therefore, whilst recognizing the distinctive nature of cultural context, and the degree of difference between countries, it is equally important to discern a number of key trends that have an element of global commonality. That there is commonality reflects a number of interdependent phenomena, faced particularly, but not exclusively, by Western economies. These pressures can be traced to a number of sources:

- The emergence of much broader international competition, and in particular the rise of economies in South East Asia with a comparative advantage in many of the manufacturing industries traditionally dominated by the West (Hay 1985).
- The increasing mobility of capital, facilitated by technological advances in communications and transport that has intensified global competition (Strange 1997).
- The dominance of a neo-liberal hegemony that has successfully promoted a free trade agenda based on economic imperatives, not social objectives (Costello *et al.* 1989).
- Demographic changes, including an ageing population, that have increased pressure on state resources and particularly the demands on pensions and health care (Bottery 2004a).

Taken together these phenomena have driven widespread economic restructuring around the world, and in turn this has driven state restructuring (Jessop 1994). Central to state restructuring has been a challenge to the welfarist principles that underpinned state policy in many Western countries in the years after the Second World War. During this period welfare systems based on principles of universalism and an explicit, if sometimes modest, commitment to redistribution had emerged and expanded. Keynesian economic policies appeared to guarantee full employment and this provided labour with the bargaining power, and the state with the resources, to confidently expand welfare provision. However, at the time critics from both the right (Bacon and Eltis 1976) and left (O'Connor 1973) questioned the long-term sustainability of this post-war welfarism in the West. Bacon and Eltis (1976) argued that the inexorably expanding state would absorb ever-increasing resources and 'crowd out' private sector investment. From a Marxist perspective, O'Connor (1973) had arrived at similar conclusions, arguing that capital required the welfare state to create the conditions for capital accumulation (notably a workforce developed by the education system with appropriate skills and attitudes), but that the rising cost of welfare provision would ultimately reduce profitability. The ensuing fiscal crisis was caused by simultaneous pressures to increase spending on welfare services such as education, whilst decreasing tax and public borrowing in order to maintain private sector profitability. In their different ways, and from quite different perspectives, a discourse developed that questioned the sustained affordability of welfare in an age of global capitalism. The discourse of 'affordability' continues to dominate welfare debates, and at least in part accounts for the increasing emphasis placed on education as investment in human capital (see Chapter 3), hence locating educational policy as supply-side driven economic policy, rather than as social, or 'welfare' policy.

Such an approach to education policy highlights a key contradiction in public policy: capital requires a labour force with appropriate skills, qualification and attitudes if it is to be competitive. There is no discernible evidence that a free market in educational services will meet this need (Bottery 2004a). However, state provision is expensive and requires public funding in a way that is characterized as creating disincentives to capital. The result is a 'funding gap' between what is required to meet capital's needs and what capital appears able to 'afford'. Efforts to square this circle have generated a kind of global economic orthodoxy in which a number of common policy trends emerge. Taken together these amount to a restructuring of state policy and state institutions with significant implications for the funding and provision of education services. Key features of state restructuring include:

- A restructuring of public services through the use of devolved management and quasi-markets, thereby securing improved 'value for money' (discussed in more detail in Chapter 5).
- Opening up areas of public sector activity to private enterprise. In some cases this represents the abandonment of public sector provision to the private sector, in many cases it takes the form of complex public/ private partnerships in which private capital is used alongside public investment (Whitfield 2000), an issue explored further in Chapter 7.
- A shift in the burden of cost from the collective to the individual whereby users of educational services are increasingly expected to purchase, or at least contribute to, what they consume (illustrated by the introduction of tuition fees for higher education in England and Wales by the UK government).
- The formation of powerful centralized inspectorates that have a role in monitoring contracts and the meeting of performance standards (Pollitt 1992), discussed further in Chapter 5.

Although these policies have been particularly prevalent in the Western economies, and most common amongst the Anglophone nations (Smyth 1993), it is also possible to discern the themes of privatization, de-regulation and an increased emphasis on markets in African, Asian and Latin and South American education policy (Burbules and Torres *et al.* 2000, Torres 2002), partly because all countries are responding to similar global pressures, and partly because of the power of international institutions such as the World Bank and the International Monetary Fund that often drive these policies. Moreover, it is important to recognize that such 'policy cloning' (Dimmock 1998) is not simply a case of Western orthodoxies being imposed on economies elsewhere, but that the new neo-liberal orthodoxy is in part a response, through imitation, to the emergence of powerful Asian economies. Furthermore, Jacques (2005: 17) has argued that as global pressures develop 'cultural traffic will no longer be one way'. He rejects the orthodox view of globalization as one that 'is overwhelmingly one of westernization' and asserts that Western cultural values in particular will become increasingly contested as Asian economies grow and Asian nations become correspondingly more confident.

It is clear that the powerful structural forces that are associated with globalization exert a significant influence on state policies in general, and on education policies in particular. These themes are developed further in the chapters in Part 2 of this volume. However, it is important to recognize the complexity of the globalization process in which global orthodoxies are not solely the product of a Western hegemony but are also, and increasingly, part of a complex mix of global influences. Furthermore, despite the emergence of clear global orthodoxies it is important to recognize the enduring influence of specific cultural contexts and the extent to which cultural factors will always mediate and shape policy at a regional and local level – the result is a rather more complex picture than is often suggested:

> . . . rather than a full-scale globalization of education, the evidence suggests a partial internationalization of education systems which falls far short of an end to national education *per se*. National education systems have become more porous in recent years. They have been partially internationalized through increased staff and student mobility, through widespread policy borrowing and through attempts to enhance the international dimension of curricula at secondary and higher levels. They have also grown more like each in other in certain important ways. However, there is little evidence that national systems as such are disappearing or that national states have ceased to control them. They may seem less distinctive and their roles are changing but they still undoubtedly attempt to serve national ends.
>
> (Green 1997: 171)

State restructuring and institutional policy development

Whilst it is essential to recognize significant differences between cultural contexts, it is also important to identify that way in which global pressures have driven state restructuring in the way described previously, and the particular way in which restructuring at a micro-level shapes policy development at an institutional level. Gewirtz (2002) has argued that in the UK the restructuring of education represents a shift from welfarism to 'post-welfarism' with a corresponding shift in institutional values. Within a welfarist regime state education was developed to shield individuals from the vagaries, and the inequities, of market forces (Marshall, 1981). Within a post-welfarist regime market forces become the driving force of the system – simultaneously intended to drive up 'standards' and ensure 'accountability' (Tomlinson, 2001). Hence educational leaders' actions are determined first and foremost by what is required to ensure organizational survival in a competitive and unforgiving market. However, it cannot be assumed that as policy develops it is implemented uncritically by those working in educational institutions. In Chapter 1 it was argued that as policy develops it is interpreted, re-interpreted and often challenged as the strategic direction of policy develops into organizational principles and operational practices. Such conflicts are more likely where the dominant values expressed within the socio-political environment are out of kilter with the values of those actors involved in implementation. Wherever there is a values dissonance there is likely to be increased conflict over policy development. However, to what extent is it possible for those working in educational institutions to challenge the values base of state policy? Just as the state may be ascribed some degree of autonomy whereby state policy does not correspond exactly to economic conditions so too might it be argued that individual educational institutions enjoy a degree of autonomy from the central state?

Gewirtz and Ball (2000) have argued that the restructuring of education has created a new managerialism in which those leading educational institutions have been compelled to forego a 'welfarist' approach to management. The pressure to perform in a market, or quasi-market (LeGrand 1990, Bartlett 1992) compels the manager to focus on performance and productivity. Educational values are forfeited as the priority is to maximize added-value. For example, according to Ironside and Seifert (1995) the public sector manager is faced with the same challenge as a commercial employer – to get 'more for less' from employees as market, not educational, priorities prevail. However, the collision of values ensures that the outcomes are neither clear, nor certain – but are often the subject of negotiation, compromise and struggle:

> The shift in values and language associated with marketization – and the construction of the post-welfarist settlement more generally – is contested and struggled over. In trying to respond to pressures created by the market, headteachers and teachers find themselves enmeshed in value conflicts and ethical dilemmas, as they are forced to rethink long held commitments.
>
> (Gewirtz 2002: 49)

How then is this tension between values resolved when the personal priorities of individuals are at odds with the dominant values expressed in the socio-political environment within which they function? Empirical studies provide a range of different scenarios in which the scope for individual agency varies considerably. Commentators who see little opportunity for those at an institutional level to shape their own policy agendas emphasize the power of external structures. They argue that those working within educational organizations have no more than the most limited capacity to develop internal policy agendas that challenge external influences – quite simply they are overwhelmed by the power of external structures whether it be the inspectorate, the local market context, or more likely, a confluence of the two. One of the clearest exponents of this perspective is Wright (2001) who argues:

> Leadership as the moral and value underpinning the direction of schools is being removed from those who work there. It is now very substantially located at the political level where it is not available for contestation, modification or adjustment to local variations.
>
> (Wright 2001: 280)

Wright (2003) asserts that school leaders may have 'second order' values (such as a commitment to team working or involving staff in decision making) that can stand in contrast to the dominance of the culture of performance, but that they are unable to challenge the first order values. First order values, in the form of system aims and outcomes, are determined elsewhere and reinforced by powerful control mechanisms that render them effectively unchallengeable. Second order values may result in internal policy agendas that appear more acceptable, but this is no more than a discussion about means rather than ends – the ends remain beyond debate. More recently Wright has argued:

> principals are [not] necessarily unprincipled people, far from it, but . . . the system in which they have to operate stipulates the overall framework, values direction and often the detail of what they have to do.
>
> (Wright 2004: 1–2)

Such an argument points to the importance of the second and third dimensions of power presented by Bachrach and Baratz (1962) and Lukes (1974) and discussed in Chapter 1. In such a scenario school leaders at best have no scope to question fundamental objectives (Bachrach and Baratz 1962), and at worst the structures of the system result in them simply internalizing the logic of the market (Lukes 1974). Wright's views are echoed by others such as Hatcher (2005) and, perhaps more guardedly, by Thrupp (2004), who both highlight the difficulties of school leaders being able to develop internal policy agendas that may be at odds with central priorities. Thrupp (2004: 8) suggests that school leaders' capacity to 'mediate' the external policy agenda is very limited, but that there may be potential for what he described as 'passive resistance' or 'unofficial responses'.

The arguments presented by Wright (2001, 2003 and 2004), Thrupp (2004) and Hatcher (2005) were in part a specific response to work by others (Day *et al.* 2000, Moore *et al.* 2002 and Gold *et al.* 2003) who have offered a more optimistic view of school leaders and argued that effective leaders can create spaces within which progressive and distinctive internal policy agendas might be developed, even when these are at odds with the demands of external structures. These contributions may perhaps be described as 'critical optimists'. They are critical because in their studies the school leaders in question were sometimes going against the grain of state policy – in summary, there was a 'values clash' between the priorities of individual school leaders and the demands of the external agenda. However, these studies suggest that school leaders were in some way able to achieve a reconciliation between their values and those of the external policy environment. This was not a simple case of lowest common denominator compromise, but a creative response to retain clear personal principles in a hostile environment.

Day *et al.* (2000) identify in their study school leaders who are either 'subcontractors' or 'subversives'. The former might be presented as the passive implementers of external agendas with internal policy agendas being entirely aligned with those generated externally. In such cases the authority of the leader is challenged by a perceived lack of independence. On the other hand 'subversives' may be equally problematic. These leaders seek to challenge external policies by deliberately undermining them. It is Day *et al.*'s (2000) contention that the subterfuge and duplicity involved in this process may similarly undermine the moral authority of the leader. Day *et al.* (2000) argue that in their study effective leaders were those who were able to 'mediate' the external policy agenda so that it aligned with the values and vision of the school. In their view it was not inevitable that this was done in an underhand manner, but rather it was capable of being achieved with transparency and integrity. These views are echoed by Gold *et al.* (2003) whose study of ten 'outstanding' school leaders highlighted an ability not simply to retain values in the face of contrary pressures from elsewhere, but to ensure that these values provided the moral compass required to guide school development:

> The school leaders in our case studies remained committed to a set of strongly held values and a simple shift from 'welfarism' to the 'new managerialism' (Gewirtz and Ball, 2000) was not apparent. This is not to say that school leaders were unaware of the need to manage resources effectively, including human resources, and of the significance of parental choice and market forces, but that they were not fundamental. They were driven by a different set of values and these . . . were based on intrinsic values and not those imposed by others, including governments. Of importance was the wider educational, social and personal development of all pupils and staff. Effective or 'outstanding' school leaders are those who are able to articulate their strongly held personal, moral and educational values which may, at times, not be synonymous or in sympathy with government initiatives or policies.
>
> (Gold *et al.* 2003: 136)

The debate about the relationship between external and institutional policy agendas, and the extent to which those at an institutional level have meaningful control over policy highlights the central issues with which this volume is concerned. In an age where the strategic direction of state policy is often becoming increasingly centralized the prospects for those working at an institutional level to shape policy in innovative and distinctive ways may look limited. However, the manner in which the strategic direction of state policy emerges in the organizational principles and practices of individual institutions is clearly complex, and this is illustrated further through the chapters in Part 3.

Conclusion

In Chapter 1 it was argued that policy needs to be seen as an expression of values, but it is important to recognize that values are operationalized in specific contexts – they are both a product of that context, and they create it (Giddens 1984). How policies manifest themselves in different contexts requires a broader understanding of the sites of policy development. Those locations, not always physical, are the spaces where actors in the policy-development process engage in the discussions, the negotiations and, sometimes, the struggles that forge policy. There are inevitably struggles, because these are disputes about values in all their shapes and forms. It is important therefore to develop an understanding of the sites of policy development that emphasizes the interdependent links between them and that highlights the importance of power in the policy-development process.

The significance of the state in the development of educational policy cannot be overstated. The influence of the state, and state institutions, in shaping the socio-political environment is profound. Voices from within the state are powerful and have the capacity to shape decisively the dominant discourses within which policy is framed and from which strategic direction emerges. One can argue that these discourses reflect the function of the state in securing economic, social and ideological objectives, and the role of the state in articulating these objectives is explored in more detail in part 2. In particular, the dominance in recent years of economic interests has had a significant impact on how education policy has been aligned with the need to develop human capital. However, the state is not simply the expression of a monolithic set of social or economic interests, formulating policy solely in the interests of a narrow elite. Consent is far more important than coercion (Gramsci 1971) and it is important to see state policy, and the discourses it develops, as sites of contestation in which different interest groups seek to assert their value positions. Understanding who has power in this process and how this power is exercised, becomes central to understanding the development of state policy and how it emerges in the form of organizational principles and practices.

Of course educational institutions come in a variety of forms, not only between countries but within countries. Different forms of ownership, governance and accountability all contribute to shaping quite distinct relationships between different institutions and the state. Policy analysis must be capable of reflecting the complexities of these cultural and institutional differences. Nevertheless educational institutions function in a context that is very largely framed by the state. Even where institutions are nominally independent of the state they operate in a context where state regulation is substantial, where state funding decisions are often crucial and where the social and political discourses that are shaped by the state have a profound influence on the policies of individual institutions. However, just as the state can be considered to have a degree of autonomy from the powerful structural forces that shape state policy, so too do individual educational institutions have relative autonomy from the influence of the state. Therefore the struggles over the shaping of policy that take place at the socio-political level do not disappear as policy filters down to the organizational and operational levels. Indeed, conflicts may increase as the values underpinning state policies and discourses may be challenged by those working at an institutional level. However, there is no more likely to be a unified homogenous response to policy within institutions than there is within the state policy-development process. As external policies are implemented in institutions, and as institutions develop their own organizational policies, actors in the process will seek to shape, and sometimes challenge, policy. Value conflicts at the socio-political level will be mirrored at the operational level, with the precise nature of these conflicts reflecting the particular configurations of power, structure and influence in each institution. Beginning to understand these processes becomes crucial to understanding how policy develops at an institutional level and provides the immediate framework within which learning takes place.

Part II

Themes in educational policy

Part II examines in detail some of the main themes that appear to shape educational policy in many countries. Almost inevitably, the themes are interconnected and often overlap but, for the purposes of clarity and relative brevity, three such themes have been identified, separated and treated as if they are relatively discrete. What might be thought of as sub-sets of themes, economic utility and citizenship for example, are considered separately within the context of the wider analysis. This part of the book concentrates on the socio-political environment and the policy context from which these themes emerge and shows how these contextual socio-political factors shape the text of policy and its strategic direction. In Chapter 3, the first, and perhaps the most dominant of the current global themes that shape educational policy is analysed, that of economic utility and human capital. It is argued that although the organizational principles and operational practices that are derived from human capital theory may take a variety of forms, the language of legitimation that emanates from human capital theory shapes educational policy in many different countries. In Chapter 4 the theme of citizenship and social justice is explored. It is argued that in many countries education is perceived by policy makers to be a major factor in determining and sustaining national identity. Education is used to foster desired images of both the nation state and the nation's citizens. The sub-theme that emerges here is that of social justice and cultural diversity. How can the fostering of a national identity be reconciled with cultural diversity in a socially just manner? In Chapter 5 the themes of accountability, autonomy and choice, often interconnected in many policy arenas, are drawn together. Accountability is linked both to pupil performance and economic utility. The devolution of choice to parents and autonomy to educational institutions is used by many policy makers to hold to account those responsible for the work of those institutions.

3 Educational policy and human capital

Introduction

In Chapter 1 it has been argued that with globalization and the associated breakdown of the ability of nation states to sustain economic nationalism, it has become widely recognized that the future prosperity of nations will depend on their ability to be internationally competitive (Brown *et al*. 1997). Bottery (2004a) has noted that globalization is not a unified and coherent movement but consists of a number of loosely interconnected global trends that appear to have a significant influence on the shaping of educational policy in many countries. The most important of these is what Bottery calls 'economic globalization' which: 'not only sets the context for other forms of globalization. Its language is also increasingly used to describe their activities – it "captures their discourses" ' (Bottery 2004b: 7).

Economic globalization has a profound effect on many countries, in part, because no other global system appears to exist which allows alternative forms of activity and organization. It also leads to an increasing emphasis on economic growth by both multi-national companies and nation states. Consequently, on the part of both private and public sector organizations, there is an increasing concern with economic efficiency and effectiveness coupled with an emphasis on the individual as consumer. This contrasts with traditional public sector values of care, trust and equity. As Bottery argues, this has a direct impact on much social policy:

> Ultimately, the dominance of this agenda leads to an emphasis on economic functionality rather than to the pursuit of things in their own right, and in so doing, undermines the intrinsic value of other pursuits.
>
> (Bottery 2004b: 7)

In this context, a set of implicit, explicit and systematic courses of action are established based on a human capital approach to education. The growing impact of globalization has forced nation states to enhance the skill levels of their labour force. In turn, this has produced comprehensive reviews of their education systems (Mok 2003). This form of globalization has important effects on education for a number of reasons:

> First, the economic imperative dominates much thinking and ... becomes a form of 'discourse capture' where radically different conceptual agendas such as those of education are reinterpreted through its language and values. Second, it affects the financial probity of nation-states and their ability to maintain adequate provision of welfare services, including that of education.
>
> (Bottery, 2000: 8)

This 'discourse capture' legitimates the social and economic values to which Kogan (1975) refers and from which the educational and institutional values and concomitant actions are derived. Increasingly, these values and actions are derived more from the economic imperative than from educational principles and procedures. This, as Kogan (1975) notes, represents a significant re-ordering of the values hierarchy on which education policy is based with those values derived from human capital theory becoming first order values while educational and personal values are relegated to the level of second order.

Human capital: an overview

Capital in all its forms is generally seen by economists as the resources available through marketized networks to individuals, groups, firms and communities, within which people are believed to act rationally and function as equals (McClenaghan 2003). Thus, if physical capital is the product of making changes to raw materials then human capital is created by changing people to give them some desired skills and/or knowledge (Ream 2003). As Schultz puts it, human capital consists of: 'skill, knowledge, and similar attributes that affect particular human capabilities to do productive work' (Schultz 1997: 317).

Human capital is the sum of education and skill that can be used to produce wealth. It helps to determine the earning capacity of individuals and their contribution to the economic performance of the state in which they work. It is usually measured by examining the level of skills and knowledge of the recipients such as members of a firm or a cohort of school pupils.

> At the centre of the policy discourse that links human capital and education is the belief that there is a paradigm shift out of Fordism towards a post-Fordist, high-skill, or knowledge-driven economy whereby investment in human capital ... constitutes 'the key' to national competitiveness.
>
> (Lloyd and Payne 2003: 85)

As Schultz (1997) argues, economists have long known that people are an important part of the wealth of nations. If this is the case, it follows that for individuals seeking access to sought-after employment opportunities, their self-interest will be served by personal investment in the acquisition of qualifications and relevant experience (Rees *et al.* 1996).

At the level of the individual, therefore, an approach to education based on human capital would indicate that:

> People invest the level of time and effort ... in education and training that their individual utility functions suggest they should ... individuals can have more than their future earning potential in mind and their utility function can be made to incorporate all conceivable benefits which could possibly be derived from investment in human capital including the achievements of particular accomplishments ... or the assimilation of particular values.
>
> (Fevre *et al.* 1999: 118)

At the national level it has been argued that educational policies developed on the basis of human capital theory may produce a greater cohesion and reduce inefficiency in the use of scarce resources (Mace 1987). The human capital approach to educational policy also works on the assumption that there is a national economic benefit to be gained from education and from having an educated and skilled work force. As Leadbetter (1999) argues, the generation, application and exploitation of knowledge is driving modern economic growth so it is necessary to release potential for creativity and to spread knowledge throughout the population. In many social systems education is regarded as the main process by which such transformations might take place, although the issues surrounding which skills and knowledge are to be acquired, by whom and who makes those decisions often lack clarification. Little wonder, therefore, that:

> Economists and other social scientists have long viewed education as the solution to many social challenges including productivity [and] economic growth ... Education is viewed as an investment in human capital that has both direct payoffs to the educated individual as well as external benefits for society as a whole.
>
> (Levin and Kelly 1997: 240)

How then, does human capital theory inform educational policy?

Human capital and education: an analysis

The impact of human capital theory on educational policy can best be identified by examining the socio-political environment which provides the impetus for policy making and from which, in most instances, the legitimation for that policy stems. The languages of legitimation used to present and justify educational policy (Bell 1989), reflect the dominant discourses within the socio-political environment. Thus, in the last two decades of the twentieth century in most pluralistic societies the discourse within the socio-political environment has been dominated by the struggle between economic individualism and social collectivism as preordinate determinants of social organization. Hence, educational policy is shaped by and located within what Taylor *et al.* (1997) term the context of the outcomes of debates in the wider socio-political environment and the language in which that policy is expressed is derived directly from its dominant discourse. Within this context, a range of social and political influences have combined to establish economic functionality as the dominant discourse. This policy text is supported by reference to individualistic languages of legitimation that underpin a belief in the efficacy of market forces as a mechanism for social organization and in the capacity of education to supply appropriately skilled labour for employment. The outcome of this, as far as education is concerned, is exemplified by the use of principles derived from economics generally and from human capital theory in particular, to legitimize educational policy and, in many countries, to underpin the use of elements of the market place to structure decision making and resource allocation.

The text of such education policy, its overall content and the strategic direction that defines the shape of policy is also derived from that wider environment. This provides part of the text for this particular education policy. It is widely recognized, for example, that in most countries where education is subject in any way to market forces, those forces do not constitute a 'free market' in the sense that total de-regulation applies. Rather, the education market is an internal or quasi-market-one in which: 'The market functions within an overall system in which the State or government retains an important role' (Tooley 1994: 156).

Where the operation of the education market is informed by human capital theory this role is to determine the nature and mix of skills and knowledge that the system is required to produce while still retaining elements of market forces such as a mechanism for resource allocation, competition between institutions and the ability of parents to exercise choice. Reliance is placed largely, although not exclusively, on the language of economics to formulate success criteria. Reference is frequently made to efficiency, effectiveness, quality, value for money, choice and economic development. Human capital theory produces, in particular, an emphasis on the inter-relationship between individual choices, the demands of the labour market for specific skills and economic growth.

Organizational principles define, for example, the limits of autonomy, the patterns of accountability and the procedures for assessment and quality control. Educational institutions must respond to the specific demands from the centre to produce particular forms of outputs in terms of students with predetermined skills and knowledge that will sustain and enhance economic development in their particular country. In order to achieve this, some form of central control over educational provision will operate. This might be based on tightly defined and rigidly assessed curriculum content and/or pedagogy, through an extensive inspection process, or a combination of both. They must also be able to demonstrate that this is what they are doing and that they are implementing national policy in such a way as to contribute to the human capital outcomes required from the education system. Here the text and the consequences of the policy overlap because pedagogy, curriculum content and forms of assessment must be appropriate for the production of these outcomes.

Linked to these organizational principles which are largely centrally determined, are the operational practices. These are the activities which contribute to the formulation of internal policies that will enable the institution to deliver an appropriately skilled and trained set of students, the day-to-day organization of schools, the specifics of decision making and the nature and extent of delegation of responsibilities. Thus, within schools the key factors in determining the nature of the operational practices and the structuring of responsibilities are the principal/teacher relationships and the arrangements for decision making in the school. Once these are established, the nature of the curriculum and its content, pedagogy and assessment, the roles of individual teachers, the mechanisms for reporting to and involving parents, the internal management of the school and mechanisms for establishing relationships with the external environment can be established (see Figure 3.1).

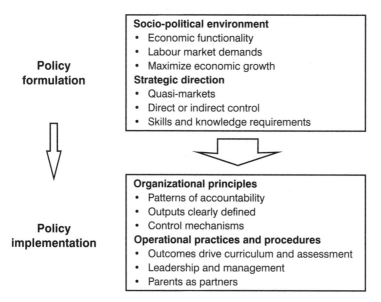

Figure 3.1   Policy into practice: human capital

There are institutional consequences of all this. As McNamara *et al.* (2000) note:

> The ideological move to construct education as a market place holds ... economic implications for schools, teachers, children and parents ... Significant ... is the necessity for schools to promote a positive image ... in terms of performance indicators of product/output.
>
> (MacNamara *et al.* 2000: 475)

The implication of this is that both students and parents are partners in the educational enterprise. As a result, parents who were once regarded, at best, as passive supporters have changed into active participants. They have now been further re-positioned as informed consumers in the educational market place. Education has become a commodity with both the individual and the state as consumer, the individual seeking to maximize personal benefit and the state seeking to maximize economic growth and development. Agbo (2004) argues that the implication of this is that the most effective route to economic well-being for any society is through the development of the skills of its population, its human capital. Consequently, education is to be regarded as a productive investment rather than merely a form of consumption or something intrinsically valuable in its own right.

The application of human capital theory to educational policy

The application of human capital theory to educational policy must be seen in the context of economic globalization to which reference was made earlier in this chapter. One consequence of the impact of economic globalization is that many nation states attempt to maximize the economic benefits that can be accrued from a system of educational provision planned to meet specific economic and business needs. Although it may not be universal in shaping the context of education policy, human capital theory is certainly extremely common as a socio-political rationale. This does not necessarily mean that it is the most appropriate such rationale or that the text of policy produced will achieve its stated outcomes. Nevertheless, the impact of human capital theory can be identified in many countries. In the USA, for example, Elmore (1988), in calling for major reforms to the American schooling system, argues that in order to sustain the present standard of living and regain its competitive position in the world economy the USA will need a better educated workforce. This implies significant changes in the relationship between schools and their wider environments, in the management and organization of schooling and the nature of teaching and learning (Murphy 1991). It also makes explicit the relationship between education and economic performance.

Similarly, in Australia, it has been seen that students' mathematical capability must be improved to enable the economy to grow and be competitive (Kemp 2000). Here it has been argued that economic rationalism based on human capital has infiltrated educational policy making to such an extent that the models and formulae of economics have replaced the values of a just, creative and humane society (Ogilvie and Crowther 1992). In New Zealand, a similar situation pertains. Here the tertiary and higher education systems must contribute to economic development by providing more graduates for science-based occupations (Gould 2001), based on a very explicit link between education and the market place: 'The value of educational qualifications does, at least in part, lie in their scarcity. Hence, education shares the main characteristics of other commodities traded in the market place' (New Zealand Treasury 1987: 33).

Here, the language of economics is used as a rationale for educational provision. In Greece the introduction of a range of new scientific programmes and new technologies is intended to contribute to the economic development of the country (Saiti 2003). Much of this provision is located within the new public technical and vocational lyceums that are intended to facilitate:

> The development of the necessary skills and abilities in order for the graduates of such institutions, through their own initiative, to properly identify and exploit the available opportunities among the technical professions in the labour market.
>
> (Saiti 2003: 35)

The basis of this approach to education is that technical education can increase the flow of skills and, by assisting people to acquire new technologies, it can enable them to adapt to new working environments. Furthermore, investment in technical education is seen to increase a recipient's contribution to the workforce and, in so doing, expand productive capacity and improve economic performance (Saitis, 1999). This is an interestingly explicit formulation of the human capital approach to Greek educational provision that finds resonance in Germany, where the emphasis is also on the link between the knowledge-based economy and the central importance of education to economic development (Bulmahn 2000).

In England, it was argued by the then Conservative Government that: 'Our future prosperity as a nation depends on how well our schools, in partnership with parents, prepare young people for work' (Department for Education 1994: 25). Here is seen a perspective based on the minimizing of state intervention which is derived from the ideas of Friedman and Friedman (1980), albeit within a very tight accountability framework. This view developed into a more explicit and increasingly interventionist articulation of the link between human capital and education after New Labour came to power in 1997. The incoming Secretary of State for Education argued that investment in human capital is essential for success in the economic future of the country and that learning throughout life will build human capital by encouraging the acquisition of knowledge and skills and emphasizing creativity and imagination (DfEE 1998a). In particular he stated that:

> Learning is the key to prosperity – for each of us as individuals as well as for the nation as a whole. Investment in human capital will be the foundation of success in the knowledge-based global economy. We need a well-educated, well-equipped labour force. Learning enables people to play a full part in their community. It strengthens the family, the neighbourhood and consequently the nation.
>
> (DfEE 1998a: 7)

At the level of higher education, a similar trend can also be found in the *Dearing Report* (1997) that argued for:

> Higher education driven by (cost-bearing) student and employer demand ... better adapted to the needs of industry, and hence the labour market. In essence this means that market mechanisms should ... tend to ensure that individuals will follow the *kinds* of courses that raise productivity
>
> (Killeen *et al.* 1999: 100)

It is in Asia, however, where the links with human capital has shaped educational policy most obviously and where the relationship between economic performance and education often find its clearest articulation. In many parts of the Asia-Pacific region the close connection between education and economic development is widely recognized and a significant number of improvement initiatives have been introduced to strengthen the contribution of education to economic growth. In Singapore, for example, the government deliberately adopted a policy of avoiding the low labour cost economy common to several of its neighbours. Instead, a policy of developing a highly educated work force was pursued:

> The political leaders saw it as their task to ensure that, as industry developed, the human capital was in place to make effective use of the physical capital. The result was a very close relationship between educative and productive systems.
>
> (Ashton and Sung 1997: 209)

Significant education reforms were introduced after the *Report of the Economic Planning Committee* (Ministry of Trade and Industry 1991) demanded substantial educational expansion and improvement to meet the needs of Singapore's economic development in a very competitive environment (Ashton and Sung 1997). These focused on three areas: the identification of those basic skills necessary for people effectively to contribute to an advanced industrial society; the development of intermediate-level technological skills; and the expansion of higher education. The specific aim of these reforms was to use a more educated labour force to establish Singapore as an economically developed nation. This policy produced a very close relationship between the government and the education system in which considerable control was exercised from the Ministry of Education over what was taught, how it was taught and how it was assessed in schools. Singapore Teachers' Union (STU) put it thus: 'The main focus of our education system was on meeting manpower needs. At the same time we had to teach values of good citizenship' (Singapore Teachers' Union 2000: 1). Here the labour market needs are presented as being value neutral, although this is far from the case, while citizenship is seen to be rooted in explicit and shared values.

In South Korea, the Presidential Council planned to introduce educational reforms with the specific intention of addressing the new challenges by manpower planning (Cheng 1999). In Malaysia the entire education system is being reviewed in order to meet the manpower requirements of the knowledge-based economy and a system of lifelong learning is being promoted to ensure that workers can continuously upgrade their skills and knowledge (*Third Outline Perspective* 2001). A similar restructuring has taken place in Israel while in the Special Administrative Region (SAR) of Hong Kong the Education Department issued a booklet in 1997 entitled, *Medium of Instruction: Guidance for Secondary Schools*. This stipulated that most schools must, from the following September, adopt Chinese as the medium of instruction while giving significant exemption for a minority of schools (Education Department 1997). Although the rationale for this appeared to be that Chinese, the mother tongue, was most appropriate for educational instruction, in fact this policy had its origins in a strongly human capital approach to education based on a utilitarian discourse about the centrality of the English language for the economic survival of Hong Kong (Choi 2003). These recent educational developments in Asia provide strong evidence to support the assertion that educational reform is the most important means of supporting the economic development of many societies (Cheng 1999). Indeed, effective schooling is often defined as that which facilitates the maximum contribution to the economy (Bell 1999a). In order to achieve this, however, educational provision, particularly in relation to curriculum content and its assessment, has to be tightly controlled and carefully planned if human capital outcomes are to be achieved.

In some countries, England for example, this control largely takes place at school level within a framework of national policy. The national curriculum, its assessment and patterns of accountability based on national inspection, publication of examinations results and the management of teacher performance provide a tight national framework within which school-level decisions are taken. In others, such as Greece, planning takes place largely at the national level but is relatively loosely controlled at the local level while in Singapore, for example, control is facilitated by the relative smallness of the country and the extent to which the government has retained central control. It tends, for schools, to take place at the level of the Ministry of Education.

It is here that the curriculum, testing, admissions and staff appointments are controlled and guidance provided on pedagogy and curriculum materials, although there have been recent attempts to devolve autonomy to schools or clusters of schools. In these examples, the strategic direction is determined nationally based on a broad policy thrust in order to maximize the benefits that may accrue from such a strategically planned system. Thus, planning based on sets of policies derived from human capital theory is used in order to try to produce relatively high-level economic benefits from the educative process, benefits that may accrue at a regional and national as well as at school and individual level.

The limitations of human capital theory

As Bowles and Gintis (1976) have argued, education policy based on human capital closely reflects the needs of industrial society for workers with particular skills and, at the same time, illustrates the role of the state in ensuring that such a workforce is available. This interconnection between human capital and educational policy, however, has its limitations. These can be found at each of the four levels of the analytical model and are sufficient to cast doubt on the efficacy of the human capital approach to education as a sufficient legitimization for the structuring of the educative process in most societies. At the level of the socio-political environment the extent to which the fundamental tenets of human capital theory pertain to the educative process is open to question. It is far from certain that there is an economic benefit to be gained from additional or specific forms of educational investment or that education does make a significant contribution to economic growth and development. As Killeen *et al.* (1999) argue, the relationship between expenditure on education and the economic performance of any particular country is largely one of correlation rather than one of cause and effect: there may well be intervening variables at work here such as investment in infrastructure or in research and development. The connections between schooling, training and economic performance are complex and by no means clear:

> Relatively successful economies may make greater investment in the education of their populations, measured by the duration and level of schooling and training but ... [this] may, at least in part, be a result, rather than a cause, of economic success.
>
> (Killeen *et al.* 1999: 99)

It is particularly difficult to establish the precise nature and value of such investments in human capital (OECD 1996a). The OECD Report argues that while educational investment does constitute the formation of capital, its value is hard to establish:

Though its [human capital] value may be appraised by the individual in whom it is embodied, it may also be appraised by others, including the policy-making members of society. Such appraisal, however, is arbitrary and subjective.

(Machlup 1984: 424 quoted in OECD 1996a: 46)

Monteils (2004) goes even further. Using data from a survey of 10 countries over a two-year period, she failed to find any positive correlations between investment in education and economic growth. Thus, at the societal level questions can be raised about the context from which such policies emerge and the extent to which education grounded in human capital theory can achieve its stated outcomes.

Similar questions can also be raised about the impact of the text of these policies. How far, for example, can and does education increase the productive capacity of individuals? Rather than generate such an increase, education may merely act as a selection device that enables employers to identify those potential workers with particular abilities or personal characteristics that make them more productive (Woodhall 1997). Even if this is not the case, education systems may not successfully produce the skilled labour force required by employers. Choice mechanisms militate against this to the extent that individual choice may be constrained by limited knowledge and resources, or available options restricted by an imperfect understanding of future skill requirements. The structuring of choice and opportunity within any society is such that a large number of factors will influence the extent to which such personal investments might take place. Individuals may choose to undertake education and training only to the degree that they are aware of both educational and employment opportunities available to them and can establish what are the required types and levels of knowledge and skills. At the same time, family support and pressure, financial resources and the limitations of realistic aspirations all operate to limit the extent to which free choice can be used by any individual to gain the maximum benefit from education (Hodkinson *et al.* 1996). However, it is not only the specific and finite access to resources that are important. The relative levels of inequality will impact on family well-being and influence the choices that are made (Wilkinson 1996). As Psacharopoulos (1986) notes, such limitations on choice mechanisms may produce results contrary to those expected by policy makers – more social science students rather than more engineers. The capacity of any society to match the human resources produced by its education systems to the demands of the labour market is, at best, imperfect and, at worst, potentially damaging to the very economies that should be sustained. As Bulmahn (2000) argues, those who deploy human capital theory as the sole or predominant legitimation for educational provision at the socio-political level, and who thus consider education from the perspective purely of national economic self-interest, will be unable to develop long-term policies for the future. In short, the free market itself may not meet the human capital needs of either developed or developing nations.

At the strategic level, economic utilitarianism based on human capital theory may not only be short-sighted, it may prove entirely counter-productive since, as Agbo (2004) argues in the case of some African countries, it can facilitate the establishment of an educated elite who are socially mobile to the disadvantage of the society as a whole or cause a society to lose touch with its cultural roots in response to a search for technology which is globally accepted. It can also have an adverse effect on the ability of nation states to compete in the global economy because it may leave a large majority of the future working population without the human resources to flourish in a global economy.

> The risks are twofold: firstly, given the time lag between entering a training programme and completing it, market demand for a particular type of training may have changed with a resulting lack of jobs. In the competitive global market, such an outcome is all too likely. Secondly, industries of today are likely to be tomorrow's dinosaurs. As a result, employer-led training schemes may not contain the vision ... required in order to maintain the high skill base necessary.
>
> (Brown and Lauder 1997: 178)

Thus, the consequences of such policies may be counter-productive. Attempts to establish too tight a focus for education or to exercise too much control of curriculum, content and pedagogy will lead to a trained incapacity to think openly and critically about problems that will confront us in 10 or 20 years time (Lauder *et al.* 1998). At the strategic level, therefore, it is doubtful if an educative process legitimated purely on the basis of human capital theory will have the capacity to produce an appropriately skilled labour force.

The organizational principles on which the relationship between human capital and education rest tend to be based on a technical-rationalist approach to education generally and to the organization of schools as institutions in particular. This gives little consideration to the benefits of education other than economic utility. As Marginson (1993) has maintained, this emphasis on economic rationalism has meant that education values have become marginalized, thus distancing education from both the social and the cultural. The application of human capital limits, therefore, the wider benefits that may be gained from a more liberally based education and marginalizes the ethical dimensions of education that might shape both the nature of educational institutions and the totality of the educational enterprise. In fact, matters related to schools as social and moral organizations, living with others in a diverse community and wider issues of social justice may be ignored in the quest for a narrowly defined form of academic attainment. Thus the social and the moral are subordinate to the economic and the utilitarian.

This failure fully to consider the wider purposes and benefits of education has allowed researchers, and more especially, politicians to deduce simplistic solutions to complex problems and to develop approaches that serve very narrow purposes based on limited and restrictive policy objectives linked specifically to economic utilitarianism and human capital outcomes.

Furthermore, the organizational principles on which the relationship between human capital theory and educational institutions is predicated – that the skills and knowledge that are required to initiate and sustain economic development are identifiable either by governments or employers, and will be delivered by educational institutions – can be challenged. It is assumed that teachers will respond to the rewards and sanctions within the organization to ensure that an appropriate curriculum is delivered and that children are either sufficiently malleable to respond to a school's organizational structure and processes, or that they understand their own self-interests sufficiently to follow the incentives created by the school (Lauder *et al.* 1998). This ignores the very tension that is at the centre of this type of education policy, between what the state might regard as economically desirable and what the individual might regard as appropriate personal development. As Entwistle (1977) has pointed out, it is doubtful if people are equipped to grapple with life's changing challenges by focusing entirely on meeting the immediate instrumental needs of the state.

A similar tension exists within many educational institutions that derive their operational procedures from organizational principles that emanate from human capital theory. These operational practices tend to be based on certainty, predictability and the operation of rules. They are often inflexible, impersonal, heavily bureaucratic, rule-bound and based on a rigid separation of responsibilities within the organization, an hierarchical arrangement of those responsibilities, and on exclusivity rather than inclusivity:

> The assumption ... is that the organization consists of separate parts bound together insofar as is necessary ... through universal rules or centralised control. Information flow and learning ... is mediated through the negotiated, rule-bound structures that make up the organisation's internal and external contracts.
>
> (Zohar 1997: 105)

Such organizations are efficient and reliable. They are ideal for a relatively stable, predictable, if competitive, environment. As long as rules and procedures are followed they operate with apparent smoothness and can give the impression of orderliness and of having an impressive ability to plan for and cope with the future. Many important processes, however, are marginalized in organizational forms based on order, simplicity and conformity where everything operates according to specific, knowable and predetermined rules and where actions are supposed to be rational, predictable and controllable (Chong and Boon 1997). Learning, therefore, is rooted in the Newtonian scientific paradigm of analysis through dissection, so that the parts can be isolated and understood. That which should be learned becomes the same as that which is instrumental. It is an individualistic process that proceeds in a linear way through analysis and the construction of generalizations based on empirical evidence. It inhibits the development of the very creativity, imaginative thinking and entrepreneurship that is often required to sustain economic development.

Where there is a high degree of standardization and inflexibility in educational systems or the institutions within them, these very systems and institutions become singularly less well equipped to prepare their students to face demands for greater flexibility and creativity (Bottery 2004b). Thus, schools cannot readily take account of forces emanating from the external environment in a period of rapid and extensive change and cannot generate the creativity and flexibility necessary to cope with such forces. Yet, it is widely acknowledged that the knowledge and skills that schools must seek to develop have to be based on creativity and innovation. Already there is a major concern in Pacific Rim countries about the lack of critical thinking, creativity and innovative skills amongst students. The lack of such skills is widely regarded as one of the contributing factors to the recent decline in the Tiger Economies (OECD 1996a). As both Ball (1999) and Bassey (2001) recognize, the over-riding emphasis of human capital theory on the role of education in contributing to economic competitiveness results in a set of pedagogical strategies linked to a narrow conceptualization of school improvement and effectiveness that ultimately are antithetical to the demands of a high skills economy. In other words, human capital, when applied to education, contains the seeds of its own failure. Thus, from a human capital perspective the management of learning becomes problematic in itself since effective learning in any school is the product of many factors. Thus Beare can argue that:

> Reductionism ... is why the curriculum is structured in the way that it is, cut up into key learning areas ... Positivism ... is why science and maths are pre-eminent in the curriculum ... Rationality ... [is why] values formation has always been an incidental rather than a central part of the curriculum ... Quantitative analysis ... The measurable is safer to handle than the intangible ... As a result the intuitive, the expressive, the unmeasurable, the subjective and the intensely personal have never found a satisfactory place in the curriculum.
>
> (Beare 2001: 39–40)

Learning is, therefore, based on reductionism, positivism, rationality and quantitative analysis. Thus, the processes of managing teaching and learning created by an emphasis on the human capital approach to education fail to acknowledge the complexity of school organization and the development of effective teaching and learning. This reductionist view of education is rooted in human capital justifications for the entire educational enterprise and focuses on improving national economies by tightening the connection between schooling, employment and productivity and by enhancing student outcomes, employment-related skills and competencies (Carter and O'Neil 1995).

The links made between educational, human and economic development, therefore, produce an excessively utilitarian approach to schooling that can lead to an inappropriate narrowing of educational objectives and processes because of the emphasis on national economic competitiveness (Kam and Gopinathan 1999). The human capital justification for the structuring of educational provision has produced an excessive instrumentalism in the curriculum:

> Instrumentalism has produced the competencies movement; it has affected the curriculum, producing concepts like 'key learning areas', as though learning is not legitimate unless it is information-driven and packaged into traditional subjects … It has driven the outcomes approach to schooling, a concentration on tests, the publication of school-by-school results and 'league tables'.
>
> (Beare 2001: 18)

These operational practices are all control devices to compel schools and colleges to concentrate on utilitarian outcomes linked to economic productivity and the demands of the labour market. Consequently younger children must become proficient in the basic skills of literacy and numeracy while their older siblings need to enhance their skills through an emphasis on information technology, science and mathematics. In tertiary colleges and universities the focus shifts to that of the knowledge-based economy and lifelong learning to respond to the changing demands of the work place (Bassey 2001). It is evident that the narrowing of the focus of education in Singapore, for example, has helped to create an education system that produces students who are excellent at passing examinations but very limited when it comes to creative thinking and the development of enterprise (Ng 1999). The STU noted that, in Singaporean education: 'The emphasis was on results. We bred a generation of Singaporeans who were examination smart … but we killed the joy of learning' (Singapore Teachers' Union 2000: 1).

Not only has the joy of learning been destroyed but here, as in other places, the sole emphasis on producing a workforce to sustain economic development is likely to lead to a trained incapacity to think differently (Lauder *et al.* 1998). The present global emphasis on developing human capital within a market or economic development paradigm, therefore, is based on a model of education policy that is deeply flawed in a number of ways. At the socio-political level, the human capital discourse of legitimation is both confused about the extent to which individuals can and do make educational choices based on human capital criteria and unconvincing about the degree to which investment in human capital does contribute to economic development. At the strategic level, the concentration on economic utility of education at the expense of its many other contributions may have adverse consequences for both society and the individuals within it. The organizational principles that shape the relationship between human capital and education produce organizational structures that mitigate against the development of the very skills that may be required to meet future economic challenges while the related operational practices lead to inappropriate forms of leadership and a reductionist approach to teaching and learning, to the ethical dimensions of leadership and the wider issues of morality and social justice at a school level. Thus, human capital as the sole legitimation for the educative process in any society has severe limitations and may be counter-productive.

Conclusion

It can be seen, therefore, that human capital theory when applied to the educative process leads to education being treated as a private consumable, a commodity or a positional good in the market place at both individual and state level (Bottery 2004b). The rationale for change and re-structuring in education is largely cast in economic terms, especially in relation to the preparation of the workforce and repositioning national economies to face international competition (Levin 2003). The impact has been significant: 'leading to changes in management processes and organization, institutional cultures (at all levels) and in perspectives on a wide range of dimensions of education from teaching and learning, to resource management and external relations' (Foskett 2003: 180).

Nevertheless, as has been argued above, human capital theory as the sole legitimation for educational policy has severe limitations such that its outcomes may be counter-productive. It has produced a situation in which education has become merely a way of increasing the value of human labour. This fails to recognize that both education and labour are more than commodities, they are value-driven social processes. The human capital discourse, therefore, requires either to be replaced by an alternative form of legitimation or a significant leavening by incorporating the key aspects of an alternative legitimation.

Education is more than the production of human capital. It is about values and beliefs, ethics, social justice and the very nature of society both now and in the future. As Hills has argued, the basis for education in the future is not:

> Facts and figures ... the explicit knowledge of the internet, the textbook or the lecture theatre because much of this is quickly obsolete and is often an obstacle to new ideas. It is the implicit knowledge gained from experience, or ... case studies, because ... these are the bases of values,

morals and character. They prepare a person for the unexpected and the difficult decision.

(Hills 2004: 27)

As will be seen in the next chapter, education also has an important part to play in developing concepts of citizenship and social justice, especially in culturally diverse societies. The relationship between economically driven educational provision, the meaning of social justice and the nature of citizenship is both complex and contested.

4 Education policy, citizenship and social justice

Introduction

In Chapter 3 the links between education policy and the economy were analysed. The global shift to supply-side economics has effectively elevated education policy to a pivotal element of economic policy with the development of human capital being perceived as central to the creation of economic growth. However, education policy has always been about much more than economic policy, it is social policy too. Perhaps, more precisely, it can be argued to have a social function – concerned not solely with matters of welfare, but with matters of ideology too. This delineation between the provision of education as economic and as social policy is not neat and tidy – the relationship is often one of interdependence. However, the focus in this chapter is on the extent to which a wide range of social values shape education policy and how education policy reflects the diverse, and sometimes contradictory, social functions associated with it.

The social functions of education policy reflect tensions and contradictions in the wider role of the state and state policy. Education has a crucial role in promoting a sense of individual and collective welfare and through this a sense of social cohesion. It also has a similarly ideological role in developing what are considered to be appropriate values in society and in establishing a sense of national identity. In short, education plays a pivotal role in developing a sense of citizenship whereby individuals take their place in their communities be that at a local, national or even a global level. However, notions of citizenship are both complex and contested (Plant 1991). They change over time and vary between cultures (Jenson and Phillips 2001). Education for citizenship focuses attention on central questions that are a recurring theme in this volume – what is education for? Who receives what, and who decides? Similarly, what does it mean to be a 'citizen', and who decides? Such questions are inextricably linked to notions of 'fairness' and, therefore, to concepts of social justice. But what is social justice, and how can the development of education policy contribute to the pursuit of a social justice agenda? This chapter explores ways in which education policy is shaped by the related notions of citizenship and social justice and the role that education plays in legitimating notions of a 'fair' society. It highlights the need to explore the philosophical and ideological arguments within the socio-political environment that influence education policy, give it a strategic direction and produce its organizational principles and operational practices and procedures. It makes the case for a multi-dimensional approach to citizenship capable of reflecting contemporary conditions and differing cultural contexts. The manner in which these shifting notions of citizenship can then shape educational policy is briefly illustrated by three international examples of policy development: the implementation of education reform in Rwanda following the 1994 genocide, the introduction of citizenship education into UK schools following publication of the Crick Report (QCA 1998), and the emergence of system-wide restructuring in Israel following publication of the Report of the National Task Force for the Advancement of Education (NTFAE 2004).

State Policy and Citizenship – shaping the discourse

In straightforward legalistic terms, being a citizen implies being a native or naturalized member of a nation state. However, the concept is broader than this and may be considered both philosophically (linked to concepts of justice) and socio-politically (Faulks 1998). These broader definitions of citizenship offer the following conceptualization:

> Citizenship is a status that mediates the relationship between an individual and a political community. It is characterised by a set of reciprocal rights, the extent and nature of which are defined through a complex set of social and political processes including: the struggle between opposing social forces, political compromise, and historical and economic circumstance.
>
> (Faulks 1998: 4)

The notion of citizenship as a series of reciprocal rights and responsibilities was central to the concept of citizenship developed by T.H. Marshall in his highly influential volume, *Citizenship and Social Class* (1950). Marshall's study of citizenship in post-war Britain argued that citizenship rights had developed in three distinct phases. First was the development of civil citizenship by which individual freedoms emerged such as freedom of speech and the right to own property. Second was the development of political rights whereby the right to stand and vote in elections provided 'the right to participate in the exercise of political power' (Marshall 1950: 11). Finally, a third element of citizenship developed, based on the belief that citizens had an entitlement to an element of social security, broadly defined – referred to by Marshall as social citizenship. Social security, in its broadest sense, was fundamental to citizenship and therefore required the provision of a range of basic social services (education and health for example). Furthermore,

Marshall argued that just as political entitlements were not the product of market exchanges (the right to vote is not purchased in a market transaction) so too should citizenship services be removed from the process of commodity exchange. Basic rights to social security should be no more dependent on wealth and individuals' market values than equivalent rights to political freedom:

> In contrast to the economic process, it is a fundamental principle of the welfare state that the market value of an individual cannot be the measure of his right to welfare. The central function of welfare, in fact, is to supersede the market by taking goods and services out of it, or in some way to control and modify its operations so as to produce a result that it could not have produced itself.
>
> (Marshall 1981: 107)

Marshall's case for social citizenship was not explicitly a theory of social justice. However, it was strongly rooted in notions of 'fairness'. Marshall was certainly not an opponent of capitalism, or the market. He accepted that the market had an important role to play in providing incentives and allocating resources, but his concern was that its inevitable inequalities might be 'excessive'. The state therefore had a legitimate role to play in both tackling unacceptable inequalities and putting in place a floor of basic rights to welfare provision that existed regardless of personal wealth. Subsequent attempts to theorize what this 'fairness' might look like are most commonly associated with the work of Rawls, whose *A Theory of Justice* (1972) sought to make an intellectual case for the 'fair' distribution of resources. Rawls' theory of social justice has had considerable influence on the development of welfare policy, particularly in the West (Angelo Corlett 1991) and may be seen as an attempt to reconcile a liberal commitment to the freedom of individuals with egalitarian commitments to a more equal distribution of resources. His approach rests on two principles; first, that each individual should have access to the most extensive range of basic liberties compatible with similar liberties for all. Secondly, that social and economic inequalities can be justified only in so far as they provide the greatest benefit to the least advantaged, and that they are linked to offices and positions that are open to all on the basis of open and fair competition. It is the first element of the second principle that is reflected in social welfare programmes that involve the state undertaking a significant element of redistribution and which has significant implications for the role of education policy in creating a more egalitarian society.

Marshallian concepts of citizenship, informed by Rawlsian principles of social justice, certainly achieved a significant degree of influence over the development of welfare policy in many western countries in the third quarter of the twentieth century. However, the extent of this influence was never global and nor has it been enduring. For example, critics from the right have challenged the notion of social citizenship and its concomitant link to policies of collective provision and universalism. More fundamentally, it has been argued that the concept of universal provision is not compatible with notions of fairness and social justice, a view most clearly expressed by Nozick's (1974) assertion that market solutions represent the only just way to allocate resources. For Nozick (1974) the market offered an objective valuation of resources and an individual's capacity to acquire resources should be linked to effort. The market was the means by which the value of resources and the value of effort were brought into alignment. Hence the state had no role to play in securing social justice, indeed attempts to do so constituted an injustice as individual liberty was violated by use of the state's coercive apparatus (Nozick 1974). From the left critics have focused on the failure of universal provision to fundamentally challenge social inequalities (Halsey *et al.* 1980; LeGrand 1982), and also the tendency to treat the notion of access to universal services unproblematically – failing to take sufficient account of how service users may be marginalized from active participation in service delivery (Coote 2000). Moreover, in recent years there has been a growing recognition of the inadequacy of Marshall's claim that the expansion of universal services would lead inevitably to a 'great extension of common culture and common experiences' (Marshall 1950: 75). Whilst this argument is understandable within the historical and geographical context within which he was writing, it has limited application for societies that were, or have become, culturally diverse and multi-ethnic (Giroux 1992). In these contexts, Marshall's dominant view of citizenship has proven inadequate in a contemporary context, unable to reconcile notions of equality and universalism with difference and diversity (Osler 2000; Olssen 2004). It also fails to reflect the growing importance of citizenship and national identity at a time when increasing cultural diversity and population movement raises fundamental questions about what it means to belong to a 'nation state'.

Marshall's contribution to the debate on citizenship was focused on Britain in the period following the Second World War. It is a product of its time and place. However, the challenges it poses, and the critical perspectives it has generated, continue to exert significant influence on the discourse within which policies relating to citizenship and equality agendas are framed, not just in the UK, or indeed the West, but globally. What does it mean to be a citizen in a particular country and how does education policy both shape and affirm a sense of citizenship?

Citizenship and education policy

Defining the relationship between educational policy and the wider citizenship agenda is particularly complex. Education is distinct from other forms of social provision because of the unique way in which it represents not only a key citizenship entitlement, but also has a hegemonic influence (Apple 2004) and its unique capacity to shape the discourse relating to how individuals define themselves as citizens. In short, education services are not only a material form of citizenship, but ideologically they help shape our conception of ourselves as citizens. Education policy therefore is both shaped by, and shapes, our sense of citizenship. Here there is a powerful duality of structure (Giddens 1984) in which education has an often explicit role in 'creating citizens'. During times of social turbulence or uncertainty, when accepted notions of citizenship are being more obviously challenged, this ideological function of education can become correspondingly more important, and this is illustrated by the three examples later in this chapter. In each of these cases, significant education policy developments arise from particular perceptions of a 'problem'. How problems are defined, and who has the power to define problems and present solutions (Lukes 1974) has important implications for policy development.

In Marshallian terms, education was a crucial component of social citizenship – universal provision of schooling represented an important citizenship entitlement, available apparently, if not in reality, to all citizens equally, regardless of their market power and material resources. Such a position reflects the period of welfare consensus in the post-war years, but which has progressively fragmented in the period since the mid-1970s. The emergence of a neo-liberal orthodoxy has posed fundamental questions about the purposes of education, and the form in which it is provided. These growing tensions about the nature and purpose of education policy reflect the values tensions that underpin policy and that shape the socio-political discourses within which contemporary education policy debates are fashioned and from which organizational principles and operational practices emerge. Within these discourses it is possible to identify four key themes that highlight the polarized nature of the values that underpin citizenship education policy and how conflicts over values shape the socio-political environment.

Citizenship and access – entitlement notions of citizenship place considerable emphasis on individuals' ability to access education services – but what can citizens expect as a right? What should the balance be between collective provision at no direct cost to the student and more privatized forms of provision and consumption in which the student pays? Marshall argued that basic rights should be removed from market exchange – to what extent is social citizenship compromised by an increasing emphasis on private sector provision, the use of market solutions and the tendency towards shifting costs to the student? Crouch (2001 and 2003), for example, argues that the drift towards privatized provision of education is likely to undermine the commitment to universal provision that is a feature of public service values. Citizenship entitlements are undermined as market-driven solutions reduce public education to a residual service for the poor: 'residual public services become services of poor quality, because only the poor and politically ineffective have to make use of them' (Crouch 2003: 11). Such possibilities raise fundamental issues relating to social justice – namely, who receives what and who pays?

Citizenship as participation – traditional views of citizenship have placed little emphasis on user engagement in the provision of services. How should users be engaged in determining policy within educational institutions? Are democratic schools based on collective participation a realistic possibility (Apple and Beane 1999), or is user engagement more effectively achieved by increasing consumer power through market solutions?

Citizenship development – education has a distinctive role in developing individuals for participation in all aspects of society as active citizens. In essence, this is about preparing learners with the knowledge and skills to be engaged members of their community, with the capacity to exert influence and agency. But these are not value neutral aspirations. What type of society are students being prepared for? What is their role within it? Who decides? Such questions highlight a tension between education for reproduction (Apple 2004) and a more radical conception of citizenship education that seeks to promote the knowledge and skills that enable students to understand, analyse and criticize, and if appropriate to challenge, society's underlying dynamics and values. These questions also raise fundamental issues relating to the curriculum – not only with regard to purpose and content, but involving wider questions about who has the authority to determine purpose and content.

Citizenship and social justice – citizenship concerns are inextricably linked to wider questions of social justice and specifically the *distribution* of rights and entitlements. To what extent are rights collective as well as individual? To what extent do rights embrace access to resources as much as more traditionally defined civic and legal rights? Liberal perspectives assert that it is for the market to allocate resources and that the role of education is to support the effective functioning of a free market. Critical theorists reject such market-driven approaches and assert that education has a key role in promoting an equitable society where equality is not conceived in narrow political terms, but in terms that embrace economic and cultural considerations as well as those relating to the distribution of power and political rights. Such approaches explicitly acknowledge the need to address structural inequalities, for example those based on class, gender and ethnicity. Equity concerns extend beyond issues of access and opportunity and are more concerned with equality of outcomes. In this world view, education has a direct role to play in not only reducing inequalities, but in tackling the sources of inequality. Here the illusion of education's ideological neutrality is expressly rejected and explicit values positions emerge more clearly. Within this approach to social justice, and drawing on the work of Gewirtz (1998) and Cribb and Gewirtz (2003) it is possible to distinguish between three different approaches to social justice each of which impinge on educational policy at both a state and institutional level.

- Associational justice – the extent to which individuals and groups are able to participate in policy-making processes. The social justice dimension places a particular emphasis on the involvement of social groups traditionally under-represented in decision-making structures such as the poor (Lister 2003) and those who have often been marginalized in traditional institutional hierarchies (within educational institutions students provide an obvious example here).
- Distributive justice – influenced by Rawlsian principles (Rawls 1972), distributive justice is concerned with the allocation primarily of economic resources across social groups, but this may be considered more widely to include various forms of capital (Cribb and Gewirtz 2003) – economic, social and cultural (Bourdieu 1997). This raises fundamental questions about the role of education as a redistributor of resources and the extent to which an explicit function of education policy is to challenge inequalities.
- Cultural justice – the extent to which all cultures within society are recognized and valued. This dimension of social justice addresses the issue arguably found most wanting as a result of recent developments in contemporary society. Again, the emphasis on social justice imputes a responsibility to challenge inequalities and to prevent the marginalization of minority cultural groups through policies and practices that privilege the majority and deny the minority. Cultural justice may be considered to require a specific commitment to challenge racism, and within the field of public policy, to challenge institutionalized racism. The implications of this for institutional policy development are explored in more detail in Chapter 8.

The tensions highlighted in the discussion of these four themes point to the contested nature of the citizenship concept, and this is often reflected in the experience of policy development in this field. Figure 4.1 provides an illustration of how citizenship policy as it is experienced in individual institutions flows from the dominant discourses relating to citizenship issues and what it means to be a citizen. However, this is not a neutral agenda and debates about issues as fundamental as this are inevitably the subject of dispute and struggle. There are tensions between social cohesion, social justice and a strategic direction that encompasses access and entitlement based on differentiation. Here the potential for policy refraction becomes clear as the consequences of the multiple interpretations of texts emerge in the form of increasingly diverse organizational principles and operational practices. These issues are pursued further in Chapters 7 and 8 which illustrate how those responsible for developing policy at an institutional level have sought to operationalize policies based on a commitment to promoting active citizenship and social justice. The remainder of this chapter is devoted to three policy vignettes that illustrate how in very diverse contexts issues

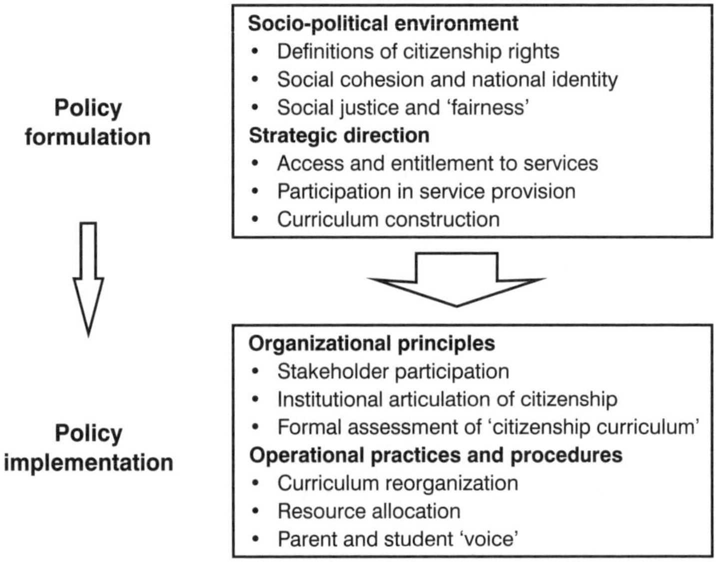

Policy
formulation

| Socio-political environment |
| • Definitions of citizenship rights |
| • Social cohesion and national identity |
| • Social justice and 'fairness' |
| **Strategic direction** |
| • Access and entitlement to services |
| • Participation in service provision |
| • Curriculum construction |

Policy
implementation

| Organizational principles |
| • Stakeholder participation |
| • Institutional articulation of citizenship |
| • Formal assessment of 'citizenship curriculum' |
| **Operational practices and procedures** |
| • Curriculum reorganization |
| • Resource allocation |
| • Parent and student 'voice' |

Figure 4.1   Policy into practice: citizenship and social justice

of citizenship and social justice have driven important educational policy initiatives.

Citizenship and social justice – policy vignettes

The following examples provide brief illustrations of how concerns with citizenship issues have exerted significant influence on shaping the formulation of education policy. There is no attempt to analyse consequences, but at this stage the aim is to provide examples drawn from diverse international contexts that highlight the links between the emergence of a socio-political discourse and the subsequent development of state policy in the form of strategic direction. In each case policy develops from the perception of a problem. In the first example, Rwanda, the nature of that problem was stark – genocide. In the example from the UK the problem is presented as one of political disengagement (QCA 1998) and is much less clear cut. In Israel there is a perception of an education service in 'crisis' (NTFAE 2004) that has failed to provide either economic success or social cohesion. Despite these different contexts, in all the cases problems emerge and are articulated by those with the power to shape and influence the socio-political discourse (Lukes 1974). Policy responses flow from the perception and presentation of these problems, and in each of these cases they draw on different dimensions of citizenship. In each of the examples the importance of cultural citizenship is apparent, and this serves to highlight the need to develop a broader model of citizenship that reflects the importance of cultural diversity. More detailed studies of policy development that explore consequences as well as context and text are presented in Part 3.

Education, citizenship and reconciliation – the case of Rwanda

Education policy in Rwanda, as in virtually all aspects of Rwandan life, is decisively shaped by events in 1994 – the year of the genocide. In 1994 nearly one million Rwandans were killed, with as many more displaced to neighbouring countries. The state-sponsored genocide by extremist Hutus was driven by ethnic divisions that were largely socially constructed during the years of Belgian colonial rule (Sibomana 1997). These divisions had been continually reinforced in the colonial and post-colonial eras by the education system, through for example the use of ethnic quotas (Shyaka 2005). In the years since 1994 the Rwandan education system has therefore confronted major challenges. The foremost priority has been to develop a sense of national unity in a country that had literally torn itself apart, and in which victims still live cheek by jowl with transgressors. Tackling such problems has taken place alongside the need to re-integrate those who had been refugees during the immediate post-colonial period, and who have been able to return to Rwanda in large numbers since 1994. All these challenges must be set within the context of Rwanda as a sub-Saharan African country beset by the chronic problems characteristic of the region – primarily dealing with the effects of poverty and the HIV/AIDS pandemic.

Given the very specific challenges facing Rwanda the priority for government has been not just to develop education to promote reconstruction, but to make education central to promoting reconciliation. Indeed, given this context, education for national unity and cohesion is seen as central to reconstruction and economic growth – 'Without reconciliation there can be no reconstruction. Reconciliation must come before anything else, because without it, nothing else is possible' (John Rutayisire, Director Rwandan National Curriculum Development Council – personal correspondence).

Since 1994, therefore, Rwanda has embarked on an ambitious programme of educational reform. This has been difficult, as any expansion in real terms has had to exceed that necessary simply to absorb refugees returning from neighbouring countries. Priority areas have been to increase participation in both primary and secondary education, with a particular focus on increasing participation rates in rural areas and amongst girls (Gahima 2005). A key challenge has been to achieve these objectives by developing teacher capacity after the genocide denied Rwanda of many of its qualified teachers. This commitment to secure expansion within the resource constraints available has been accompanied by a programme of substantial curriculum reform focused on reconciliation and driven by a number of institutions established following the genocide, notably the National Curriculum Development Centre, the National Examinations Council and the General Inspectorate (Rutayisire *et al.* 2004).

Curriculum reforms to support reconciliation in Rwanda have focused on creating and resourcing a curriculum that confronts the issues raised by the genocide and helps young people formulate for themselves their attitudes to those events and their consequences. At the centre of this initiative is the development of a new history curriculum, which it is acknowledged has a key role to play in developing a sense of national unity.

> This is important because education is seen as a major tool for transmission of values and socialization towards national identity. Education is vital in social and political reconstruction in that schools can also be arenas in which children learn to think critically about a range of view points. Primary and secondary education in war-torn countries has the potential to be an important resource, not only for economic development, but also for the pursuit of conflict resolution and social reconstruction. It is for this reason that the government of Rwanda has recognised that schools help shape the collective memory of the nation, remould social identity, and can encourage cross-ethnic affiliation.
>
> (Rutayisire 2004: 12)

A complementary approach has been to place a high priority on a skills-based programme of civic education in addition to curriculum reforms described above. It is argued that only by actively engaging citizens in political processes designed to tackle the problems confronting all Rwandans will a sense of social unity emerge. However, this must mean developing policies appropriate and relevant for Rwanda's unique context.

> Civic education strengthens democratic political culture to promote acceptance by both citizens and political elites of a shared system of democratic norms and values, and to encourage citizens to obtain knowledge about their system of government and act upon their values by participating in the political and policy process, with a potential to shape the democratic skills, values and behaviours of ordinary citizens. This leads to political tolerance or willingness to extend procedural liberties such as free speech and association to popular or disliked individuals or groups. This has long been viewed as essential for a stable and effective democratic system.
>
> (Rutayisire, 2004: 10)

This initiative seeks to develop the civic-political dimensions of citizenship through providing the skills for citizens to engage in local political processes; in so doing it draws on both associational and cultural forms of social justice.

Education for democracy – the case of the UK

What might broadly be called 'citizenship' education first formally appeared in the curriculum in England and Wales following the introduction of the 1988 Education Reform Act and the implementation of the National Curriculum. Prior to that, teachers enjoyed significant teacher autonomy in relation to curriculum matters (Lawton 1980) and although citizenship-related issues were often covered within the curriculum, this was often in diverse, and sometimes serendipitous ways (QCA 1998). The National Curriculum introduced the notion of 'cross-curricular' themes, including citizenship, health education and economic awareness, that were intended to permeate the curriculum, rather than be taught as discrete subjects. In National Curriculum terms the cross-curricular themes were non-statutory – this immediately presented problems. The 1988 Act introduced a 'league table' driven accountability model, based on pupil performance in statutory subjects. In such circumstances schools inevitably focused on what their organizational success depended on, and paid only lip-service to those aspects of the curriculum considered to be non-essential.

The commitment to citizenship education grew in large part in response to concerns that 'citizens' in general, and young people in particular, were disengaging from formal political processes and institutions (QCA 1998). Although there was evidence of interest in 'public issues', activity in these areas was often being conducted through non-mainstream channels – with young people in particular appearing to be more comfortable operating within less formal organizations in wider civil society, rather than within state institutions (Beck 1992). The concern was that a sense of alienation from formal political processes may in turn threaten more fundamental aspects of the body politic. This was the articulation of the problem that formed the backdrop to subsequent policy development, and is exemplified by the following assertion: 'We should not, must not, dare not, be complacent about the health and future of British democracy. Unless we become a nation of engaged citizens our democracy is not secure' (Lord Chancellor, quoted in QCA 1998: 7).

These concerns reflect the discourse from which the Crick Report (significantly titled *Education for Citizenship and the Teaching of Democracy in Schools*) was published in 1998. Following publication of Crick, citizenship has become more established within schools, partly because it is now a statutory order and partly because its delivery by schools has come under greater scrutiny from the Inspectorate. In the Report it is argued that education for citizenship and democracy is so central:

> that there must be a statutory requirement on schools to ensure it is part of the entitlement for all pupils. It can no longer sensibly be left as uncoordinated local initiatives that vary greatly in number, content and

method. This is an inadequate idea for animating the idea of a common citizenship with democratic values.

(QCA 1998: 7)

Reflecting the nature of some of the issues identified above, Crick argued that traditional conceptions of education in political literacy were inadequate – rather citizenship education needed to be conceived of more widely. Hence its proposals focused on promoting citizenship through encouraging social and moral responsibility, and community involvement, as well as political literacy. However, the more complex challenge for Crick was to tease out the implications of a 'common citizenship with democratic values'. Marshall's notion of citizenship was predicated on a perception of a largely homogenous nation, with social divisions dominated by class (Marshall 1950). As indicated, there was clearly an expectation that those divisions would diminish as the provision of social entitlements reduced inequalities and contributed to more common experiences and common culture. However, twenty-first century Britain looks very different to post-war twentieth-century society. Most significant is the increasing cultural and ethnic diversity, particularly in urban areas. How far is it practicable, or desirable, to define citizenship in terms of a 'common citizenship'? Crick's approach was to not only recognize the diverse nature of contemporary British society, but to seek to weld this, in part through citizenship education, into a unified national identity.

A main aim for the whole community should be to find or restore a sense of common citizenship, including a national identity that is secure enough to find a place for the plurality of nations, cultures, ethnic identities and religions long found in the United Kingdom. Citizenship education creates common ground between different ethnic and religious identities.

(QCA, 1998: 17)

The publication and subsequent implementation of the Crick Report illustrates how a discourse emerges based on a perceived need to restore, or re-articulate, a sense of Citizenship for 'New Times' (Andrews 1991). However, the difficulties inherent in this task soon become clear because these are highly contentious issues. Crick recognizes diversity, but seeks to mould this into a sense of shared national identity. Such an integrationist approach (Parekh, 1991) does not pass without challenge. One response is to reject any notion of multi-culturalism and to see education as a means of re-asserting traditional mono-cultural values (Hall 2004). An alternative approach questions the extent to which Crick's aspiration is feasible, or desirable. Olssen (2004), for example, argues that Crick pays insufficient attention to cultural diversity and in trying to forge a shared identity is largely concerned with trying to 'fit' minority communities into a common, majority culture. Such issues highlight the tension between universalist conceptions of citizenship in an age of multi-cultural diversity, tensions which are illustrated in different ways in the next example.

Developing citizenship and national identity through education – the case of Israel

Israel is a modern nation state, formed in 1948 following the re-drawing of the world map after the Second World War. It has always been a country of diverse ethnic groups, containing not just differences between Jews and Arabs for example, but significant differences *within* the Jewish population, for example between Sephardi (Eastern-origin) and Ashkenazi (Western origin) Jews, and between immigrant Jews (for example, those from the former Soviet Union) and native Israelis. These ethnic differences are overlaid by significant differences based on social class. Not only does Israel lack the historical roots of a common culture, but it continues to evolve rapidly due to continued immigration on a significant scale. Even its borders are not clearly defined with obvious conflicts over the status of the Occupied Territories. For Israel there is a perception that fundamental issues of national security depend on developing a strong sense of national identity, within the context of a highly heterogeneous society. Israel aspires to a strong unified sense of national identity in a society that is not only culturally and ethnically diverse, but in some senses deeply polarized. This desire for national unity has traditionally been reflected in a highly centralized education system that has placed a premium on seeking to transmit a powerful sense of nationhood through schooling.

However, despite the appearance of centralization, the reality has often been very different. A feature of the Israeli school system is the high number of parochial schools that receive large sums of public money, but have little accountability to the public system (Gibton 2004). There is often a failure to comply with government regulations, including those relating to the curriculum. Hence a system that presents as centralized is often in reality highly fragmented, with religious groups exploiting the fragile nature of Israeli coalition politics to assert their independence regarding school governance and accountability.

The outcome of this state of affairs has contributed to a sense of crisis in the Israeli education system with a concern that Israel's economic prosperity, its national security and its very sense of nationhood are all threatened by an education system that fails to achieve its aims. In particular, there is a concern that the prevalence of huge social inequalities, that are also reflected in the education system, threaten to undermine the prospects for promoting national identity and social cohesion. The National Task Force for the Advancement of Education (NTFAE) (commonly known as the 'Dovrat Report') refers to the 'largest socio-economic disparities in the world' (NTFAE 2004: 2) and highlights the significant differences in educational outcomes based on the ethnic and social class divisions identified above. There is a clear recognition that within the school system the Arab population suffers 'considerable discrimination' (NTFAE 2004: 4). The Report argues that the scale of the crisis requires a root and branch reform of the education system, in order to ensure that education continues to be 'the cornerstone of society, the basis for culture and national unity' (NTFAE 2004: 1). It goes on to highlight the central role of education in securing important social objectives:

> In its goals and actions, the education system should reflect the essence of the general and national culture for which the society strives, so that the citizens of the future will have both cultural depth as individuals and a shared intellectual world. Education is the basis for molding citizens and imparting humanistic and democratic values: cooperation and social solidarity, consideration for others and contribution to others, justice, and equal rights
>
> (NTFAE 2004: 1–2)

The key recommendations set out in the Report largely mimic the educational restructuring that has dominated countries like the UK for many years (Gibton 2004). It highlights the global spread of educational policies (Green 1997), but raises important questions about the efficacy of policy-importation (Dimmock 2000). The NTFAE recommendations focus on extending the very modest policies of site-based management that Israel had already established. However, these are to be developed with much more apparent autonomy for schools, coupled with increasing accountability based on bench-marked performance data. The reforms are overwhelmingly managerialist in their perspective (Gibton 2004). It is striking that a report that proposes root and branch reform of the education system in order to improve its performance and effectiveness has almost nothing to say about the curriculum and pedagogical matters, but is focused almost exclusively on accountability structures.

A key concern of the Report is to improve the performance of the education system, and it is considered that this is fundamental in order to secure economic competitiveness. The development of human capital is seen as central in an economy with limited natural resources. However, citizenship issues also drive the reforms as there is an explicit recognition of the need for the education system to:

> deepen its students Jewish identity – to consolidate the conceptual core that constitutes the underpinning of the nation's presence in its land, the national home of the Jewish people, which is a Jewish center and a focal point for identification for all of world Jewry.
>
> (NTFAE 2004: 5)

The desire to promote a sense of national identity of the type described above, whilst for example, supporting the right of the Arab population 'to preserve its cultural and social identity' (NTFAE 2004: 6) raises important questions about the Report's recommendations. For example, the drive to 'improve performance' leads the Report to recommend decentralizing structures and placing more emphasis on individual school and headteacher accountability. Yet in important respects this undermines the desire for coherence. Will a more fragmented system make it more or less difficult to promote the national values identified above? The Report refers to strengthening the role of the Core Curriculum across all schools – introducing both centralizing and decentralizing pressures. Similarly, the Report identifies inequalities in society as a major threat to national unity, and yet proposes to reduce gaps by creating a more decentralized and differentiated system. The intention is to link funding of public schools more explicitly to compliance with state regulations. However, this may simply propel some schools into the private sector, creating a more hierarchical system. Tensions therefore emerge between the desire to decentralize accountability and the need to take actions that can reduce, not increase, inequalities. As the case studies in Part 3 reveal, reconciling these tensions may not always be easy.

Conclusion

It is important to recognize that education policy has many important functions and is driven by many pressures. Globalization and the increasing demands of international competition, have emphasized the central link between educational policy and economic considerations. However, this is never a crude relationship. Not only are economic pressures complex and sometimes contradictory, but so too are the social functions of education. Education systems have never developed purely in response to the needs of capital and economic considerations, but are rather the product of struggles in which wider social forces have asserted their rights to welfare as an important citizenship entitlement (Gough 1979). Hence social pressures in education policy can be both progressive and reactionary – challenging, or reinforcing, the *status quo*.

It is important to recognize the link therefore between education policy and differing, and shifting, conceptions of citizenship. Such a connection is always likely to be complex. Sharp ideological differences relating to the nature of citizenship and the linked theme of social justice ensure that such conflict is an ever-present feature at all levels of the policy development process. Education policy on citizenship goes to the heart of core values relating to the nature and purpose of education. Contestation at the level of the socio-political environment therefore becomes mirrored at the strategic, organizational and operational levels as policy progresses from formulation to implementation. These tensions become more significant in an age of rapid technological and social change which challenge commonly held assumptions about existing notions of citizenship and what it means to be a citizen. Economic pressures have challenged the concept of universal welfare provision, and have increasingly privileged privatized solutions to welfare problems. It remains to be seen whether education provision based on principles of choice and diversity can be reconciled with the citizenship concept of entitlement and equity of access. Universalism in its wider sense has in turn been challenged by the increasingly diverse nature of contemporary societies, particularly in terms of culture and ethnicity. Within such a context, tensions emerge between universalist, and uniform, provision and the need to respond to the differing demands and aspirations of diverse communities. Questions are also posed by the changing expectations of citizens regarding issues of access, participation and accountability. Conceptions of citizenship that emphasize access to education services but fail to address issues of participation and accountability provide only a partial picture of citizenship. The citizenship agenda is about developing individuals as active agents of change, not simply assuming that the users of services are passive recipients of producer determined product. Accountability issues are therefore inextricably linked to issues of citizenship and these themes are discussed in more detail in the next chapter.

5 Accountability, autonomy and choice

Introduction

In previous chapters the inter-related themes of economic utility, citizenship and social justice have been examined. Educational institutions are now, more than ever before, required to produce students with the appropriate skills and capabilities to match national priorities. Education also is now seen to be important in developing national identity, citizenship, social cohesion and social justice. As Scott (1989) has pointed out, many governments across the world are now increasingly exercising their right to determine the broad character of the schools and colleges that they support so that they contribute fully to the goals that have been established. This right on the part of government derives from the extent to which, because the system is largely supported by public funds, education must be accountable for the use of those funds.

In order to meet these demanding challenges, significant changes have taken place in many education systems, not the least of which has been the introduction of a range of measures designed to hold those institutions to account for the contribution that they make towards meeting national priorities through the performance of their students. The third main theme that will be considered, therefore, is that of accountability.

In this chapter an examination of market accountability will be followed by a consideration of choice as a mechanism for holding schools to account. The involvement of parents and other stakeholders is such a significant element in establishing accountability within the sphere of educational policy that it is treated separately from the more general discussion of choice and market accountability. This is followed by an analysis of accountability through performance management, while accountability through decentralization and site-based management will be discussed in the penultimate section.

Both the concepts and the mechanisms for accountability as they relate to education have moved a considerable distance from that posited by Sockett (1980) who argued that that accountability had both a simple and a complex meaning. At its most basic, it implies being obliged to deliver an account as well as being able to do so. In its more complex form, accountability can mean responsibility for adherence to codes of practice rather than outcome. Following the emergence of alternative discourses in the socio-political environment based on neo-conservativism and neo-liberalism, the nature of accountability has changed (see Olssen *et al.* 2004). Now teachers and schools can be held to account both through the control mechanisms of the state such as inspection processes and the procedures within the school such as performance appraisal. Both teachers and schools are accountable to parents and to the state, although in the case of tertiary institutions the accountability is more likely to be to the state and, perhaps, to the students. The educational institutions and the individuals working within them are now held accountable for student performance and for the contribution to national priorities or performance targets. Thus, accountability is: 'the submission of the institution or individual to a form of external audit [and] its capacity to account for its or their performance ... accountability is imposed from outside' (Scott 1989: 17). This change in the strategic direction of education policy has thus brought about radical revisions in both the organizational principles on which accountability is based and the operational procedures through which accountability is delivered.

There is, however, a contradiction here for teachers who have to operate as professionals within an organizational framework. A measure of autonomy is required for practitioners to be effective while the school remains accountable for their performance. Is it possible to establish a balance that ensures that guidelines concerning performance can be applied while a degree of professional autonomy is retained? Edwards (1991) notes that that accountability leads to control while autonomy and choice foster the release of human potential. This is a dilemma already encountered in the discussion of the human capital approach to educational policy in Chapter 3. This is not to say that all modes of accountability are linked to or derived from the direct link between education and the need for nation states to further develop their economic capacity, but, within the prevailing discourses, the economic imperative does exercise a very powerful influence over the requirement to hold educational institutions to account. The issue here is about organizational principles that shape the forms that the accountability mechanisms take and the impact of the application of their concomitant operational procedures.

Leithwood *et al.* (2002) suggest that there are four different approaches to accountability that have been adopted by New Right governments. These are market approaches, management approaches, decentralization and professional control through site-based management. The common thread that binds these together is: 'A belief that schools are unresponsive, bureaucratic, and monopolist ... Such organizations are assumed ... to have little need to be responsive to pressure from their clients because they are not likely to lose them' (Leithwood 2001: 47).

It was as a direct result of this perceived unresponsiveness that attention turned to the creation of the use of market forces for holding educational institutions to account for their performance: 'Educational markets operate within an institutional framework and the government's job is to design the framework ... If this framework is designed with care and concern markets can be allowed to work their wonders with it' (Chubb and Moe 1992: 10–11).

Market accountability

It has been argued in earlier chapters that globalization has had a significant impact on the formulation and implementation of the education policies of many nation states and has often resulted in a shift in emphasis from policy related to provision, to policy concerned with regulation. The economic challenges posed by the weakening of national boundaries have made it more difficult for any nation state to sustain the high cost of welfare provision, including that of education. Consequently, it has been argued that the most effective mechanism to match the state's capacity to provide welfare services with the requirements of the economy was through the operation of a market within which individuals can exercise choice by acting as rational consumers in pursuit of their own economic interests. The socio-political philosophy for this approach to policy was based on the work of right-wing political economists such as Friedman and Friedman (1980) and Hayek (1973). Their argument, that market forces were the most appropriate way of allocating resources and structuring choices in all aspects of human endeavour including social and educational policy, provides the context within which the text of policies on accountability and their strategic direction have been established in many countries. Such policies are often interpreted as enabling the state to restrict its role to that of a regulator of the market place by identifying the strategic direction in which provision should be focused. As Ball (1993) observed, the intended consequence of such policies is that collective, bureaucratic controls, structures and relationships will be replaced by market forces, performance management and with individualistic and competitive relationships.

The market driven approach to accountability in education policy found one of its most coherent forms in England and Wales in the 1990s. It was translated into a strategic direction and a set of organizational principles and operational procedures for public sector institutions by Joseph (1976) and Scruton (1984); the main tenets on which this legitimation was based were:

- the absolute liberty of individuals to make choices based on their own self-interest;
- the freedom of individuals to exercise such choices without being subject to coercion from others;
- the freedom to choose being exercised daily through spending choices rather than every 5 years through the ballot box.

Competition becomes the motive force for policy implementation and through it improvement in the nature and quality of service was to be brought about as the family becomes a unit of economic consumption, its members making choices about products and public sector institutions behaving as firms seeking to maximize both profits and market share. This was articulated in a government document, *Choice and Diversity* (Department for Education and Science (DfES) 1992). *Choice* was to be exercised most fundamentally by parents over where and how children were to be educated within a policy framework based on the provision of a range of different types of schools to which parents might choose to send their children. Parents who sought to transfer their children from one school to another had the ability to influence school budgets since funding follows pupils. Thus, it was in the interests of all schools to compete against other schools to recruit and retain as many children as is possible. The assumption was that the more successful schools would attract more pupils. Those schools that did not reach an acceptable standard and were therefore deemed by the Office for Standards in Education (OfSTED) to be 'at risk', were liable to be closed if significant specific improvements were not made within one year. The mechanisms for holding schools to account: inspections, publishing results, annual reports and meetings, were all determined by national policy but were operationalized at a local level.

While the text and context of educational policy is derived from the language of the market place, choice provides the organizational principle that underpins the operational procedures. When associated with a pluralist schools sector, choice provides a range of different opportunities for parents to act as consumers of education. Some of those opportunities involve working within the public, state-provided system of education, others depend on private sector funding while yet others may be a combination of public and private sector initiatives. At the same time, mechanisms for facilitating choice have placed an emphasis on decentralization in countries as diverse as Sweden and China while in other countries, New Zealand for example, choice is located within a framework of a centralized curriculum and state inspection and accountability processes. The operation of the voucher schemes in parts of the USA is underpinned by the view that teachers alone should not determine the goals of education:

> Teachers will indeed become accountable but not to publicly appointed bodies or other professionals in public authorities. The head and staff will have their behaviour conditioned by the degree of success that they achieve in attracting pupils.
>
> (Kogan 1986: 53)

Similarly, the School Management Initiative (SMI) in Hong Kong also reinforces the market accountability model, although to a lesser extent than in the UK. According to Wong (1995), the recommended introduction of an annual School Plan and School Profile was to enable consumers to exercise their choice in deciding which schools would best satisfy their needs. The *Hong Kong Education Commission Report of 1997* (ECR7) emphasized the link between accountability and school autonomy. One of the recommendations involves 'allowing school management greater autonomy in general administration, finance and personnel matters but at the same time requiring a higher degree of accountability for school performance' (*Education Commission Report Number 7* 1997: xii).

As Apple (2000) has argued, accountability through choice in this context requires that all people should act in ways that maximize their own personal benefits and thus, for neo-liberals, consumer choice becomes the guarantor that education will become self-regulating. In some countries, the USA for example, this can lead to an emphasis on standardized testing and in others to a national census-based student assessment system such as that in Uganda (Riley 2004). In other states, Singapore and England for example, it leads to league table ranking of schools on the basis of which those parents with the resources to exercise choice are able to select schools for their children. In Singapore the emphasis is on achievement rather than choice, on standards rather than autonomy but accountability remains important (Kwong 1997). Here accountability is expressed both through the implementation of system-wide educational reforms and through the part played by education in forming, maintaining and improving the economic infrastructure of Singapore. Yip *et al.* (1997) highlighted this when referring to 'the pivotal role of education in the task of nation building and in fashioning the vibrant Singapore economy with a competitive edge in the world market' (Yip *et al.* 1997: 4).

Bush and Chew (1999) also recognized the tight coupling of economic development and education that is a feature of Singapore. They note that over the last three decades of rapid economic growth school and tertiary education has frequently been restructured and expanded in response to the changing human resource requirements of an increasingly diverse economy. The education system has been harnessed to achieve national development goals. This is both a highly developed form of accountability and a recognition of the importance of educational standards that goes beyond narrow market perspectives.

Such accountability mechanisms are based on a policy text that treats policy making as uni-linear whereas it is far more complex and is subject, in a pluralist society, to diversity, influence and re-interpretation at every level. Consequently the organizational principles produce specific policies that are difficult to operationalize because they are based on a central control model that is inappropriate for educational institutions in many societies. The implications for education are considerable. As McNamara *et al.* (2000) note:

the ideological move to construct education as a market place holds both systemic and economic implications for schools, teachers, children and parents and may involve policies ... based on 'educational' values being marginalized in favour of those lodged in market imperatives. Significant amongst them is the necessity for schools to promote a positive image of themselves in terms of performance indicators of product/output.

(McNamara *et al.* 2000: 475)

Furthermore, there are four important limitations to the market model of accountability and choice. First, how far is it appropriate to identify the parent as the sole customer of the education service, especially as education is largely funded from tax revenue, is largely state provided and the provision exists within the context of a national framework of priorities? Secondly, in spite of attempts to develop choice within the system, the education system's position as a near-monopoly supplier is maintained by a range of regulations, including compulsory attendance at school, the qualification and registration of teachers and, in many countries, a national curriculum. A free market would require all these controls to be lifted. Thirdly, the option of establishing co-operation and collaboration that could achieve similar ends is often ignored, although this is being re-established as part of the policy agenda in English schools. Schools are now encouraged 'to choose to establish new partnerships with other successful schools, the voluntary sector, faith groups or the private sector' (DfES 2001a: 44).

Even where such collaboration is attempted, such as in the case of schools involved in Education Action Zones discussed in Chapter 7, for example, residual competition between institutions makes collaboration difficult. While the curriculum may be national, the localization of accountability is such that the onus is on individual schools to achieve successful implementation and to accept the consequences if they fail. These reforms in most countries are introduced on the grounds of quality and efficiency, improving pupils' learning while both obtaining a better return on money spent and ensuring that national labour requirements are met.

Fourthly, accountability based on choice emphasizes the role of the parent as customer choosing between service providers. Both accountability and choice will be maximized without the need for political interventions. The argument is that if the customer can be placed in a direct relationship with the supplier of the services they seek, then a self-regulatory market can be allowed to operate. As Gewirtz *et al.* (1995) warn, however, the limited capacity of many parents to avail themselves of the opportunity to send their children to any but the nearest school could mean that exercising such choices could become a middle-class form of engagement with the mechanisms of choice and accountability. This gives rise to what Stoll and Fink call the quality-equity paradox:

> The reforms ... have been made in the name of quality and efficiency. They provide the rhetoric of equity but fail to accommodate the changing nature of society. Indeed many changes tend to be ways for the 'haves' to escape from the 'have nots' ... various choice and voucher initiatives ... gain favour with the affluent but ignore the impact of post-modernity on the least empowered elements of society.
>
> (Stoll and Fink 1996: 7)

This unequal distribution of resources within most societies means that the opportunity to make such choices is also unequally distributed. As Davies (1990) warns, parental choice through the direct application of market principles is an uncertain avenue to equality of provision since the ghettoization of schools could be reinforced where choice is exercised on grounds of social and racial prejudice. Accountability based on choice may, therefore, have adverse effects: 'There is growing evidence that the quasi-market in education is leading to greater inequality between schools and greater polarization between various social and ethnic groups' (Walford 1996: 14).

This resonates with the argument developed in Chapter 4 about citizenship and social justice involving both contributing to society and benefiting from being a citizen in equitable ways. Such choice mechanisms may jeopardize the ability of public schooling to provide equal opportunity for all students since unequal resources produce situations in which good schools attract the advantaged students while other schools are left with rejected students and thus result in socially unjust provision (Gaskell 2002).

The role of parents is central to this policy mechanism. Sallis (1988) argues that schools and parents must be accountable to each other for their contributions to a shared task; 'true accountability can only exist in an acceptance of shared responsibility for success at the level of the child, the school and the service' (Sallis 1988: 10).

Parents and accountability

The development of an accountable relationship between schools and parents as individual family units representing their child or children at a particular school represents a complex and sometimes controversial aspect of state intervention in education. In 1994 a survey of teachers and school principals in Hong Kong conducted by the Education Department indicated a positive belief in the desirability of having effective home-school cooperation. There was far less agreement about the nature of that partnership or about empowering parents to hold teachers to account. The Hong Kong School Management Initiative, introduced in 1991, encouraged schools to adopt new management practices in order to improve the quality of school education (Cheng 1999). Teacher and parent members were added to School Management Committees in Hong Kong (Wong 1995). Ng (1999) noted that the intention of the policy change was for parents to play a much more proactive role in supporting learning and in facilitating the attainment of a much higher level of academic performance by their children. However, only a few schools in Hong Kong permitted parents to become partners in the process of determining school policies (Ng 1999). Nevertheless, the Hong Kong Government remains committed to the promotion and strengthening of this form of accountability and has recently produced the School-based Management Consultative Document to pursue this further. Part of the title of the document indicates the strategic direction in which the policy is intended to move: *Transforming schools into dynamic and accountable professional learning communities* (Education Department 2000). This links initiatives to involve parents and teachers jointly in school management closely to the improvement of pupil performance and the development of high-quality school education.

Parent-school relationships in Hong Kong are similar to those in many other countries although as Cheng (2002) points out, in many Asian countries parental involvement in schools is a difficult issue because many of those countries lack a culture to support such involvement. There has developed a growing awareness of the importance of involving parents, not least because they can share the management responsibility, monitor teachers and monitor school operations (Cheung *et al.* 1995). A similar awareness has recently emerged in Israel where there is a highly centralized education system with limited private school alternatives. Here there is a move towards local school autonomy in an attempt to meet the diverse needs of parents (Goldring 1997). Furthermore: 'Parental and community involvement is often perceived as distrust of teachers and principals. To involve parents can be perceived as a loss of face among professionals' (Cheng 2002: 110).

The most radical forms of parental partnership in education can be found in charter schools in the USA. Here parents are able to found or co-found schools that are governed by boards of directors composed primarily of parents (Yancey 2000). These were not to be independent schools in that they charged fees but they were to be free from regulations that governed other schools. They were, in effect, mini school districts (Amsler 1992). By the early 1990s more than 20 states had introduced or were introducing legislation to establish charter schools (Sautter 1993). In New Zealand the Picot Review (1988) required every school to produce a charter while in Australia, reforms in Victoria and New South Wales introduced similar changes. Increasingly these agendas are driven, not by parental interests, but by a focus on accountability through pupil achievement.

The importance of parents in the education of children has long been recognized in England. The 1986 Education Act (DES 1986) increased parental representation on governing bodies and required the governing body of each school to deliver an Annual Report for parents and to hold an Annual Meeting of parents to discuss that report. By implementing such operational procedures the education system in England may have moved further than in many other countries to establish mechanisms through which parents can hold the school to account and, conversely, can be held to account by the school. This relationship has now been formulated into a formal agreement. All schools are required to prepare a written home-school agreement and associated parental declaration (DfEE 1998b). These agreements must contain a statement of the school's aims and values, a definition of the responsibilities of both the parents and the school together with a statement of the school's expectations of its pupils. Once the agreement is finalized, schools 'invite' parents to sign it. The rationale for this development is children are more successful when schools and parents work together and this can happen most effectively if parents know what the school is trying to achieve and, therefore, can hold the school to account.

Within this context, however, contradictions are inevitable. For example, many parents focus on what is best for their own children. What matters is what their children are doing as individuals. Teachers, on the other hand, are more concerned with aggregate notions of school performance and improvement and attempt to involve parents in the quest to achieve improved aggregate goals. Thus, the impact of this is that accountability mechanisms through parental involvement are largely concerned with individual self-interest, on ensuring that cost is minimized and benefit is maximized from the services available. As was seen in Chapter 4, this interaction between accountability and benefit is often expressed at a societal rather than an individual level. In many nation states the close connection between education and economic development is widely recognized and a significant number of improvement initiatives have been introduced to strengthen the ways in which educational institutions are more accountable for their role in meeting economic priorities. School–parent accountability, therefore, is grounded in relatively high level benefits, benefits that may accrue at a societal, regional, local community or even school level. Most parents, however, are concerned about the benefits that pertain to their own children as individuals within the education process. It is for this that they wish to hold schools to account rather than for their more general performance and the achievement of targets. Such different perceptions of accountability lead to significant tension.

The implication of the operational procedures that produce these accountability mechanisms is that both students and parents are partners in the educational enterprise. As result, parents have been re-positioned as informed consumers in the educational market place. Thus, the nature of this accountability relationship has now shifted from parents as passive or active consumers to one in which the total mobilization of resources in support of the educative process has sought to transform parents into productive partners who accept a responsibility for the success or failure of the educational enterprise. This involves the sharing of both responsibility and risk. The responsibility for the attainment of pupils in any school, both collectively and as individuals, has now been broadened to include parents.

This, of course, shifts some of the burden of responsibility away from teachers. Nevertheless, the extent to which schools succeed is determined by measures of pupil performance and it is against such measures that schools will be held to account. As Crozier (1998) says of the relationship between parents and schools:

> part of this process of accountability is a device for surveillance. However, this surveillance is not one-way: as well as the accountability of teachers through surveillance, school relationships have been underpinned ... by some form of surveillance and social control of pupils ... and parents.
> (Crozier 1998: 126)

Such accountability requires teachers to persuade parents to adopt the school's definition of what it means to be a 'good' parent and a 'good' pupil. Thus, while parents and external educational agencies may monitor the work of teachers, teachers are using home-school partnerships to monitor parents who monitor pupils. This can often mean the imposition of an entire value system on parents for, as Crozier (1998) notes, this accountability procedure is carried out through a process of teacher domination and on the basis of the teachers' agenda.

Perhaps to counteract this domination of the agenda by educational professionals, more recent British government publications such as *Schools Achieving Success* (Department for Education and Skills (DfES) 2001a) have re-defined accountability in a number of ways to further facilitate consumer choice. In New Zealand, similar partnerships are envisaged between schools, colleges, stakeholders and local communities to reinforce existing accountability mechanisms (Government of New Zealand 1998). The extent to which such partnerships are beneficial and sustainable will be considered further in Chapter 7. Linked with this process of establishing such operational procedures for accountability are the organizational principles of modernization through deregulation such that central to achieving both a more accountable school system and the improvement of educational standards is: 'The confident, well-managed school running its own budget, setting its own targets and accountable for its performance' (DfES 2001a: 63).

Accountability and school-based management

School-based management is closely linked, through the language of markets and choice to organizational principles grounded in notions of autonomy. The argument here is that institutional accountability can be strengthened if decisions are made by those inside schools and colleges rather than by national or regional officials because those on the ground are best able to make decisions about appropriate provision of education and the allocation of resources. They can then be held to account for those decisions and greater effectiveness through greater flexibility and more efficient deployment of resources achieved (Caldwell and Spinks 1998; Bush 2002).

In Hong Kong the introduction of school-based management was legitimized by rhetoric of raising standards:

> The School Management Initiative (SMI), embarked upon in Hong Kong in March 1991, is a major restructuring of the operations of secondary and primary schools, with the belief that greater self-management can enhance school performance. With self-management, schools are more free to address their own problems, and ... manage changes and routines in a controlled manner.
>
> (Wong *et al.* 1998: 67)

Similarly, the Schools Excellence Model (SEM) in Singapore is based on devolving power over some aspects of resourcing and pedagogy to the institutional level, but within a framework of competition between schools. SEM was specifically intended to improve student performance: 'once people in the school setting are motivated and sensitized to the drive for excellence ... then organisational excellence will be eventually achieved, thereby schools will become agents for continuous improvement and innovation' (Mok 2003: 357). This model is based on the development of internal accountability mechanisms that can enable the school to respond to the national quality framework. The SEM schools do have powers devolved to them, but they are not, in any way, autonomous.

As Caldwell (2002) points out, self-managing schools may not necessarily be autonomous, since autonomy implies a degree of independence that is not provided within education systems in which central control is exercised over curriculum content, pedagogy and outcomes. Schools can be considered self-managed when 'there has been decentralized a significant amount of authority and responsibility to make decisions related to the allocation of resources within a centrally determined framework of goals, policies, standards and accountabilities' (Caldwell 2002: 35).

Karstanje (1999) warns, however, that while decentralization may shorten the distance between the policy makers (government) and the policy implementers in the schools and colleges, it may not mean that the institutions gain more autonomy. Such autonomy may shift to the regional level and make little difference to schools. Deregulation, however, does lead to an increase in institutional autonomy if the effect of the deregulatory process is to shift the locus of decision making to the institution. The effect of both deregulation and decentralization is often to shift the financial responsibility and risk away from governments and towards the institutions that, through deregulation, become accountable for managing their own resources (Karstanje 1999). This can be seen as 'a deliberate process of subterfuge, distortion, concealment and wilful neglect as the state seeks to retreat ... from its historical responsibilities for providing quality public education' (Smyth 1993: 2).

Alternatively, deregulation can be interpreted in a more limited way as a selective withdrawal on the part of the state from areas in which it has had difficulty succeeding such as providing equality of opportunity through education (Nash 1989). It may also be an example of the way in which the state seeks to cut expenditure on public services during a period of economic stringency induced by global pressures. Whatever the interpretation adopted, school-based management can be seen as a technique for shifting accountability away from government to individual institutions, especially as the failure of individual schools and colleges can then be attributed to poor leadership and resource management at local level. Deregulation 'will give ... schools greater freedom and less central control. New freedoms ... will enable them to meet the needs of pupils and parents more effectively' (Blair 2001: 44 quoted in Caldwell 2002: 36).

With freedom comes responsibility because self-managed schools, while able to set their own targets, will be accountable for their own performance within an established national accountability framework (DfES 2001a). As Bullock and Thomas (1997) point out, the main thrust of deregulation in England is to give schools control over spending priorities while, at the same time, autonomy over what is taught has been severely reduced:

> The autonomy of schools has been enhanced in the area of control over human and physical resources but control over deciding what is taught has been reduced by the national curriculum. Accountability has been altered and the role of the professional challenged.
>
> (Bullock and Thomas 1997: 52)

They note that a similar pattern can be found in Chile, China, Poland, Uganda, the USA and in most countries where resource management has been devolved to schools. In Australia, many schools have had devolved to them the responsibility for utilities, buildings, flexible staff establishments and appointments but this has been associated with greater centralization of both curriculum and assessment while in New Zealand:

> Accountability has been strengthened by performance monitoring against ... objectives established by the school and the government. The efficiency of the change depends on the 'market' benefits being generated. The impact of the reforms may threaten equity.
>
> (Bullock and Thomas 1997: 54)

The autonomy of schools has increased over the management of the operational grant at the expense of greater central control over the curriculum. There is a paradox here, because:

The decentralization of control over resources can be viewed as consistent with the 'market' principles ... Yet the centralization of control over the curriculum would appear to be ... more consistent with the principles underlying planned economies.

(Bullock and Thomas 1997: 211)

The link between these two apparently contradictory sets of policies lies in the need to challenge the autonomy of the educational professionals and to link the outputs of schools more closely with the perceived needs of the economy. The centralized elements of educational policy in most countries serves to enable the mechanisms of choice to operate and to allow government agencies to hold schools to account for their outputs. At the same time, those within the schools are accountable to government and stakeholders for the management of resources to achieve those outputs: 'Under a system of school-based management, accountability for student achievement rests squarely with the individual school' (Carlson 1989: 2 quoted in Murphy 1991: 5).

Accountability in self-managing schools, therefore, is concerned both with student performance and with institutional efficiency:

creating more efficient and cost effective school administrative structures is a ... central goal for devolution. Typically, this goal is pursued through the implementation of an *administrative control* form of site-based management that increases school-site managers' accountability ... for the efficient expenditure of resources.

(Leithwood 2001: 49 original emphasis)

Although the idea that locally managed schools could be run more economically is critical to this approach to accountability, there is no evidence that this is the case (Gaskell 2002). The nature of the work carried out by headteachers and their staff, however, will change.

Fergusson (1994) has identified the extent to which the accountability for resource management can result in headteachers becoming embroiled in administration and performance management at the expense of educational matters. Heads no longer have the responsibility for formulating broad educational policy within their schools. Instead, they are responsible for implementing a set of policies the strategic direction, organizational principles and operational procedures of which have been externally imposed. Their role is restricted to achieving the outputs determined by those policies. Consequently, unless resources are diverted away from teaching and learning and into administrative structures, there is, on the part of heads and senior staff in schools:

a radically increased emphasis on budgetary considerations ... and less attention to providing leadership about curriculum and instruction

... As an approach to accountability, site-based management is wide spread ... Considerable evidence suggests, however, that by itself it has made a disappointing contribution to the improvement of teaching and learning.

(Leithwood *et al.* 2002: 858–9)

Here the operational procedures are producing outcomes that are contrary to both the articulated strategic direction for these policies and to the very language of legitimation of the policies themselves.

The contradictions here are obvious. The administrative demands of site-based management take senior staff away from the arena of professional practice while the requirements of performance management, as will be shown in the next section, are predicated on authority and hierarchy. Little wonder then, there is an increased emphasis on leadership training, the creation of national standards for school leaders and even pre-appointment certification such as the National Professional Qualification for Headship in England and Wales. A common theme in such training is that leadership and accountability should be dispersed throughout the organization. This, of course, means that if accountability as well as leadership is so distributed, then performance management within the organization becomes a key element in the accountability process. Thus, as Ball (1994) argues, the doctrine of site-based management can be seen as one in which surveillance of school work and holding staff to account is conducted by heads and other senior staff as part of a process of carrying out the intentions of central government. Self-managing schools, like all other modern organizations, can be seen to be concerned with regulation and surveillance, either explicitly or implicitly (Foucault 1997).

Accountability and performance management

The most significant distinction between accountability through market forces, parental choice as an accountability mechanism and accountability through performance management is that market forces and parental choice focus largely on external accountability while performance management is concerned with the internal processes of the school. Accountability based on control at school level through performance management focuses on the extent to which teacher and pupil performance can be managed, progress towards relevant targets can be observed and reported upon and on the extent to which progress towards specified outcomes can be measured. Thus, the main formal characteristic of accountability through performance management is that of a managerial hierarchy concerned with organizational effectiveness and efficiency in terms of achieving targets that are institutionally driven by appraisal and improvement planning. This goes beyond accountability by a public comparison of performance through league tables and published inspection reports that underpin market accountability and accountability by parental choice. For example, teachers are required by contract to perform tasks set by the headteachers and determined by national targets, and subject to review. Teachers are held accountable by the headteacher for their work. This is supported by a technical-rational view of both the curriculum and the nature of teachers' work. It assumes that the curriculum consists of a series of specific outcomes and the pedagogy to achieve them: 'Outcomes remain the focus, but they are now constituted as targets and benchmarks, rather than just comparisons with other institutions' (Fergusson 2000: 208).

Once this framework of targets is established, it then becomes the function of the teacher to use the prescribed pedagogy to ensure that the outcomes are achieved. Accountability thus becomes a matter of assessing how successfully teachers have deployed the relevant pedagogy based on the testing of pupil performance. This, in turn, is based on performance management: 'It is necessary to establish clear organizational goals, agree to the means of achieving them, monitor progress and then support the whole process by a suitable system of incentives' (Normore 2004: 64).

The framework for such performance management in schools in Britain is enshrined in legislation that requires headteachers to ensure that all teachers are appraised on a regular basis (DfEE 2000). This performance review must take place annually and must result in the setting of at least three targets, one of which must relate to pupil performance. Typically the other targets are linked to professional development and management responsibilities. The performance targets are, in Britain at least, part of a wider agenda of target setting that culminates in the school development or improvement plan that sets targets for pupil performance within the school – based on nationally determined priorities and standards. Lay governors are responsible for ensuring that these targets are achieved but it is the headteachers and subject leaders who are directly accountable for the performance of teachers and pupils.

In parallel with this emphasis on managing the performance of teachers is the requirement to set standards of performance targets both for individuals and schools. One of the main management functions of senior staff becomes the monitoring of performance towards achieving both individual and school targets. *The Teaching Standards Framework* (DfES 2001b) states that subject leaders must provide management to secure high-quality teaching, effective use of resources and improved standards of pupil learning, a clear indication that middle managers are now responsible and accountable for performance management. This has implications for workloads, professional relationships and for the training of middle managers. Although in Britain promotion to senior management positions in schools is now subject to more specific criteria and certification, the current level of provision for middle managers in Britain, at least, is regarded as wholly inadequate (Busher and Harris 2000). In the further education sector in Britain, Gleeson and Shain (2003) note that performance management has substantially redrawn the lines of responsibility and accountability which have led to greatly increased regulation of professional workers and intensification of workloads.

Similar performance management measures can be found elsewhere. Tomlinson (2000) identifies a number of performance management mechanisms in the USA, many of which link pay to the appraisal of performance, although these tend to be whole school rather than individually-based rewards packages. In Hong Kong schools, there is what Wong (1995: 521) describes as 'a bureaucratic system of staff appraisal based on a managerial model'. Here the principal is accountable for the work of the school and has authority to discharge that accountability. In New Zealand, a rigorous performance management system based on teacher appraisal has been in place since 1997. This is predicated on the view, expressed by the New Zealand School Trustees Association (NZSTA) that:

> Performance appraisal ... is a tool by which the board can measure whether the objectives set for the school are being met. Through performance appraisal, the board and the principal can ascertain whether the elements of the job description, the performance objectives, and the outcomes ... take both the individual and the organisation forward.
>
> (NZSTA 1999: 7)

Concerns have been expressed about the extent to which this process has been used to implement accountability mechanisms: 'What has happened in New Zealand is that the accountability edge and thus the organisational demands of performance appraisal have insidiously been increased' (Middlewood and Cardno 2001: 12).

In countries such as Israel, Japan and the USA the process of assessment before moving on to the next career stage is well established (Middlewood 2002) while appraisal for school principals in Singapore focuses largely on their accountability for school improvement (Chew 2001). As Beare argues: 'Assessing performance is normal practice these days ... for accountability purposes, for efficiency, and for explaining and keeping track of how resources are used' (Beare 2001: 170).

For the first time middle managers in schools are involved in the process as appraisers, thus reinforcing the hierarchical management emphasis of the process. Information from the performance review statement can be used to inform decisions about pay and promotion to the extent that up to the normal pay threshold teachers can expect an annual increment if they are performing satisfactorily. Double increments for excellent performance would need to be justified by review outcomes. Performance review is used to inform applications by teachers for promotion to the upper pay spine while it also forms part of the evidence which can be used to inform decisions about awarding performance pay points to eligible teachers (DfES 2001c).

In some places, Connecticut and Queensland for example, this extension of the responsibilities of middle managers into the realm of performance management has been accompanied by a more rigorous control over criteria for entry into teaching and the extension of provision for teacher professional development (Leithwood *et al.* 2002). In Germany, a similar extension of performance management can be found in some universities. Here attempts are being made to strengthen central control within institutions by introducing accountability through contracts (Kreysing 2002). In one university this is based on a process of contract management between the central board and the faculties as operative units based on negotiated objectives which the heads of faculty or of the units within the faculty are responsible for achieving within a specified budget over a pre-determined time period, after which progress is reviewed. This is having a significant impact on the lines of accountability within the university.

The impact of such performance management mechanisms on the leadership role of the headteacher is extremely significant. In essence, headteachers are in danger of ceasing to be senior peers located within professional groups and are becoming distinctive actors in a managerialist system, in which the pursuit of objectives and methods which are increasingly centrally determined is their main responsibility. They must account for the deployment of those methods and the achievement of those objectives and, at the same time, ensure the compliance of their teaching staff (Fergusson 1994). Such accountability mechanisms must establish clear aims and acceptable criteria that are relevant and can be applied in educational institutions. At the very least, it is necessary to establish a balance between performativity and the legitimate professional concerns of teachers. There is evidence that this is not the case. Performance management as a form of accountability is widely seen as disempowering the professional domain within educational institutions at the expense of a strengthened management domain (Normore 2004). As Simkins (1997) argues, such accountability cannot be achieved through professional collaboration and accountability and will, therefore, generate tensions between heads and staff. In the USA, performance management schemes have been abandoned and, in Britain, the recently introduced appraisal process linked to performance-related pay has not been successful:

> Not only did the scheme ... fail to meet its own aims, but it has a number of shortcomings ... Respondents are unclear about targets and standards and are unclear about the extent to which performance can be measured on an individual basis. Moreover, there are clear concerns about the equity of such a system.
>
> (Farrell and Morris 2004: 101)

Furthermore, even where additional payments are made through the award of performance-related pay, there is significant evidence that such awards made a difference to what teachers do (Wragg *et al.* 2005).

Policy
formulation

Socio-political environment
- Economic utility
- Value for money

Strategic direction
- Market accountability
- Direct and indirect control
- Autonomous schools

Policy
implementation

Organizational principles
- Choice and autonomy
- Individual and institutional performance
- Targets/outputs clearly defined

Operational practices and procedures
- Performance measures identified
- Monitoring mechanisms established
- Monitoring information published

Figure 5.1   Policy into practice: accountability

Conclusion

It has been argued in this chapter that the imperative for a skilled work force and a competitive economy that has emerged from the socio-political environment over the last two decades has provided the context for the development of rigorous accountability mechanisms within education systems across the world. This movement has been reinforced by an emphasis on individual and institutional performance and the need for financial stringency in the public sector in many countries. These factors have combined to provide a strategic direction that can lead to the integration of accountability into the organizational principles and establish a range of operational mechanisms for holding educational institutions to account (see Figure 5.1).

Green (1999) notes, however, that while there is clear evidence of the impact of common global forces on education systems in Asia, Australasia, Europe and North America and convergence around broad policy themes such as decentralization and accountability: 'This does not appear to have led to any marked convergence in structures and processes' (Green 1999: 6).

Nevertheless, there is a discernible shift towards establishing and maintaining accountability procedures in many countries (Glatter 2003). Consequently, those within education in many states operate with multiple levels and senses of accountability that often co-exist in a confusing manner (Ferlie *et al.* 1996). Market accountability has been reinforced by the ability of parents to make choices and to determine where their children shall be educated. Schools have to be responsive to client needs if they are to thrive in this new market-led climate. In other words, schools will only provide the outcomes and services and meet the standards required of them when they are directly accountable to their clients, in the form of parents, through choice mechanisms and to the state through specified performance targets. Although the strategic direction of many of the accountability processes considered here is relatively uniform, it is evident that the organizational principles and the operational practices may differ widely. As Sui-Chu has recognized, if it is to form a significant element in accountability, parental choice and involvement in Asia generally and Hong Kong in particular:

> should take a broader view of parental involvement, to encompass both home-based and school-based activities. It appears to be more feasible, under Asian culture, to move parental involvement … from the traditional home-based form to less rational school-based form according to teachers' attitudes and zones of acceptance.
>
> (Sui-Chu 2003: 72)

Thus, while public accountability remains powerful, its nature is influenced by local policies and practices such as the extent to which teachers can influence and re-shape policy implementation and the locating of additional powers at school level.

Some argue that one of the consequences of the emergence of these forms of accountability is that professional accountability, where it existed, has reduced in importance because of government imperatives for higher standards and parental demands for greater responsiveness (O'Neill 1997). Professional norms often have to be subjugated to these public and market pressures, not least through processes of performance management although, as Middlewood (2002) argues, this has also produced an increased emphasis on and extended opportunities for continued professional development. Such opportunities, however, are often closely linked to the achievement of overall national priorities (Bolam 2002). Perhaps the issue here, however, goes beyond the specific nature of teaching as a profession and the nature of professional values and services. Given that the main features of professionalism are not immutable, it is evident that the nature of teacher professionalism, in the UK at least, has been transformed in conjunction with a growing emphasis on economic performance.

There is also evidence of growing central control over both processes and outcomes in schools and colleges. Much depends on where the balance lies between achieving outcomes through specific processes and the wider continued professional development of teachers and lecturers. How far, for example, does appraisal focus on targets and outcomes and how much attention is given to professional development? If the emphasis is on outcomes, then accountability and performance management are providing the organizational principles for control through accountability. This is less likely to be the case if the operational practices focus on professional development. Much of the evidence in this chapter suggests that the greater concern is with targets and outcomes. Consequently, questions are raised about the nature of teacher autonomy, responsibility and accountability (Hoyle and John 1995). It would be a mistake not to recognize that:

> The more complex a professional activity becomes, the more policy interventions have to take into account the views of practitioners and leave space for local adaptations ... practical problems cannot be solved for the institutions by central regulations.
>
> (OECD 1996b: 11)

A further consequence of such accountability rooted in the legitimation of economics and the market place, has been the diminution of the weight attached to concerns for equality of opportunity, equity and social justice. The operation of market forces through competition and choice may lead to some schools being over subscribed and others having rapidly declining enrolments, resulting in a climate that is not conducive to setting and achieving reasonable educational standards for those students that remain. The use of raw measures of pupil achievement as the sole or even the major criterion for making judgements about the efficacy of the performance of schools may also result in discrimination against certain groups of pupils, those with special needs or whose first language is not the language of instruction in the school, for example. The exercise of parental choice may exacerbate these inequities. As Goldring (1997) notes in the cases of both Israel and the USA, one effect of parental choice mechanisms is that students from more affluent families are more likely to leave their neighbourhood schools in favour of magnet schools that cater for students with special skills. In Britain, Gewirtz (2002) has pointed out that one of the most significant weaknesses of accountability based on market forces, choice and performance management is that such mechanisms merely tend to produce a redistribution of students amongst schools and colleges without addressing the root causes of educational under-attainment and issues of equality of opportunity and social justice. There is evidence from both Australia and France that professional families are far more active in exercising choice than are those of manual workers (Hirsch 1997). How to address such issues at institutional level is often left to individuals within schools and colleges who have to implement and often mediate the impact of educational policy on both teachers and students. The three chapters in the next part of this book each explore different aspects of local interpretation and implementation of policy.

Part III

The impact of educational policy

It was argued in earlier chapters that much educational policy can be seen as a response to the impact of globalization, human capital theory, concerns about citizenship and mechanisms for holding to account the leaders and managers of educational institutions. These themes, together with advances in information technology typify the wider socio-political environment and provide the context within which the text and strategic direction of much educational policy is located. In this third part of the book, the impact of some of these themes is explored at the local and institutional level. Here the strategic direction, the implementation of the text of policies and organizational principles and operational practices on which policy implementation depends will be explored. In Chapter 6 the implementation of school-based strategic planning in two policy contexts is discussed. There is a particular focus on how leaders at an institutional level seek to mediate national policy agendas in order to develop institutional policies relevant and appropriate to their specific context. In Chapter 7 there is an analysis of the impact of a particular policy, that of Education Action Zones (EAZs), which links the themes of economic utility, citizenship and social justice. In this instance a high-profile national policy was introduced in order to tackle perceived problems of educational underachievement in socially disadvantaged areas. The chapter explores how a policy principally concerned with issues of social justice is legitimated within a discourse dominated by economic considerations and the need to develop human capital. In Chapter 8, citizenship, cultural diversity and social justice are explored in the context of multi-ethnic school leadership. Here there is a particular focus on how school leaders have sought to develop institutional policy agendas that reflect the diverse nature, and needs, of their school communities. Research in this chapter illustrates how institutional policy agendas could be both supported and constrained by policy discourses emanating from the state and how leadership often involved simultaneously maximizing the opportunities, and minimizing the threats, presented by state policies.

6 Policy, strategy and leadership

(with Daniel Chan)

Introduction

The strategic direction of much of education policy is frequently justified not as an end in itself, but as a means to enhanced economic development leading to a more competitive economy, greater productivity and increased wealth. The essence of such a policy, as far as it impacts on schools is that:

> The effectiveness and efficiency of schools will be improved ... as they become more strategic in their choice of goals, and more ... data-driven about the means used to accomplish those goals. The approach encompasses a variety of procedures for 'strategic planning' ...
>
> (Leithwood *et al.* 2002: 861–2)

More broadly, investment in education and training is believed to provide the key both to national competitiveness and social cohesion, often by seeking to improve the performance of schools through setting targets for pupil achievement and requiring schools to develop improvement plans to meet those targets. These targets are intended both to provide the organizational principles and to inform operational practices in schools. Frequently, however, the setting of such targets does not take into account contextual features such as the socio-economic mix of the school, its funding or teachers' skills and experience (2004). It is claimed, however, that: 'Even schools suffering from high levels of deprivation can achieve genuine improvements through careful rational planning' (Hargreaves and Hopkins 1994: ix).

This implies that school context is irrelevant and that if development planning is pursued vigorously, then school improvement will be the inevitable consequence. The strategic direction that emanates from such a discourse is based, as was seen in Chapter 5, on accountability and target setting linked to a rational planning process. Such organizational principles and operational practices are not confined to secondary schools and tertiary institutions which tend to have a direct link to the labour market. They also impact directly on schools in the primary sector. The emphasis here is on primary schools because there is a dearth of research evidence on similar processes in secondary education. What work there is tends to be largely descriptive and prescriptive (Jayne 1998; Davies and Davies 2005). This chapter explores the consequences of such policies on primary schools and examines the role of school leaders in implementing and mediating such policies. It is based on an analysis of how far primary school headteachers in England and Hong Kong adopt rational planning and to what extent they seek to modify both the process and the outcomes. These two countries are chosen partly because they represent different policy contexts. Each of these education systems has also been subject to significant changes in its wider socio-political environment over a similar time period, resulting in a shift of strategic direction and changes in organizational principles and operational practices.

Educational policy and schools in Hong Kong

The Hong Kong education system has its origins in British colonial rule and, therefore, shares some features with its English counterpart. The differentiation between primary and secondary schooling in Hong Kong is broadly similar to that in England although English children begin and end their primary education a year earlier than their Hong Kong counterparts. Both education systems have been subject to significant reform in recent years. In each country, 1997 proved to be a watershed for changes in the socio-political environment and, therefore, for educational policy. In Hong Kong what has been termed 'The First Wave' of educational change (Cheng 2000) had its roots in the *Llewellyn Report* (Llewellyn *et al.* 1982) and the subsequent establishment of the Education Commission that produced six reports between 1984 and 1996. The strategic directions articulated in these reports:

> had their roots in the assumption that policy makers could establish best practices to enhance effectiveness ... to solve major problems for all schools ... they were generally characterised by a top-down approach with an emphasis on external intervention.
>
> (Cheng 2000: 23)

These changes focused on more effective teacher training, the use of new technology, revisions to the curriculum (Education Commission 1984) and improving language teaching (Education Commission 1996). Measures were also introduced to enhance teacher quality (Education Commission 1992). A major policy thrust revolved around curriculum development through establishing a Curriculum Development Institute and the refinement of assessment by introducing a set of attainment targets at key stages in children's education (Education Commission 1990). In 1991 the School Management Initiative (SMI) was announced.

In 1997 Hong Kong became a Special Administrative Region (SAR). New sets of policies followed. Although this was a significant change in the socio-political environment, the strategic direction of educational policy remained largely the same, at least for the time being, as did the organizational principles. These included developing school-based management as the major approach to enhancing effectiveness and quality assurance in education. Operational procedures were based on upgrading primary teachers to graduates, implementing a long-term information technology strategy and working towards the target of whole-day schooling for all primary students (Education and Manpower Bureau 1997). The link between the two policy phases was provided by the SMI that established a school-based approach to school management (Cheng and Cheung 1999):

> The School Management Initiative ... is a major restructuring of the operations of secondary and primary schools with the belief that greater self-management can improve school performance ... Consequently, the development of an annual school plan is one of the changes required under SMI.
>
> (Wong *et al.* 1998: 67)

By 1999 a new vision had been established for Hong Kong education which was intended to cultivate good learning habits in children, lay the foundations for lifelong learning of students and prepare them for the building of a civilized society based on learning, a sense of social responsibility and a global outlook (Education Commission 1999).

There was a distinct paradigm shift, not in strategic direction but in organizational principles, between the first and the second waves of reform. The first wave depended on a resource-based top-down approach which largely ignored school needs, while the second wave put a strong emphasis on a school-based approach that recognizes the importance of institutional level planning and management (Cheng 2000). While school-improvement planning became an important operational practice in the implementation of education policy in Hong Kong, the central organizational principle was the devolution of responsibility to schools. It was strongly promoted through the SMI and, more recently, through school-based management (Advisory Committee on School-Based Management 2000) which is integral to the reforms included in the *Education Commission Report Number 7* (Education Commission 1997). At the same time, the importance of the role of the headteacher in the planning process and in establishing the organizational conditions within which effective planning can occur has been recognized. Teachers and members of the school community are now also expected to be actively involved in both planning and school-level decision-making:

> Each school needs the capacity to manage its own affairs. Local governance gives a school direct access to the expertise of its key

stakeholders – the sponsoring body, the head teacher, teachers, parents, alumni and members of the community. Above all, it empowers the stakeholders themselves to work effectively for the educational welfare of the students under their care.

(Advisory Committee on School-Based Management, 2000: 1)

The School Development Plan (SDP) was introduced in September 2000. Every new school was required to produce a plan that:

set out specific targets for implementation, school budgets, performance indicators and means of evaluating progress during the school year. The planning on various school programmes should reflect the development priorities of the school. The Plan should be endorsed by the SMC, signed by the school head and sent to the School Senior Development Officer (SSDO) by October every year. The school should also make the Plan available for the perusal of parents, teachers and members of the public.

(Education Department, 2000: 1)

Thus strategic planning for school improvement has become the linchpin of education policy in Hong Kong. In so doing, policy implementation in Hong Kong has tracked similar developments in England.

Educational policy and schools in England

It can be argued that planning in schools in England has evolved through at least three different forms in the last three decades. Each of these forms may be regarded as strategic although each has a different emphasis.

- Planning at LEA Level. Before 1988, planning was largely the province of local education authorities (LEAs). It consisted of staffing and resource management, allocating pupils to schools, seeking to match available places to projected pupil numbers, and, latterly, in-service provision.
- School Development Planning. *The Education Reform Act* (DES 1988) linked new patterns of accountability to the school development plan. The responsibility for ensuring that the National Curriculum was taught and tested and for the deployment of resources rested with governing bodies on which parents and representatives of the local community were in a majority. Inspectors from the Office for Standards in Education (OfSTED) were required to make a judgement about the management of the schools through the quality of the school development (OfSTED 1992).
- School Improvement Planning. The emphasis on strategic planning in schools continued under the New Labour Government after May

1997. This policy was firmly located within the context of school improvement articulated through a centrally determined set of targets. The interpretation and use made of strategic planning by New Labour was different from that of the out-going Conservative administration. Improving pupil and teacher performance were both central to this policy agenda and head teachers were responsible for policy implementation: 'Specific targets that inform the strategic planning in individual schools are set in conjunction with LEAs but must move towards those set nationally'.

(Bell 2002a: 411)

As this target-driven approach to educational planning has been pursued, a significant change has occurred both in the nature of development planning itself and in the relationship between the state and schools. It is no longer sufficient for staff in schools to set their own targets and to be accountable for achieving them. School targets must now be derived from national achievement targets for similar schools and incorporated into the School Improvement Plan (DfEE 1998c; DfEE 1998d). This plan must focus on strategies for bringing about curriculum change that will lead to improvements in pupil performance in line with national targets determined by policy makers at the centre (DfES 2001a). This approach to planning is similar to that advocated in Australia and elsewhere: 'The principal must be able to develop and implement a cyclical process of goal-setting, need identification, priority setting, policy making, planning, budgeting, implementing and evaluating' (Caldwell 1992: 16–17).

The essential features of planning in schools

Strategic planning as conceptualized in the previous section is essentially forward-looking, based on environmental scanning; proactive in the sense that the school will recognize opportunities and take advantage of them; creative so that present practice can be improved upon; and holistic by dealing with all the school's operations, not just teaching and staffing (Fidler 1996). The process is also outward looking, positioning the school, college or university in relation to the external environment, particularly its competitors (Lumby 2002). Such strategic plans will be based on evidence from: 'The external environment (both now and future predictions); the internal strengths of the organisation; the prevailing organisation culture; the expectations of stakeholders and likely future resources' (Fidler 2002: 616).

These plans may focus on different planning horizons or time scales including the very long term (what will life be like in the future); long-term desirable developments (10 years); medium-term (five year) plans and short-term (three year) institutional development plans (Fidler 2002). Strategic planning is: 'a list of actions so ordered as to attain over a particular time period, certain desired objectives derived from a careful analysis of the internal and external factors likely to affect the organisation' (Puffitt *et al.* 1992: 5).

Strategic planning has become closely associated in management terms with the rational expectations of those who wish to direct and shape an organization towards specific ends. It is a top-down process that develops from analysis through the identification of objectives or targets and actions to achieve them. It is predicated on being able to predict the environment and on the ability to exercise sufficient control over the organization and its environment to ensure that planned outcomes can be achieved by the deployment and redeployment of available resources. Its essential purpose is to assess the environment in which the organization operates, forecast the future for the organization and then to deploy resources in order to meet the predicted situation (Whipp 1998). Van der Heijden and Eden (1998) term this the linear rational approach to strategic planning which separates thinking from action and proceeds by analysing the evidence, choosing the best course of action and implementing it which approximates to school development and school improvement planning as it is required of English primary schools. Such a plan:

> follows a logical sequence; takes into account the external and internal environments and the stated aims of the school; priorities, targets and success criteria are set within these aims; action plans make targets operational; and subsequent reviews lead to annual up-dating and rolling forward plans ... for the next few years.
>
> (Wallace and McMahon 1994: 25)

This chapter explores how far such a linear rational approach to school improvement planning can be identified in primary schools in Hong Kong and England, using seven key questions:

- What do heads understand by strategic planning?
- To what extent do values and vision shape the planning process?
- Over what time period do heads believe it is feasible to plan?
- How far do audits of the internal strengths and weaknesses of the school play a part in formulating strategic plans?
- How far do factors in the external environment play a part in formulating strategic plans?
- What is the relationship between strategic planning and the organizational structure and culture of the schools in this study?
- What barriers to strategic planning do heads identify?

Strategic planning in primary schools

The answers to the questions posed above are provided from data gathered from a series of interviews with primary school headteachers in England and primary principals in Hong Kong (Bell and Chan 2004). The schools for which these heads are responsible are not intended to be a random sample of all primary schools in those countries. They form an opportunity sample based on schools that have taken part in previous studies. However, while each school is in some way distinctive, taken together they share many characteristics that are typical of most primary schools in their respective countries. In England, School E1 is a junior school in a small town that has just emerged from special measures. The head [EH1] has now been in post three years. This is her second headship. School E2 is a junior school in another town in the same area. It has also emerged from special measures. The head at this school [EH2] was appointed from within the school three years ago and this is his first headship. School E3 is a community primary school in a large town hit by the closure of its major industry. The head [EH3] has been in post 22 years. School E4 is also a community primary school and is in a newly created village. The head of this school [EH4] was appointed two-and-a-half years ago, nine months before the school opened. It is her second headship. In Hong Kong, school H1 is the afternoon session of a bi-sessional school. The headteacher [P1] has been in post 22 years. School H2 is the morning session of a bi-sessional school in a public housing estate. The headteacher [P2] has been in post since 1989 and this is his second headship. School H3, on a public housing estate, has recently moved from being bi-sessional to whole day. The headteacher [P3] has been in post eight years and this is her second headship.

What do heads understand by strategic planning?

The heads in this study saw planning as a set of complex operational procedures. Many of them use graphic metaphors to identify and cope with the complexities of the strategic planning processes. The head of School E1 saw it as:

> being on a dance floor. You know you have been involved with various partners but the perception from the balcony might be entirely different so you know that the key factor is being in both places at the same time or moving between the two … It is incredibly complex.
>
> (EH1)

The head of School E4 also recognized the complexities involved in strategic planning:

We have to get used to the idea of not being able to have a set strategy, of it not being a linear process in effect and being able to try to work out when and how you jump even before the ideas are past their sell by date.

(EH4)

In Hong Kong, one headteacher found that he had to deal with immediate problems before even considering longer-term planning:

I realized soon that the school was a mess, the corridors and the staircases; even the classrooms were so dirty. ... However, I told myself that even if it was a 'decayed apple'; I have to live with it. I would first concentrate on problem shooting so as to pave the way for further development. If you can't fix the problems that you are facing, how can you talk about future development?

(P1)

Similarly, P2 found himself having to make difficult decisions about priorities:

I needed to be more patient, more tolerant and more receptive because I did not have a strong team of teaching staff. I knew that I needed to work on people first, because, if a school is a place for learners, it should accommodate pupils, teachers, and also parents.

(P2)

The headteacher of School H3 felt herself being exposed to tensions from five different perspectives, cultural, political, technical, human resources and educational:

Culturally, the existing teachers were very much worried about the change from half-day schooling to whole-day schooling, ... Politically, the teachers were very disappointed by the SMC's decision to ask them to secure their post by re-submitting their applications to the 'new school'. Technically, nearly 80 per cent of the existing staff were computer illiterate. Human resources-wise, middle management had become a vacuum as a result of the merger and the retirement of the eight senior teachers. Educationally, the culture of professional development was not there.

(P3)

In order to cope with these difficulties she had to re-shape her entire approach to planning in her school. In School E2, is a response to equally difficult circumstances, the head compared his planning process to steering his boat down a river: 'It's the river and rocks and banks and stuff. The skill is actually in knowing what's coming and navigating a course' (EH2).

In School E3, the head also saw planning in nautical terms:

> We are on deck but from time to time the leader has to climb the rigging and just keep a view on what's happening whether it's outside the school, nationally, whether it's about politics ... then you are back down in the engine room because this sailing boat has got an engine and it's just helping to keep the wheels oiled.
>
> (EH3)

This is similar to what Mintzberg (1995) terms seeing through, looking ahead, behind, above and below, beside and beyond.

Strategic planning in these schools is difficult and demanding. The heads used a variety of metaphors to try to understand it and cope with its demands. They did not necessarily subscribe to the linear-rational models of strategic planning identified in Figure 6.1. The procedures are more complex. There is no doubt that, for all of them, their process was informed by a commitment to a strongly held set of personal and professional values and beliefs about their own role and about the nature of their educational enterprise (Bell 2004b).

To what extent do values and vision shape the planning process?

Heads paid significant attention to the place of values in shaping both their approach to strategic planning and the content of the plans themselves (Chan 2002). Some headteachers stated these quite explicitly:

> Education is about people ... Individuals are the key building blocks in this school. We care and value every individual who comes together to make a team ... Helping every individual to develop their potential is a never ending job in this school.
>
> (P1)

Developing individual potential was also identified as important by the head of School H3: 'We aim to build a learning organization that emphasizes on self-managing and self-learning. The main purpose of strategic planning is to enable all our students to be self-respecting, self-managing, self-discipline, self-care, and self-learning' (P3).

A third placed a similar emphasis on developing the whole child:

> The main purposes of strategic planning is to build on the strengths of this school to develop and to enhance the effectiveness of 'management and organization', 'teaching and learning', 'support for students and school ethos', and 'students' achievement'. We aim to provide whole

person development that embraces the six virtues of moral, intellectual, kinaesthetic, societal, aesthetic, and spiritual, for our pupils with due emphasis on our core value of 'Not to be served but to serve'.

(P2)

Here is an example of culturally distinct values shaping the planning in this school. In other schools the espoused values emerged as a consequence of debates and discussion:

How did we arrive at those values? After discussion and debate about what the school should be about, what is important and it was a whole school approach, all staff were party to that ... We send questionnaires to parents. Our parent governors talked to people ... We asked the children.

(EH2)

The values that informed the planning in School E4 were also the product of consultation and debate, but this time the whole staff, all children, all parents and governors and representatives of the local community were involved. As a result the head could say: 'We knew where we were when we set up the school ... We had a vision and an idea of what parents wanted from us ... We had our key values' (EH4).

The core values in other schools were arrived at in somewhat different ways. One head, for example, was aware that she had a vision based on core values but it was essentially her vision rather than a shared one:

I have a very strong vision for my school and maybe it's not communicated to all staff as well as it should be. My deputy knows what it is. My senior management, particularly one senior manager who has worked with me in two schools know exactly what I am after. I have been developing my vision and making the vision not mine but everybody's vision for the school ... The vision is now going out to children. We have asked the parents their views on it.

(EH1)

This experience is shared by many heads who have to reconcile their own strongly held professional values with the need to enable a variety of stakeholders to help to shape the values of a school and the vision that is derived from those values. This gives rise to the other major dilemma that faces many primary school heads: 'We are all different you cannot offer all things to all people ... so everything we do is in terms of valuing people' (EH3).

This reveals a paradox encountered by most of the headteachers in this study; how to reconcile the competing interests of different stakeholders in the school. Their commitment to educational and personal values provides one way forward while their capacity to exercise their authority to influence both the nature of the debates within their schools and the outcomes of those debates provides another possible solution.

Over what time period do heads believe it is feasible to plan?

It was clear that for most of these heads, strategic planning in primary schools was significantly different from most of the approaches identified in the literature reviewed above. In particular, few of the heads planned over what Fidler (2002) described as the long or medium term, that is over three to five years. School H1 was an exception to this: 'In view of the future development of this school ... I think it is time for us to have a longer term planning strategy which can guide the development of this school for the next ten years' (P1).

More typically, however, heads saw the planning horizon in much shorter terms. As the headteacher of School E2 put it:

> We started off very grandly, thinking about getting to grips with what the school development plan needed to be. We tended to make it too detailed. We tried to do a five-year plan but couldn't get to five years. It just became impossible. We are trying for a two-year one now.
>
> (EH2)

In School H3 the head adopted a slightly more flexible approach: 'I shall start with a one-year plan, followed by a three-year plan that serves to cope with the new ... educational initiatives' (P3).

In School E1:

> We have three sets of planning. We have long, medium and short. The long-term plan is the curriculum map which is an overview of all the year's work and it's a really good one.
>
> (EH1)

The curriculum map referred to here is intended to eliminate the overlap between curriculum content that previously occurred as children moved through the school and to provide a basis for children's continuity and progression through the national curriculum. It provided a basis for planning work on a termly and weekly basis.

In School E4, which at the time of interview was in its second year, the head planned over the short term: 'The very first plan was about nurturing. What did we want to grow? How we went about it? Who did we want to involve? The second year plan was layered on the first' (EH4). In some ways this is a special case because the school is new, but the time period described is similar to that in other schools. This is hardly long term in the sense that it is normally used in the literature and it is certainly not what

Davies and Ellison (1999) term futures thinking. Nevertheless, it is based on a real attempt to give systematic consideration to the future and it has much in common with the planning horizons adopted by the other heads. Most regarded a year as the longest period of time over which planning was feasible or desirable.

How far do audits of the internal strengths and weaknesses of schools shape strategic plans?

Almost all the heads in this study relied heavily on internal audits to help to identify and justify planning priorities and to facilitate the implementation of those plans. In School E1 the planning process was informed by a review of teaching methods and materials:

> We reviewed what we were doing, looked at good practice, all the things you would do to make decisions ... We knew we had to adapt because our standards were low ... Very shortly teachers were saying we can't operate like that. We started very quickly to adapt the materials we had and add the dimensions that we needed to it.
>
> (EH1)

In England, OfSTED inspections could play a similar role. For example, following the OfSTED inspection in School E1 the head recognized that action had to be taken so she implemented a process of:

> Looking at the children. Looking at where they come from and where they are meant to be going. We found they were standing still ... but it's not like that any more ... The idea is that all children who are potential level 5s get there and that our border level 4s become level 4s. Those who are not level 4 get a good grounding to continue their education ... This helps with our planning.
>
> (EH1)

In Hong Kong, school inspection is very infrequent and, therefore, plays a less significant part in developing the internal audit process. None of the Hong Kong schools in this study had ever been inspected by the Quality Assurance Inspection process. Heads here tend to rely on their own resources for this: 'I knew that the school would undergo what Cheung and Cheng called the three strategic stages: defrost, change and enforcement. The environmental analysis has actually accelerated the defrosting' (P3).

The environmental analysis here included a self-analysis of the principal and a questionnaire for the staff in order to produce a thorough internal audit of the school's present position and future development. Heads in both the Hong Kong and English primary schools conducted similar internal audits that shaped both the short- and longer-term plans for their schools.

How far do factors in the external environment play a part in shaping strategic plans?

The educational reforms that have been introduced in Hong Kong, particularly the School Management Initiative with its emphasis on curriculum development and school-based management, have caused at least some headteachers to use external initiatives to drive internal improvement (Chan 2002). The head of School H2, for example, used two approaches, the Purpose, Process and Product model and the Walking by Two-Leg approach in conjunction with university colleagues. In so doing, he illustrates the extent to which heads are subject to the tensions generated by the need for change and the requirement for stability:

> What the '3P' reminds us is to begin any action with the end in mind. It can be either used in planning and curriculum development, or even in daily routine work. Whenever we discuss issues in respect of strategic direction, teaching and learning, staff development or resources management, we will critically reflect on whether the process and the expected outcome are in line with the original intention or purpose, and vice versa. The 'dual strategy' serves to strike a balance between maintaining the basic competency and aiming at the development of creativity and other higher order thinking skills among our students, if it is used in curriculum planning and development. Our dual strategy is to emphasise the core competencies through traditional method on one hand, and developing the creativity through dissolving the stereotype or traditional thinking on the other.
>
> (P2)

As was seen above, each English school should have a set of specific targets derived from national priorities and benchmarks and agreed with the LEA. When asked how far externally set targets influenced the school's planning process, the head of School E1 remarked that the targets set in conjunction with the LEA inspector were realistic and largely matched the results of the school's own analysis. Not all heads took such a sanguine view of external targets. The head of School E3 dismissed them as having no value, especially for a school in which a significant proportion of the children have profound learning difficulties. He recognized that there was a danger that such targets: 'Can become rigid and ... seen as an opportunity to narrow that range of attainment that these youngsters are performing at in a uniform system of target setting' (EH3).

The reaction of heads to another possibly significant set of external influences on their school's strategic planning, namely government policy documents, was equally mixed. In School E1 the head used *Excellence and Enjoyment* (DfES 2003) to shape what happens in the school:

> I sent away for enough copies for all my staff and governors ... My literacy co-ordinator went on a course last week. She was the only one [on the course] to have been given it to read and discuss ... I am afraid I don't agree with everything in it and nor does my deputy because it's only if you can get the money that all those things in that book work.
>
> (EH1)

Where they were influential, these government documents were almost as influential as parents on the planning process. Headteachers in both England and Hong Kong were very conscious of the importance of involving parents. For instance, P2 was very proud of having almost 40 parents who act as helpers in his school every day.

In School E1 the parents were consulted about the content of the strategic plan while in School E4 the parents were involved in drawing up the plan from its inception. Largely, however, consultation with parents was restricted to asking for comment on decisions already made. Few heads made any systematic attempt to identify parental opinion more widely or in a rigorous way. One head reported that he closely monitored parental satisfaction: 'All teachers are required to tell me of any concerns or any changes including new babies. I monitor complaints and issues parents bring to me' (EH3).

He seems to be encouraging his staff to work within a customer-provider relationship rather than one of partnership. This head is trying to keep informed about the potential demand for places in his school and the customer-satisfaction of parents but no more than that. Scanning the external environment is not a significant part of the strategic planning process for most heads although they do make use of government policy documents. Earlier studies of the extent to which headteachers monitored the parental aspect of their school's external environment found headteachers taking a similar stance. Even the most outward looking heads did not establish rigorous and systematic procedures for scanning their environment or collecting information about their market place (Bell 1999b; James and Phillips 1995).

What is the relationship between strategic planning and the organizational structure and culture of schools?

The values on which strategic planning was based in each of these schools may not always have been articulated clearly and shared widely, but there is no doubt that all the heads in this study wanted to involve their entire staff in the planning process and to empower them to deliver the agreed outcomes both through utilizing the expertise of the staff and facilitating their further professional development. All three schools in Hong Kong focused on school-based curriculum development (SBCD) in subjects and topic work within a cross-curricular framework. Collaborative cultures were established and enhanced by co-operative teaching, peer lesson observation, action research and creating lesson preparation time for teachers in each year level.

In School H1, in-service days are held once a month to discuss the SBCD, seminars are conducted about new issues in curriculum development and pedagogy and intensive training provided for middle management focused on the knowledge and skills of the Subject Panel and enabling them to work collaboratively with teachers to formulate the strategic intent for their subject areas. As a result, senior staff and subject co-ordinators have become powerful allies of the headteachers and form part of a set of corporate alliances in negotiating over school policies (Chan 2002). This, however, is usually based on the headteacher's own perception of the school's culture. As the head of School H3 puts it: 'My vision is to build a self-managing and self-learning culture that incorporated a cyclical process of environmental analysis, planning, action and progress assessment' (P3).

This focus on dispersed leadership finds expression in the way in which all the heads talked about the culture of the school, differentiating in quite specific ways from the school's management structure. This was particularly true in the English schools emerging from special measures:

> I am not a teaching head … mainly because I had so many issues throughout the school … We have a senior management team and that is made up from year leaders and … my SENCO.
>
> (EH1)

This head then commented that her deputy had worked with all subject leaders: 'Empowering them to understand their areas and to take responsibility … teachers are now working together and not working in an isolated way' (EH1).

Obviously, this shift from isolation and an inability to take responsibility to collaboration and empowerment is an important cultural change both for this head and for the school and those who work in it. This head also noted the importance of school culture: 'It's about providing a safe environment … I see myself providing support for the people that I employ to do a job … It's about creating the right atmosphere' (EH1).

This emphasis on atmosphere and the role of the head in creating an appropriate atmosphere in which strategic planning could be attempted in a collaborative way was an important theme for both the English and Hong Kong heads in this study. They were at pains to differentiate between management structures and the cultural aspects of their school's organization. Typical of their comments was that made by the head of E3:

> Management to me is about having systems in place … You can have the greatest systems in the world, a beautifully managed school where nothing happens … What I try to do in my school is to create an

atmosphere and ethos where people are valued, where there are high expectations.

(EH3)

One head noted the difficulties encountered in trying to create a collaborative culture in a new school where staff have been recruited from a range of different contexts:

> They are used to being in a school where somebody else has done the policy and you ... follow it. This idea that we have to share our ideas and share our thinking and make a commitment ... has been quite challenging for people.
>
> (EH4)

Nevertheless, this head had a clear view of the culture that she wanted to establish in the school. She wanted it to be based on learning communities:

> Learning communities in the classroom, learning communities among teachers, learning communities in the broadest sense ... based on reflectiveness, resourcefulness, repositioning, accepting challenge, looking at opportunities, problem solving, creativity. Exactly what I want for the children as well.
>
> (EH4)

The Hong Kong heads had similar aspirations but move forward in somewhat different ways. They used opportunities for professional development with their teachers to facilitate such cultural changes. This was possible because the teaching loads of Hong Kong primary school teachers are significantly lighter than those of teachers in English primary schools. In both the English and the Hong Kong schools, although the value of structures was acknowledged, the organizational cultures of these schools were seen as particularly important in enabling the planning process.

What barriers to strategic planning do heads identify?

The difficulties encountered by most of the heads in both countries clustered into three main areas, namely resources, staffing and their inability to predict the future. Several heads commented on the impact that budget uncertainties had on their capacity to plan over a period longer than a year. Typical of their comments was this: 'If I had a budget over a longer period we could plan more strategically. Not knowing what resources you have got makes planning difficult' (EH1).

It is not just the knowledge about future funding that is of concern. The level of funding is also a problem: 'Primary education is not funded as well as secondary ... if I had secondary funding I could create a different school' (P3).

In England, EH3 echoed this view and, like many of his colleagues, tried to overcome their budgetary problems by bidding for external funding. This is becoming increasingly important but it brings its own difficulties. The uncertainty of the process makes strategic planning difficult (E2) and an unsuccessful bid can mean: 'We were going to do things all in one go; instead we are going to have to do it in little bits ... The time scale has changed' (EH1).

If anything, staffing issues cause even more difficulties for heads than budgets. One head described his staffing issues over a year as: 'a roller coaster ride ... I cannot get a Year 6 teacher ... I am also short of a literacy co-ordinator' (EH2).

The same head noted that sometimes staff changes can be positive but they often create difficulties for strategic planning. This view was echoed in the views of the head of School E3 who remarked that he had lost 13 staff over the past few years, many of whom had left on maternity leave and not returned:

> We have to be continually moving on and evolving and dealing with staffing difficulties ... Who is going to know what circumstances arise ... what you do will depend on who is going to come in and this affects your major decisions.
>
> (EH3)

In Hong Kong, similar difficulties of finding suitable staff emerged: 'Most of the existing teachers did not have a degree ... most of them were approaching their retirement age' (P2).

The difficulties of predicting an uncertain future play a major part in shaping the approaches of these heads to strategic planning: 'Because circumstances change in schools ... you can have a view about what might be happening next June but the nonsense is in expecting it to happen' (EH3).

Some heads perceived their schools to be subject to significant and uncontrollable external forces:

> Even now we cannot predict what external forces are going to be pulling you in this way or that. We can't say we are definitely going there or doing that. This does not help us to be strategic.
>
> (EH2)

The head of School H2 made a similar point: 'We felt frustrated that whenever the government implemented a new initiative, they tended to throw away all the old stuff so as to convince others that the new initiative is good' (P2).

This leads many heads to adopt a flexible, perhaps incremental approach to the planning processes in their schools:

In terms of how you can predict the future and the areas that are going to be of interest, I don't think you can … As a result the greatest strength we have developed is the flexibility to deal with different situations.

(EH3)

Small wonder, therefore, that one head described her attempts at strategic planning as: 'Being on the edge of chaos' (EH4). The world of the primary school is complex and, in many ways, uncertain. Predicting the future, responding to changes emanating from the external environment and coping with internal changes is extremely difficult. Such relatively small organizations do not have the capacity to respond easily to changing circumstances.

Conclusion

There are some interesting similarities between the socio-political environments from which educational policy is derived in Hong Kong and England. In both countries, major changes in the political context took place in 1997 bringing a new government to power in England and introducing a new form of government to Hong Kong. These changes produced a shift in the dominant discourse that shaped educational policy. In Hong Kong there was a move away from the former colonial-based forms of legitimation towards a discourse that was more appropriate to Hong Kong's new status as a Special Administrative Region, including the introduction of Chinese as a medium of instruction so that schools could prepare students to play a greater role in the wider Chinese society. In England, the discourse of the market economy became subservient to a discourse based on linking education more closely to producing a skilled labour force, while, at the same time, fostering a concept of good citizenship. Thus, in their different forms, the themes of economic utility and citizenship provide a context and set the strategic direction of education policy in both countries. The organizational principles of these policies are based, in different ways, on accountability, which provides much of the text for policy implementation. In England, school inspection and the publication of results within the context of authority devolved to schools. In Hong Kong, autonomy has also increased at school level but here accountability operates more directly through the mechanisms of parental and community pressure. It is at the school level, however, where the impact of education policies can most clearly be seen. It is here that the operational practices are applied and, in some ways, mediated by heads as they seek to reconcile the competing demands of policy and practice.

Heads in this study recognized both the dominant discourses that set the context for the education policies that they were to implement and understood the strategic direction of such policies. Many of them, however, either rejected all or some of the organizing principles and concomitant organizational practices or sought to re-define them in the light of their consequences for particular schools. This is reflected in Figure 6.1 where

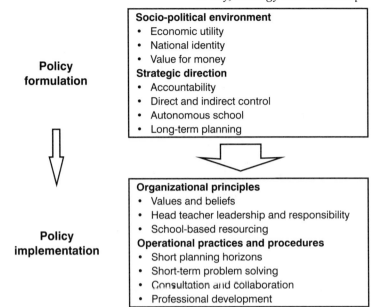

Policy
formulation

Socio-political environment
- Economic utility
- National identity
- Value for money

Strategic direction
- Accountability
- Direct and indirect control
- Autonomous school
- Long-term planning

Policy
implementation

Organizational principles
- Values and beliefs
- Head teacher leadership and responsibility
- School-based resourcing

Operational practices and procedures
- Short planning horizons
- Short-term problem solving
- Consultation and collaboration
- Professional development

Figure 6.1   Policy into practice: strategic planning

there is a clear disjunction between the strategic direction and operational practices and procedures adopted by most of the heads in this study. This disjunction can be explained by the extent to which heads re-interpret policy in the light of their own values, beliefs and understandings. These heads deploy a range of complex metaphors in order to understand the nature of planning and to cope with the demands that it places upon them and their staff. They only plan strategically, however, insofar as they seek to establish relatively limited but coherent and agreed plans for their schools. They often act as 'subversives' rather than passive implementers of policy (Day *et al.* 2000). They want to ensure that: 'Everybody is singing from the same hymn sheet' (EH1). As Lumby (2002) argues, strategic planning is, at least in part, a bureaucratic process, especially when it is required by national policy as it is in both England and Hong Kong.

Nevertheless, as one of the heads notes, primary schools are not bureaucracies, they tend to be loosely coupled systems. The degree of autonomy enjoyed by the staff of these primary schools, the complexity of both the teaching and learning technology and the relative unpredictability of the external environment: 'renders problematic any simple translation of intention or instruction into planned outcomes ... It is particularly hard to relate specific management activity to improvements in teaching and learning' (Lumby 2002: 95).

Thus, although the strategic direction and organizational principles of policy are recognized, the operational practices and procedures are mediated by heads in the light of their own perception of the socio-political environment. Perhaps for this reason heads tend to minimize the importance of the structure of their schools, and place greater emphasis on their collegial and collaborative cultures. Perhaps this tendency was stronger in the larger schools in Hong Kong where staff were expected to take part in professional development as part of their working day, rather than in twilight hours as in England. Such findings are not limited to the primary school context.

Cowham (1994) identified a move from a politico-bureaucratic organizational structure to a culture based on collegiality in a large further education college in England as it struggled to meet the Further Education Funding Council's planning requirements and Weeks (1994) sees a similar trend in secondary schools while noting that headteachers and their senior management teams must ensure that: 'The culture of the school [is] able to accommodate ... change ... and be ... committed to the idea of encouraging initiative and embracing failure' (Weeks 1994: 262).

Espoused values play an important function in determining the content and nature of the plans produced in these primary schools. They reflect the dominant values espoused by the schools and the philosophy that shapes their approaches to teaching, learning and professional relationships (Hopkins *et al.* 1996). As Hargreaves (1995) suggests, cultures have a reality-defining function that enables staff in schools to make sense of their actions and their situation. In this study, heads and their staff deliberately re-shaped their cultures by facilitating planning based on a recognition of the importance of developing collaborative procedures in which people are valued for what they can contribute and in which ideas can be shared, innovations tried out and where blame is not apportioned. The focus is on coping with short-term or immediate problems but the process is informed by strongly held values:

> You don't do it because somebody else has told you to or has said that it is a good idea. You have to do it because you believe in a school ... It is all about honesty and integrity.
>
> (EH4)

These values tend to provide both a foundation on which planning can be based and a rationale for decisions that are made and priorities that are identified. This depends on establishing and sustaining a culture of inquiry, a commitment to collaboration and to continued professional development (Southworth 1998). In these schools, therefore, while the need to establish a strategic direction is accepted, the organizational principles and operational practices are re-defined to reflect the contexts of the schools themselves. This certainly applies to the planning horizons adopted.

If, as Fidler (2002) suggests, strategic planning is a long-term process that might involve planning horizons of five years or even longer, then the plans that are formulated by the heads in these primary schools are not strategic. The time scale over which heads in this study plan is very short term and even here their approach tends to be based on a degree of flexibility. Thus, the organizational principles and operational practices are mediated by context. The plans themselves are based on a series of internal audits, both self-initiated and driven by external processes. These focus on both pupil and teacher performance. There is, however, no evidence of the careful analysis of external factors that Puffitt *et al.* (1992) argue should form the basis of such strategic plans. The wider policy context of education and beyond is of marginal influence, especially in England. In both Hong Kong and England, heads tend to adopt a reactive and responsive stance rather than a proactive, anticipatory one to matters that emanate from both the national and local environment. Even some elements of the internal environment, budgets and staffing for example, are treated in this way over relatively short periods of time. The organizational principles emanating from the socio-political environment are, therefore, either being rejected or mediated in these schools.

To the extent that these planning processes are based on an attempt to identify priorities that will be achieved over time and that resources are mobilized to achieve those priorities, the planning process can be considered to be strategic. With one exception, planning in most of these primary schools tends to consist of rather small and marginal changes made and evaluated over a very limited planning horizon. The plans in these schools do contain specific school development and improvement objectives, relate to agreed targets and focus on children's learning. Strategic planning in these schools, however, is not target driven in the sense that targets are interpreted rigidly, exclude the interests and skills of teachers and learners and prevent a true educational experience from taking place (Bottery 1992).

So short is the time span and so marginal many of the changes that are achieved by planning in these schools that the model of school improvement planning adopted approximates more closely to what Van der Heijden and Eden (1998) call the evolutionary/incremental approach than it does to the more structured, longer-term, linear-rational model of strategic planning that is envisaged in the expectations of the DfES and the Education Commission. Evolutionary/incremental planning is based on the view that the large number of variables with which schools have to cope make it too complex to identify and implement a specific set of actions and it is more effective to make and evaluate a number of small changes over a shorter period of time. This contrasts with the linear rational approach which assumes that schools have the capacity to achieve organizational goals through a rational process that begins with analysis and proceeds in a linear way to implementation over a three- to five-year time span. This policy text implies that planning and implementation are orderly and sequential and that schools can be shaped and controlled in order to avoid the unintended consequences of change while realizing strategic objectives. It also assumes a measure of control over both the internal and external environment to which primary schools cannot aspire. The inability of heads in these schools to control or even predict their available resources with any degree of certainty of a period of more than a year has almost inevitably created a situation in which their planning is short term and highly flexible. To that extent, planning in these primary schools is not strategic but incremental.

It is possible that headteachers could develop a form of planning based on a longer-term time horizon. The evidence here, however, is that they do not do so either out of choice or because the perceived barriers to such planning are too great to be surmounted or to be worth attempting to overcome. Instead, school structures and, more particularly, their cultures are re-shaped to facilitate this form of evolutionary incremental planning. What emerges here is a flexible and collaborative planning process that is more suited to primary school context than is the linear-rational strategic planning which is implied in the text of much educational policy. In implementing the plans the heads and their staff do seek to reconcile the need to achieve agreed objectives with available resources but the planning procedure seems to be less of a process of matching resource capabilities to priorities than a matter of shaping priorities to identify what can and should be achieved within available or uncertain resource parameters. American high school principals report a similar lack of capacity and resources:

- They ought to do long-range thinking but frequently they had to make short-range, even instantaneous decisions.
- They ought to be innovators; but they were maintaining the *status quo*.
- They ought to be champions of ideas; in reality they were masters of the concrete, paying attention to detail.

(after Leiberman and Miller 1999: 40)

To some extent, the picture of planning that emerges here is one of:

Coping with turbulence through a direct, intuitive understanding of what is happening in an effort to guide the work of the school. A turbulent environment cannot be tamed by rational analysis alone ... Yet it does not follow that the school's response must be left to a random distribution of lone individuals acting opportunistically ... Strategic intention relies on ... vision ... to give it unity and coherence.

(adapted from Boisot 1995: 36, quoted in Caldwell and Spinks 1998)

Much of what the headteachers in this study do is very short term and often does not rest on a strategic direction or even a strategic intent. Instead, these heads tend to see planning as either interim problem solving or a basis from which to cope with the unexpected, unpredicted and unpredictable. In the case of both Hong Kong and England, such an emphasis was evident from the start of their tenure of office as headteachers. They focus on the internal school environment and demonstrate a relatively low understanding of or concern with wider external issues. They believe that they have almost no capacity to control or influence that environment. These heads do recognize, however, that strategic plans go awry if the future turns out differently from the way that planners expect. Indeed, heads recognize that many of the problems associated with strategic planning in their schools: 'are not differences in goals or even personalities, but differences in assumptions ... about probable futures' (Hampden-Turner 2003: 119).

The heads in the schools represented here, therefore, do not, on the whole, believe that careful planning gives them control over the future, their schools and the people within them or over the environment in which their schools operate. At best, therefore, planning in these schools consists of: 'Incremental adjustments to environmental states that cannot be discerned or anticipated through a prior analysis of data' (Boisot 1995: 34).

The complexity of the issues facing these schools is reflected in the nature of the barriers to strategic planning identified by the headteachers. These barriers appear to be a product of the size of the schools, the nature of their resourcing and the perceptions held by the headteachers of the external environment. The environment is seen as unpredictable and uncertain in a number of ways, not least of which are the changes in educational policy that seem to occur with singular regularity. This is regarded by some heads as a direct result of the discourses in the socio-political environment that continually produces changes in the text of education policy. The consequences of this and of the organizational principles are not always recognized by policy makers. For example, in England the schools are resourced over a very limited time period, typically one year only. Their budgets are small and most schools do not have the resource flexibility to build a contingency fund sufficient to be confident of funding plans in a second year. Consequently, heads are reluctant to commit their schools to strategic plans that span a longer period. This is partly because staffing losses or changes in such relatively small organizations can have a significant impact on the school's capacity to develop in specific areas. In Hong Kong, schools are relatively well staffed but other resources are often not available. Teachers are not always well trained and may not even have suitable qualifications for the posts that they hold. These factors combine to make long-term planning extremely difficult. Therefore, on the evidence of the data collected from these schools it is apparent that primary headteachers and school principals and their staff are not attempting to espouse a model of planning that even approximates to long-term strategic planning. Instead, they are re-defining the nature of planning to meet the needs of their own circumstances. This re-definition is an example of the importance of values and the extent to which power is used to translate values into operational practices. It also illustrates the contested nature of planning both at school level and between the school and the state, a theme that is examined further in Chapter 7.

an important policy change in this key area. Although New Labour remained committed to parental choice, and it appeared ambivalent at best about the role of LEAs, it nevertheless identified LEAs as having an important role in supporting schools facing challenging contexts (DfEE 1997). However, a feature of New Labour's claim to radicalism was that the centrepiece of its attack on educational underachievement in disadvantaged communities was to take the form of local bodies established effectively *outside* the traditional LEA framework.

EAZs were established in areas of social disadvantage with the express aim of raising levels of student attainment in these areas, and combating the high levels of social exclusion. However, the socio-political discourse within which EAZ policy developed is complex and it is important to identify a number of policy themes, not all of which sit comfortably together. The strategic direction of the policy was driven by a desire to tackle specific problems in areas of social disadvantage, and involved a redistribution of resources to these areas to achieve this. Some element of distributive justice was clearly central to the EAZ project (see Chapter 4); indeed Dickson and Power (2001) have argued that EAZs represented one of only a few explicitly redistributive New Labour educational policies. However, the underlying approach to social justice was conceived in a number of specific ways. For example, tackling social exclusion involved tackling young people's exclusion from the labour market – widely regarded as one of the key organizational principles of EAZs. However, in an economy with low unemployment this was not presented as an economic problem (lack of appropriate jobs), but as an individual's problem (lack of appropriate skills or, more significantly, attitudes). One of the strategic directions of policy that emerged, therefore, was to increase the 'relevance' of the curriculum through a vocationalization of the curriculum. This not only showed a clear link between the social justice agenda and the development of human capital, but also the way in which problems of social (in)justice were capable of being conceived and legitimated in terms of 'social problems'. Hence the response to social exclusion, with all its associated social problems, was to be largely addressed through improving 'employability'. Social and economic objectives were not presented as a polarization between either/or options, but as two sides of the same coin. Tensions between first order values based on economic and social imperatives were reconciled as complementary rather than contradictory.

Developing the strategic direction of policy

Although the objectives of Zone policy were largely the traditional concerns of the social democratic left, the intended radicalism of EAZs was to lie not so much in their objectives, as in the processes by which these objectives were to be achieved. Hence as the strategic direction of policy emerged from within the socio-political environment the novel nature of EAZs became

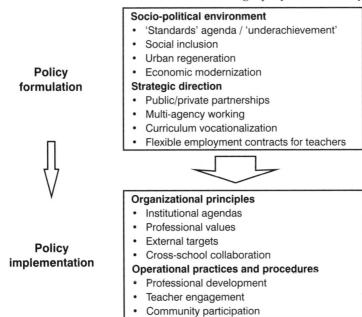

Figure 7.1 Policy into practice: the EAZ experience

more apparent. Dickson and Power (2001) have identified a number of characteristics of EAZs that distinguish them as quintessential examples of 'third way' thinking – a commitment to multi-agency working, a blurring of the public-private divide and a desire to engage local people in more participative forms of educational decision making (Dickson *et al.* 2001). To this list might be added a fourth, namely a commitment to use EAZs as laboratories for the promotion of greater service diversity, particularly in terms of curricular provision. Figure 7.1 illustrates how a strategic direction emerged from the 'third way' discourses that shaped EAZ policy and that subsequently developed organizational principles and practices that are discussed later in this chapter.

A number of specific features of EAZs were intended to support these new ways of working. For example, EAZs not only needed to enlist the support of the business community to raise funds, but were also expected to ensure that representatives of local business occupied key positions on the body established to exercise strategic and operational control over Zones – the Action Forum. Each Action Forum included a range of stakeholder representatives, with an initial expectation that the governing bodies of Zone schools would cede some of their powers to the superordinate body. LEAs were to be represented on the Action Forum, but their presence was certainly not intended to dominate, quite the opposite. This was in keeping with New Labour's commitment to avoid traditional statist solutions, and instead to look towards new forms of governance.

It was anticipated that EAZs would contribute to the promotion of diversity within the state system in two important respects. First, schools within the Zones were afforded the opportunity of flexing elements of the national curriculum, effectively allowing some students to opt-out (or be opted-out) from the curriculum that elsewhere was a statutory requirement. Here the intention was the promotion of a more vocationally orientated curriculum for those students apparently disengaged from schooling. Secondly, Zones were given powers to vary teachers' national pay and conditions, for example allowing them to vary the requirement that 'directed time' (the time a teacher can be directed in their activities by the headteacher) must not exceed 1265 hours over 195 days in any one year. This requirement was seen as an obstacle to generating the sort of flexible working required to deliver key Zone initiatives (after-school activities and summer schools for example).

EAZs were therefore central to New Labour education policy. Not only were they intended to spearhead the attack on underachievement in areas of social disadvantage, but they were also intended to act as a role-model in the distinctive third way thinking that would ensure that New Labour was seen as a modernizing, radical and reforming administration. EAZs were intended to put the 'new' into New Labour, and in so doing demonstrate that the government could tackle 'Old Labour' concerns in novel and innovative ways. However, by March 2000, despite their flagship status, it was clear that political interest in EAZs was waning. The Minister for School Standards berated Zones for a lack of progress in raising attainment (Mansell 2000), whilst Inspectors indicated that Zones had failed to be sufficiently innovative (Mansell and Thornton 2001). At this point the flagship began to show signs of sinking, and as indicated, within five years of their inception EAZs no longer existed with their original name, or in the form originally envisaged. Does this mean that EAZs must be viewed as a policy failure – another frustrated attempt to lever up educational standards in communities where educational achievement has been hugely constrained by social disadvantage? Or is it possible to identify a more positive legacy in which the impact of the Zones continues to shape the development of public policy in this area?

Developing EAZ policy – from formulation to implementation

From the outset EAZs experienced difficulties as the broad conception of policy presented in policy texts began to be implemented at a local level. Within governing party circles the strategic direction of policy was becoming clearer but the multiple interpretations of policy texts provided early evidence of policy refraction. Moreover, as these ideas appeared to become more diffuse, and as the policy moved towards local implementation, this lack of clarity emerged in the form of several policy tensions, subsequently identified by the government's own inspectorate as obstacles to policy implementation. It was noted that Zones:

> ... got off to a slow start and showed limited progress initially, generally because they had plans that were too ambitious, too diffuse, or insufficiently focused on the difficulties faced by schools in their area.
>
> (OfSTED 2003: 14)

The issues raised by this observation point to a number of policy tensions, and these were often evidenced in the case-study EAZ. For example, a clear tension existed within the organizational principles and the text of the policy, between, for example, the role of the EAZs in raising 'standards' in the form of public examination results and the commitment to promoting social inclusion. Schools indicated they were under considerable pressure to improve pupil attainment in national tests but that resources allocated for this purpose might be at the expense of initiatives designed to encourage social inclusion. The need to deliver results highlighted a further tension – that between short-term success and long-term gain. Schools, and the EAZ itself, were under considerable pressure to deliver results quickly, and to demonstrate progress against measurable targets. Political imperatives demanded that investment would secure a demonstrable, and quick, return. Yet teachers interviewed in the case-study Zone argued they were seeking to reverse the effects of decades of industrial decline and deprivation. Expectations of rapid improvements were considered unrealistic and unreasonable. Moreover, they were counter-productive as they sometimes deflected resources from the longer-term projects that interview respondents argued often had the best chance of achieving sustained improvement.

This latter point highlights a further tension between national and local policy agendas. This tension in part derived from the government's desire to articulate its policy messages to a range of different audiences. Hence the presentation of EAZ policy to the media often highlighted its most novel, and hence headline grabbing, aspirations – curriculum restructuring, private sector involvement and radical changes to the school day were all trailed. However, schools were less interested in a novelty agenda they often saw as irrelevant, but were more focused on the operational practices required to progress the initiatives to which they were committed, but which they had lacked the resources to deliver. Within the case-study Zone this manifested itself in a sense of disconnection between the Zone and schools – the initial objectives of the Zone to demonstrate impact through 'Big Ideas' did not fit with schools' priorities, and what headteachers and teachers identified as the necessary action to make progress in difficult circumstances. These problems were compounded by the creation of organizational structures in the Zone that appeared to exclude teachers from a dialogue about the development of policy (Price Waterhouse Coopers undated) – a policy outcome perceived by some to be deliberate (Jones and Bird 2000). Theakston *et al.* (2001) argue that government rhetoric pitched at the public through the media served only to raise teacher anxieties, which in turn impacted negatively on how the Zones developed. As such, it highlights the need to see the development of a policy as the development of a plurality of policies (see Chapter 2) in which ostensibly the same policy may be repackaged for different audiences and purposes. It also explains why the development of policy presented in Figure 7.1 appears to lack coherence.

These policy tensions resulted in considerable resistance to the EAZ policy at both national and local level in some quarters. Ball (1987) identifies three circumstances when change is likely to be resisted: when it threatens vested interests relating to working conditions; when it undermines professional status and self-perception; and when it presages changes in working practices on ideological grounds. EAZs could be said to provoke resistance on all of these counts. For example, in the case-study Zone, fears that the school day would be lengthened, and holidays shortened, challenged teachers' vested interests as they saw a potential worsening of conditions of service. Teachers' professional status and self-perception was undermined by the view expressed by some teachers in the case-study Zone that being part of an EAZ would label their school as 'failing'. Finally, significant ideological interests were challenged. Trade unions and political organizations, in particular, articulated a number of concerns relating to the increased role of the private sector, and a concern that 'restructuring the curriculum' would result in a more vocationalized curriculum in working-class communities, hence creating an academic/vocational divide based on social class lines (STA, 1998). This congruence of oppositional interests often created a difficult environment in which EAZs might develop as local decisions about whether or not to seek Zone status were subject to negotiations between coalitions of interest groups (Bacharach and Lawler 1980).

This latter point highlights the importance of local context. This is further illustrated by an analysis of this case-study Zone. The case-study Zone was centred in a Midlands town that had grown around the development of the steel industry, but which 20 years later was still experiencing the social fallout of rapid de-industrialization in the 1980s. By any measure of social deprivation this was an impoverished community and teachers in the five secondary schools and 21 feeder primary schools in the town faced a range of complex social problems. The research presented in this chapter is based on interviews with a range of staff conducted in all the EAZ secondary schools and seven of the primary schools. A specific feature of this locality was the highly fragmented and hierarchical nature of secondary education in the town. The town featured virtually every school type that had emerged over the previous years following government policies to create 'diversity' and tackle 'failing schools'. The town's school profile was dominated by a City Technology College, a type of school that had been established after the passing of the 1988 Education Reform Act (DES 1988). These schools function independently of the Local Education Authority, receiving direct grant funding from central government plus private sector support. Some of the other schools in the area had embraced the autonomy agenda of the 1980s and had sought one of the various categories of 'specialist' status that these reforms provided. Favourable funding, and a favourable image, had resulted in favourable market conditions for all these schools whose status as 'over-subscribed' allowed them to use an element of student selection to reinforce their privileged market position. At the other end of the spectrum, two secondary schools faced substantial problems with empty places and a disproportionate number of students with special educational needs. The consequence of this was a sharply polarized local market in which, for example, a school identified as amongst the best performing state schools in the country was located no more than 2 kilometres from a school identified as amongst the worst performing state schools in the country. The result was a significant 'Balkanization' of schools (Hargreaves 1995), with intense competition, and little co-operation, between secondary schools in the town ('The secondary heads hadn't talked to each other for years' was how one primary headteacher described the situation). This scenario not only generated huge inequalities in apparent performance between secondary schools, but made it difficult for local primary schools to manage efficient transition arrangements to secondary school.

Policy implementation – from organizational principles to operational practice

At the same time as government ministers appeared to be losing interest in the EAZ experiment, there was a change of leadership in the case-study Zone and a new Director was appointed. The change of Zone direction that followed this appointment, coinciding with a shifting emphasis in national policy, proved to be significant. As policy makers appeared to lower their expectations of Zones' capacity to innovate, the focus in the case-study Zone shifted towards new priorities. The Zone itself re-organized, placing schools (represented primarily, but not exclusively through their headteachers) at the centre of its own policy-making processes. A Headteachers' Planning Group was established and from this six project groups were formed, based on priorities agreed across all the participating schools. The immediate outcome of this was to increase significantly teacher engagement with the Zone and its processes of policy development. The focus of these groups reflected largely mainstream priorities such as literacy and numeracy and illustrates the extent to which national policy discourses had become internalized by those working in schools. However, this also ensured there was now an alignment between Zone priorities and school priorities, and as a consequence attitudes towards the Zone began to shift. Research conducted in the case-study Zone following this re-organization found that teachers spoke much more positively about the impact of the EAZ.

Zone resources were focused on the secondary school sector as this was identified as the area of greatest need. Initially there was some evidence that resources within the Zone had been distributed between schools inequitably. Individual schools had sought to maximize their share of EAZ resources and some schools were apparently more effective at this than others. The result was often that resources went where they were needed least. As a more collaborative structure was created not only did it result in increased transparency regarding resource allocation, but it also developed an understanding between schools about the basis of that allocation. Whilst resource decisions were now largely taken by the Headteachers' Planning Group, these still very much reflected national targets and priorities. Within the secondary sector there was little dissent for the pursuit of these objectives. Secondary schools are well used to trying to maximize examination performance, especially within the context of a now well-established league table culture and the shift in the Zone's style seemed to 'fit' with what these schools were already trying to do. However, at primary level there was evidence of a greater dissonance between Zone initiatives, the 'standards agenda' and the individual priorities of teachers. One project designed to provide 'bridging materials' to help students make the transition between primary and secondary school, and to be commenced in the period immediately after students had completed their national tests (SATs) was received thus:

> Rightly or wrongly students and teachers feel that after SATs they want to do all the things that have been put on hold whilst preparing for the SATs – the fieldwork, the topic work etc. After SATs there is a feeling of relief and when these [bridging materials] came in my first reaction was 'hang on'. This is when we do lots of things, like geographical work, which goes by the board in Year 6. It's alright saying the curriculum should be 'broad and balanced', but life ain't like that in Year 6 – not if you want the results. Why can't we do some fun Maths? Let them enjoy their time. Do we really want to be constantly 'challenging'? Everyone has been busting a gut up to SATs.
>
> (Teacher – primary school)

Despite the clear focus within the secondary schools on raising student achievement in national examinations, it is important to note that during the period of the research student performance at aged 16+ did not rise appreciably. This again echoes wider EAZ evaluation data that suggests that academic attainment, especially in the secondary sector, showed limited improvement in the Zones (OfSTED 2003). Perhaps the key issue was the significant difference between individual schools highlighted earlier and the difficulty in closing the performance gap between the highest and lowest performing school. The higher performing schools in the town seemed able to consolidate or improve their performance, whilst those schools with the lowest results either failed to improve, or indeed declined further, despite receiving additional resources through the EAZ. This suggests that Zone support was insufficient to overcome some of the wider structural disadvantages these schools faced in the local market. In particular, it points to the additional difficulties experienced in raising standards of academic performance in schools where intakes are skewed towards high levels of both special educational needs and poverty (Edwards and Tomlinson 2002).

Whilst it was possible to identify a clear focus for those initiatives clustered around raising 'standards', it was more difficult to establish a similar attitude and approach towards the promotion of social inclusion. Schools seemed to have internalized the discourse relating to 'standards' much more clearly. Despite widespread disagreements about the efficacy of national testing the presentation of test and league table data is clear. There is no such clarity about the concept of social inclusion (Lunt and Norwich 1999). This may simply reflect the fact that social inclusion is a more slippery concept, more difficult to conceptualize and, consequently, to operationalize. It may also reflect the priority of social justice issues relative to standards objectives, with the latter accorded more status and a higher priority than the former. As a consequence, social inclusion was often conceived of in very narrow terms – reducing truancy rates, or the numbers of pupils excluded from school. This was understandable given the pressure on Zones to demonstrate impact by achieving progress against measurable targets. In this sense social inclusion objectives were no less driven by performance targets than those based on academic attainment (see Figure 7.1). In both cases, educational aspirations were presented as, or indeed reduced to, the meeting of externally generated performance targets. However, within some schools the inclusion agenda was viewed more widely. Whilst resources were often focused on high-profile issues, for example small groups of students at high risk of permanent exclusion, there was evidence that many teachers viewed any initiative as inclusive if it engaged all students, and especially those often marginalized within the school system. These differences in operational practices generated tensions within schools as teachers sought to reconcile the desire to be inclusive with the target of raising examination performance. Progress on these issues, and the improvement in teachers' attitudes towards the EAZ, followed the Zone's re-organization so that its policy agenda was driven by those working in schools, rather than the Zone's Action Forum. Broadly, influence on the policy process had been inverted at a local level – flowing from schools upwards, rather than from the Forum downwards.

In terms of the innovative elements of EAZ policy as originally conceived, it was much more difficult to demonstrate progress. Specifically, in those areas that defined the EAZ initiative as leading edge New Labour policy, there appeared little evidence of significant achievement. For example, a key organizing principle of EAZ policy was to develop a multi-agency approach to tackling social problems. Many headteachers were able to describe a range of contacts they already had with local agencies and service providers in the town. Where there were specific circumstances, for example relating to special needs provision, schools appeared to have well established links with a wide range of agencies. However, it was difficult to identify any significant initiative where either existing links had been expanded, or new links had been forged, as a direct consequence of Zone activity – 'from where I'm sitting the Zone hasn't done anything in that area' was the comment of one deputy headteacher. However, respondents were keen not to apportion blame. There was simply a recognition that 'joined up working' required a substantial investment of resources across a number of public service agencies that were already operating at full capacity, and therefore under considerable pressure. In the absence of adequate resourcing, and in the context of high stakes accountability, schools made the strategic choice to focus on those areas of their activity that had the most impact. EAZ membership brought additional funding, but not sufficient to make a significant breakthrough in the area of multi-agency working. As a consequence, there was little evidence of multi-agency working at the operational level (although there was more evidence of this at a strategic level, for example through the active involvement of the local council on the Zone's Action Forum and Executive Board).

A similar pattern of progress emerged in relation to other key EAZ themes such as increasing the role of the private sector, the development of new forms of school governance and the use of new flexibilities in terms of both the curriculum and terms and conditions of employment. Some of the schools within the Zone already had significant contacts with local businesses and used these creatively to support the curriculum. At a strategic level there was business involvement in the Zone through representation on the Action Forum (chaired by a representative of the local business community). However, it had not been possible to find large-scale business finance to support the Zone, nor was it possible to identify at a school level any significant initiatives that increased business involvement in the curriculum. There was no enthusiasm in the Zone's schools to use their EAZ status to vary national curriculum arrangements. This difficulty in promoting private sector involvement in EAZs was not unique to the case-study Zone. Indeed, this pattern appeared to be the rule rather than exception (Hallgarten and Watling 2001; Carter 2002), and perhaps suggests that those who foresaw EAZs as the harbingers of a Gradgrind curriculum for the working class and the portent of a new role for the private sector were unduly pessimistic (STA 1998). It is the case that increasing vocationalism and privatization (DfES 2004) are very much features of the current educational scene in England, but in both cases this appears to be almost despite, rather than because of, the contribution of EAZs. In a similar way, the government appears determined to challenge the concept of national pay and conditions for teachers; however, it has had to look to alternative ways of achieving this after EAZs showed no enthusiasm to use the powers they had to vary terms and conditions.

The apparent lack of progress in the case-study EAZ on many of these high profile issues was not peculiar to this Zone. As already indicated, many of the difficulties faced by this Zone were reflected in the experience of other EAZs (OfSTED 2003; PriceWaterhouse Coopers undated; Carter 2002). It is almost certainly the case that this failure to demonstrate results on these 'flagship' issues is what in part contributed to a governmental loss of confidence and interest in the EAZ policy at a national level. However, what was clear from research in the case-study Zone was that from the perspective of those at the point of delivery this was not a policy failure. Quite the reverse, those working in schools demonstrated high levels of support and commitment for the policy – unusually so given attitudes to externally imposed policies. Support for the Zone appeared to stem from the way in which it contributed to school improvement, not simply in terms of helping teachers and others achieve better results (for which there was limited evidence), but in the way that it built the capacity to support further improvement (Hopkins *et al.* 1996). How this was achieved in the case-study Zone requires further analysis.

Analysing the policy consequences

It was undoubtedly the case that a key determinant of the shift in attitudes towards the EAZ resulted from the change of key personnel in the Zone, from its re-organization and from the way in which Zone priorities were re-established. Initially there was some hostility towards what was perceived as a top-down and often irrelevant agenda. In its early stages the pressure to be seen to generate radical and innovative approaches was strongly evident in the Zone. The result was a disengagement between the Zone organization and its member schools. Quite simply there was a break in the policy chain between schools and the Action Forum, the Zone's central strategic body. This may have been indicative of a more substantial break in the link between schools and the national policy agenda. Several headteachers reported that Zone priorities were out of kilter with their own, and Zone strategies were not relevant to what schools were trying to achieve. The collective experience of teachers, who in many of the town's schools were doing spectacular things in very difficult circumstances, appeared to be at best under-valued, and at worst ignored.

Following re-organization of the Zone, this situation changed significantly. Headteachers became central to its policy-making process. At this point, not unnaturally, there emerged an alignment between Zone priorities and strategies and the priorities and strategies of those working in schools. Interviewees indicated this provided a powerful lever to support improvement. It appeared that schools often had clear strategies to raise attainment, and to tackle difficult social issues, but had lacked the resources to deliver these. The re-organized Zone was able not only to provide the resources to make things happen, but also to facilitate collaborative working between schools which offered the possibility of generating additional benefits. The re-organized structure was also successful in distributing this leadership role beyond headteachers and engaging others in shaping Zone policy. Many interviewees spoke positively about how they were involved in shaping key Zone decisions or how Zone-supported professional development had allowed them to take on a more significant role within their institutions (Harris 2003). It was also important to recognize that this participation extended beyond teachers and included those working with teachers. The Zone therefore appeared to create a space in which fundamental questions could be asked:

> The Zone has provided time out for groups of classroom assistants, teachers or co-ordinators, or whatever, where they've had to ask themselves very specific questions, which they may not want, or even have time to ask, or may not even have thought was an appropriate question normally. So you may take a group of Classroom Assistants out and say 'What are you trying to do?' – something as simple as that; and because they see themselves as implementers of somebody else's planning they never ask themselves that. So that impacts on how they support or teach children. Those kind of scenarios have been really powerful, and they've addressed a lot of issues about co-operation and the dissemination of good practice.
>
> (Headteacher – primary school)

However, this type of involvement was not without a workload cost and an emerging issue from the research was how schools sought to balance the desire to participate in the Zone with the lack of available time. Generally, there appeared to be high levels of goodwill because teachers could largely exercise their own professional judgement about how and when to participate in Zone projects. This notion of professional control was crucial to many of the Zones' achievements. However, it was not achieved unproblematically as illustrated by the following interview evidence:

> The ideas are all good … but we need to pace them better. Sometimes the Zone feels as though it's 'over there' – people go to a meeting and do something then suddenly some new initiative appears in school. Better channels of communication need to be developed with classroom teachers and these need to be two-way.
>
> (Union representative – secondary school)

This perhaps suggests that issues of ideological opposition (Ball 1987) had largely dissipated as those involved in the Zone were able to shape its activities to be more in line with their own convictions. However, workload issues remained a problem, even when substantive concerns about local variation to pay and conditions were removed.

One key way in which teachers and others working in schools were engaged in the Zone was through the patterns of collaborative working that emerged, and which appeared to be the key factor in determining the EAZ's impact. Time and time again, interviewees identified increased collaboration as the primary benefit of the Zone, and the key issue that had driven improvement initiatives. To appreciate why this was valued so much it is important to reflect on the local market conditions that had emerged within the Zone following the 1988 Act reforms, the result of which was to fracture secondary schooling into a highly competitive market. Differences compounded by inequitable funding regimes and hugely complex admissions procedures had resulted in a system that some described as comprehensive in name only.

A recurring theme from the interviews was how the re-organized EAZ was promoting genuine collaborative cultures *within* schools, *between* schools and *across phases*, that is between primary and secondary education (Stevenson, 2004). The catalyst for this may well have been bringing the headteachers together through the Headteachers' Planning Group in the re-organized Zone. This was not without its difficulties initially, but over time increased collaboration resulted – 'we've not just been nodding at co-operation, but there has been some really rigorous discussion' commented one headteacher.

Teachers cited a wide range of benefits from developing collaborative structures that involved them in working with colleagues from other establishments – planning was reduced as work was shared, teachers learnt from good practice elsewhere, problems were shared and common solutions developed, teachers were enthused by what they saw working elsewhere and finally there was a greater understanding of the problems faced in different schools. The benefits of this collaboration are perhaps best expressed in teachers' own words in response to the interviewer's invitation to identify the main benefits of the Zone:

> Collaboration. Collaboration between the schools. It's funny, I came from another local authority where they had these structures. The Heads worked together and the subject leaders worked together. I came here and there was none of that. It was like a desert. I thought 'where's my support here?'. At least we've got that now, but it needs developing.
>
> (Deputy headteacher – secondary school)

> In the meeting of minds that is the Key Stage 4 group we have from there developed other groups – Maths, Science and English – and they have meetings, and that means, for example, that our Science department talks to the Science departments in other schools in the town, which they didn't do before. They did twenty years ago as it happens, but they haven't done for a number of years – they are now. That is a significant benefit for the school in the longer term. Teachers face the same sort of

problems across the town .... If someone else is doing something well it has to be a good thing if we can learn from that.

(Leadership group member – secondary school)

These teachers' voices help to explain the link between increased collaboration and the drive to achieve the Zone's objectives. There was little research evidence to suggest that the collaboration was contrived (Hargreaves 1995), or that people felt coerced into a form of co-operation they did not wish to participate in. On the contrary, in many ways interviewees presented this sort of collaboration as a type of liberation from the years of professional isolation that flowed from inter-school rivalry in the town. Again, the emphasis on collaboration within and between schools is not unique to the case-study Zone, but is also highlighted elsewhere (OfSTED 2003). It is important to note therefore that although EAZs may have made little progress in developing the type of partnerships originally envisaged by policy makers, for example in the form of multi-agency working or public-private partnerships, they clearly made significant progress in developing partnerships between schools. Some have argued that this unintended consequence of the EAZ initiative may yet be its most enduring legacy:

School staff rejoiced in opportunities to do things which they themselves had identified as their priorities. Local people addressed local agendas, owned and clung to them despite centralising tendencies. Few other structures have been as effective at making autonomous schools work together, including across phases.

(Wilkins 2002: 9)

Whilst evidence from the case-study Zone accords with many of the points identified by Wilkins (2002), it is also important to sound a loud note of caution. Collaboration between Zone schools was visible on an unprecedented scale when compared to the situation that prevailed before the Zone's re-organization. However, the evidence also suggested that the terms on which such collaboration took place were highly circumscribed. Where 'win-win' initiatives were identified, in which all schools stood to gain from each others' participation, it was possible to identify considerable and productive collaborative activity. However, this was still collaboration between competitors and the impact of competition was never far from the surface. For example, one secondary school in the research project refused to reveal strategies it had for recruiting primary pupils at the age of transfer, for fear of losing 'competitive advantage'. Certainly teachers and students interviewed were acutely aware of the hierarchy of schools within the town, and the competition between them – 'It's competition "Big Time"' was how one teacher characterized it. The chief consequence of this competition appeared to be the skewed intakes between schools, measured not only in terms of academic attainment, but also in terms of pupils with special needs, including behavioural difficulties.

The problem of skewed intakes was seen as a major obstacle to progress in those schools faced with the most challenging contexts. Moreover, the view was expressed by some interviewees in both primary and secondary schools that this inequality was not simply a product of parental choices, but subtle selection procedures operated by some secondary schools during the admissions phase. What followed was a virtuous circle of improvement for those dominant in the market and a constant battle against the odds for the rest. Gorard *et al.* (2002) have argued there is no widespread evidence of schools in these circumstances facing a 'spiral of decline'. Research in this study could not prove a spiral of decline, but it does suggest it as a strong possibility. Certainly the imbalanced intake appeared to make it appreciably more difficult for these schools to demonstrate significant progress. It almost certainly explains why these schools found it so difficult to raise pupil attainment in examination results (a point acknowledged by the Audit Commission/OfSTED report (2003) investigating the impact of school place planning on academic standards and social inclusion). In these schools teachers valued highly the increased collaboration with their neighbours, but the type of collaboration they craved was more substantial. What they sought was agreed changes in Admissions Policies that would allow for a more balanced student intake across the town's secondary schools. This was identified as a key factor in facilitating improvement in those schools facing the most challenging circumstances. It was also considered an unrealistic objective. At this point the aspirations of social justice seemed unable to overcome the more powerful pressures of a high stakes standards agenda driven by competition and market forces (Stevenson 2003b).

Conclusion

Education Action Zones consummately illustrate the complexity of the policy development process. EAZs were intended to be the flagship education policy of a radical and reforming government that had indicated that education was its number one policy priority (DfEE 1997). However, within five years EAZs had come, and gone. Although there is some limited evidence that schools in EAZs may have improved more rapidly than others there is only very limited evidence to demonstrate a link between Zones and improved academic performance (OfSTED 2003). Problems of underachievement in these areas are long-term problems, that require long-term solutions and significant, focused, investment – a point made prominently by Her Majesty's Chief Inspector (Bell, D. 2003). Moreover, EAZs cannot claim to have acted as a role model for radical 'third way' thinking in policy development. Many of the policy's most radical elements came to nought and the New Labour government has had to look elsewhere to promote its agenda of diversity and privatization. However, whilst the radical private partnerships that were envisaged did not materialize, it is the case that many EAZs had considerable success in creating professional cultures of collaboration (Hargreaves 1995). Indeed, in some contexts EAZs may have provided the mechanism by which the teaching profession began to rebuild the partnerships between schools that years of market competition had dismantled.

Staff, teachers and support staff, were working together within and between schools. They were not simply learning from each other, but drawing on each others' experience to provide solutions to common problems. This was not a case of the 'good school' showing the 'failing school' how to do it – but all schools working together and learning from each other. This appeared to provide a powerful model of staff development that not only increased professional self-confidence, but had the potential to impact on morale and, in turn, staff retention (a crucial issue in this locality). It would appear too that this experience from within EAZs of staff and schools working together has found favour in official policy. Despite the demise of EAZs, it is possible to discern a clear shift in subsequent policy discourses, with an increasing emphasis being placed on ensuring that schools work together and pool their knowledge and expertise. There is now a raft of important policy initiatives (Glatter 2003) that depend on, and encourage, schools to work together – 'The education service must adapt and change to meet our goals, working in collaboration, not competition' (Labour Party 2003: 9). This represents a highly significant shift in public policy and points to what may yet be the long-term, and positive, legacy of EAZs.

However, herein lies the contradiction at the heart of New Labour policy. EAZs that have worked well, such as the one in this case-study, have illustrated the power of collaboration, not only *within* schools, but crucially, *between* schools. Zones have helped to create the conditions in which longer-term school improvement becomes possible. EAZs are not the only examples of how schools have benefited from increased collaboration, but they have provided a powerful model that has clearly influenced subsequent policy. Despite their demise, this lesson may yet prove to be their most enduring legacy, and their most important long-term contribution to tackling underachievement in socially disadvantaged areas. However, the equally clear lesson from the case-study Zone was the difficulties of effective collaboration co-existing within a wider context of continuing competition. In such a competitive market, collaboration could only go so far. Competitive practices that favoured some schools and disadvantaged others were strongly entrenched – with no incentive on the part of those most advantaged to change. As a consequence, some schools faced enormous challenges and this almost certainly impacted on their capacity to deliver significant and sustainable improvements. Given the concentration of students in these schools from the most disadvantaged backgrounds, these are precisely the students that the EAZ project was intended to support. Raising their academic performance would have made a major contribution towards the government's objective of raising attainment in Zone areas. However, despite the Zone's best efforts, poverty and structural inequalities between schools appear to have provided almost insurmountable obstacles. Evidence from this research suggests that if the positive lessons from the EAZs are to have an impact on raising attainment and promoting social inclusion then they must be combined with a much more realistic long-term perspective; must be part of a wider and more ambitious anti-poverty programme, and must be prepared to challenge the educational market that ensures the most disadvantaged schools face the least favourable conditions for improvement.

8   Citizenship and social justice

Developing education policy in multi-ethnic schools

Introduction

In the previous chapter the focus was on a single national policy initiative, and how this developed as it was implemented at a local level. Whilst there was clearly significant flexibility in the policy, which was reflected in the way it was developed, the EAZ initiative remained a distinctive and discrete policy project. In this chapter the focus shifts to policy in its more amorphous and less tangible form – policy as the promotion of values. The concern here is how policy is developed at an institutional level in pursuit of equity objectives. In this instance the specific focus is on equity in the context of cultural and ethnic diversity. The material draws on a research project conducted on behalf the National College for School Leadership that investigated the contribution of school leaders to the development of successful multi-ethnic schools in a number of Local Education Authorities in England (Dimmock *et al.* 2005). The central concern is with the development of policy at an institutional level and how policy is shaped not only by national discourses and the values and priorities of key people in the institution, but also by the specific local context of the institution. Policy at this level represents a complex mix of influences in which the institutional priorities of school leaders are not only shaped by national policy debates and initiatives, but crucially by the influence of local community factors and the role of the wider market within which the schools function.

The chapter begins by exploring a specific period, and within that a specific incident, in England in the mid-1980s. At this time the publication of a major report (Swann 1985) decisively shaped the discourse relating to education in a multi-ethnic society. Individual institutions were faced with developing organizational procedures and practices within the context of the socio-political environment at the time. However, as the example illustrates, tracing the links between the socio-political environment and institutional policies is seldom clear cut. Tensions and contradictions at the socio-political level are reflected at the institutional level, and institutional policies as texts may bear little relation to the intentions of the texts' authors. Confusion, uncertainty and conflict at a national level in turn have implications for the development of policy at a local level. The main interest in this chapter is on the present and a concern with the development of policy at the level of the individual school. The study draws on recent research to explore the ways school leaders committed to principles of racial and ethnic equality are able to work within the wider national discourse. This is an area of increasing interest to school leaders around the world, but to date has been the focus of limited research (Reyes *et al.* 1999). The evidence illustrates how those working in multi-ethnic schools are able to exploit the opportunities provided by national discourses and to develop policies at an institutional level that take national policy agendas into new territories. However, the evidence also illustrated how policy at an institutional level was shaped by the institution's specific local context and that in some cases this could exert significant negative pressure on schools' capacities to promote an equality agenda. Specifically, the impact of the local market and role of parental preferences could generate a tension in which the promotion of a commitment to race and ethnicity equality might be penalized in terms of parental choice and pupil numbers (Gewirtz *et al.* 1995). Such situations created ethical dilemmas for school leaders in which the ability to operationalize egalitarian values could be compromised by the need to pursue strategies geared to success in the market.

Schooling in a multi-ethnic society: linking the socio-political environment and organizational practice

In 1985 the government in the UK published, with grudging enthusiasm (Tomlinson 2001), a major report into the education of children in a multi-ethnic society, under the title *Education For All* (Swann, 1985). This report traced its own origins to concerns expressed in the late 1960s by members of the West Indian community about the relatively poor performance of children from West Indian backgrounds in UK schools (Stone 1985; Carter 1986). The report confirmed the relative under-performance of West Indian origin school students and, most significantly, identified racism within the educational system as having a 'pervasive influence on institutional policies and practices .... [and] which can be seen as the major obstacle to the realisation of the kind of society we have envisaged here' (Swann 1985 :8). Significantly, it recognized that 'institutional racism' was 'just as much a cause for concern' as the prejudiced attitudes of individuals (Swann 1985: 29). It defined institutional racism in the following terms:

> We see institutional racism as describing the way in which a range of long established systems, practices and procedures, both within education and the wider society which were originally conceived and devised to meet the needs and aspirations of a relatively homogenous society, can now be seen not only to fail to take account of the multi-racial nature of Britain today but may also ignore or even actively work against the interests of

ethnic minority communities. The kind of practices about which we are concerned include many which, whilst clearly originally well-intentioned and in no way racist in intent, can now be seen to be racist in effect, in depriving members of ethnic minority groups of equality of access to the full range of opportunities which the majority community can take for granted or denying their right to have a say in the future of the society of which they are an integral part.

(Swann 1985: 29)

The corollary of Swann's analysis was to present a raft of recommendations which might broadly be described as providing a basis for a multi-cultural and anti-racist education, the central argument of which related to the need for these principles to underpin the education of *all* children, not just those from minority ethnic backgrounds, or those in multi-ethnic schools. Swann explicitly rejected providing a policy blue print, but argued that schools needed to develop their own organizational practices and procedures that were informed by multi-cultural and anti-racist principles.

In the year after Swann's publication, on Wednesday 17 September 1986, a 13-year-old Bangladeshi boy was killed in the playground of the school in Manchester where he was a pupil. Ahmed Ullah had been murdered by a fellow student. The background to the incident clearly indicates that Ahmed was the victim of a racially motivated attack. Ahmed Ullah's murder was an event that immediately exposed the fault lines in UK race relations, which often appeared to be at their sharpest in the education system. Ahmed was a pupil at Burnage High School (BHS), located on the south side of Manchester. Manchester is a typical industrial city in the north of England. Its importance to the world cotton trade attracted significant immigration into the textile industry and as a consequence the city was, and is, characterized by considerable ethnic diversity. However, by the mid-1980s Manchester was suffering the consequences of substantial de-industrialization – problems compounded by unsympathetic national economic policies designed to facilitate economic capital's rapid restructuring (Green 1989). The local context therefore was of a community in flux, coming to terms with both cultural diversity and economic change, the latter often experienced as significant and painful dislocation. As a consequence, BHS faced considerable challenges in the form of the complex social problems generated by economic decline.

One of BHS's responses to reflecting its local communities' needs, and the problems they faced, was to introduce specific policies promoting multi-cultural and anti-racist education in much the way envisaged by the Swann Report. However, as Ahmed's murder testified, such policies were clearly ineffective. Following Ahmed's death an inquiry into racism and racial violence in Manchester schools was established, chaired by Ian Macdonald QC. The report that followed (Macdonald *et al.* 1989) provides a comprehensive background to a tragic event – it continues to offer a fascinating account of the complexities of policy development in a single institution, the power of micro-politics (Hoyle 1982) and the serious consequences of errors and misjudgements in the policy process. Sadly it serves as a reminder of the inadequacy of seeing policy purely as *product*. Policy analysis must shed light on the *process* of policy development if it is to illuminate the realities of practice.

At BHS there was clearly 'a policy', but what was apparent was the almost total disconnection between the policy as a text, the policy as it was experienced by students and staff at BHS and the consequences of that policy. Although the Macdonald report provides a number of explanations for this, at its centre was the considerable tensions amongst the staff regarding the policy, its aims and its implementation. This was about a clash of values, as a consequence of which policy development was buffeted between conflicting sub-groups within the institution. Powerful coalitions formed within the school that challenged and undermined the policy text. At times, policy development appeared to be a conflict between those with authority, but little influence, pitted against sections of the staff with little authority, but substantial influence (Bacharach and Lawler 1980). A range of responses were discernible in this clash of values – to embrace diversity, accept it reluctantly, deny its existence or reject it? However, in many ways these simply provided a mirror to tensions and contradictions within national state policy with a continuum of responses offering a choice between assimilationism and integrationism (Parekh 1991).

Since Swann, these tensions within state policy have formed a continuous backdrop against which schools have had to develop their institutional responses to meeting the needs of their student populations. However, the socio-political environment is never static, but is continually shifting in response to local, national and international factors that shape policy. More recently, in the UK for example, debate has been decisively influenced by another racist murder, of student Stephen Lawrence, in 1997 and the subsequent publication of the Macpherson Report (1999). The issues raised by this incident were not specifically about the role of educational institutions; however, the report's identification of institutional racism as an important contributory factor in the failure to secure justice for Stephen echoed the conclusions of Swann 14 years earlier, and have had an important influence in shaping policy across all public services. One obvious manifestation of this has been the amending of the Race Relations Act (RRA) 1976 (now the Race Relations (Amendment) Act 2000) which now places a duty on public sector institutions to positively promote race equality. Here it is possible to see the interplay of a number of factors – government policy objectives (for example relating to social inclusion), high-profile national events, influential reports and national legislation – all shaping the socio-political discourse. How though have these shifts in debate, and in public policy, played out in schools? The experience at Burnage High School in 1986 highlights the need to develop a better understanding of policy development at an institutional level, and to assess how discourses framed at a national level influence and shape the experiences of individuals studying and working in schools and colleges. This is the focus of the following section.

Leadership and policy development in multi-ethnic schools

Those working in schools operate within the context of the shifting discourses and policy contexts identified above. As argued in previous chapters they are both the recipients, and developers, of policy. However, these distinctions are not neat and tidy and it is vital to see both elements as part of a single process in which those working in schools must simultaneously make sense of external policy agendas, respond to factors specific to their own context and develop and pursue their own agendas. At any one time there are likely to be tensions within institutions, between institutions and between institutions and the demands of the external environment. School leaders are at the interface of these tensions – accountable within, and without, the institution and often the link between external and internal policy agendas. The research that informs this chapter is based on a research project in five case-study secondary schools identified as being successful in meeting the needs of their multi-ethnic communities. The research sought to establish what leadership behaviours were a feature of these schools and the specific contribution of the headteacher to the school's success. How was policy developed within the school and how was internal policy development shaped by the external agenda and local context? Table 8.1 is a summary of some of the key characteristics of the case-study schools.

Participating schools were recommended by their Local Education Authorities on the basis that they were considered effective multi-ethnic schools that had significant success in meeting the needs of their school community. Academic success was important, but was just one issue considered in identifying the sample.

In this volume it has been argued that it is not sufficient to see 'policy' purely in terms of texts, or statements, or pronouncements, but that policy is a much more complex process by which resources of power are mobilized in order to operationalize values. In all the case-study schools headteachers frequently articulated their leadership role in terms of their values – values were rarely implicit, but were often explicit. There was a clear recognition that leadership was about the operationalization of values – highlighting the intimate link between leadership, strategy and policy. This intangible way of articulating policy and its link to values was expressed in the following terms:

> When I arrived it was an ethos of control and I wanted to turn it into an
> ethos of respect and equality of opportunity. I think whatever systems
> and structures you have the key way that you do that is by leading by
> example, and making it clear that you are what you preach. You put into
> action ways of working which are based on your fundamental principles

Table 8.1 Case-study schools – details and background

School	Location	Ethnic profile	Background information
S1	Northern industrial city	93% Black and Minority Ethnic (BME) – very largely a single ethnic group.	A mixed local area, but where school choices are heavily influenced by ethnic issues resulting in skewed ethnic profiles in schools.
S2	Midlands town	24% BME – heterogeneous BME profile with significant East European element.	Poor placement in academic league tables. The school has surplus places and is the subject of a re-organization.
S3	Large Midlands city	85% BME – heterogeneous BME profile.	Popular school in largely established local community.
S4	Midlands city	55% BME – heterogeneous BME profile, established local population.	Popular, over-subscribed school.
S5	Outer London District	75% BME – heterogeneous BME, significant proportion of recent immigrants.	Well-regarded local school. Over-subscribed. Changing ethnic profile based on recent arrival of new ethnic groups.

of human beings – that is why I am a head, and that is what I have tried to do while I've been here.

(S3, Headteacher)

In this instance the limitations of policy fiat as a means of securing change, so graphically illustrated at Burnage High School, is recognized. Here policy is shaped by the actions of individuals, which in turn is driven by the values of those seeking change. How policy becomes operational is hence a much more complex process than the production of a policy statement.

Within the context of multi-ethnic schools, values that were associated with social justice had a high profile. Values most frequently cited related to equity, fairness, respect and tolerance. However, the precise way in which these values were interpreted, articulated and implemented could differ significantly. There was, for example, a common commitment to the principle, strongly associated with the comprehensive schools movement, of valuing all students equally and in the case-study schools it was difficult to challenge the sincerity or the passion with which those views were held. A feature of the headteachers in these schools was that their commitment was not rhetoric or lip service, but a deeply held conviction that they were able to convey to colleagues, students and the wider community. However, valuing all students equally in a context of cultural diversity, often compounded by economic and social disadvantage, rarely equated with 'equal treatment'. Principles of both distributive justice and cultural justice clearly came into play (see Chapter 4). For example resource decisions that allocated resources in favour of particular disadvantaged groups could be controversial. Whilst not always contentious, such decisions sometimes met with resistance from staff and/or parents. Similarly, different views were expressed about the extent to which cultural diversity might be explicitly recognized. A feature of the case-study schools was the way in which the variety of ethnic cultures represented in the school was highlighted and celebrated. Valuing diversity by recognizing it, and celebrating it, was a common feature of many of the schools in the projects. However, whilst some schools confidently highlighted their diversity, others were less confident, arguing that differential treatment generated division. In one school, located in an area that had witnessed significant inter-ethnic tensions, the headteacher maintained a strong policy of equal treatment for individuals, whilst recognizing the need to respond to the specific needs of collective groups:

> It's absolutely essential to keep sight of the individual, because even within minority ethnic groups there is huge diversity. You cannot lump groups together. As a school we've learned you've got to treat people as individuals. There are sensitivities about how you put in place strategies for groups, when actually you are concerned about individuals.
>
> (S2, Headteacher)

In this instance policy options are clearly shaped by the local context and leaders are faced with the challenge of reconciling individual values positions with the prevailing local circumstances – a dilemma illustrated in more detail later in this chapter.

The strong and explicit values base expressed by leaders in the case schools could be seen as the central element of their leadership. These values then informed institutional policy in several key areas – these are identified in Figure 8.1. In each of these key areas institutional policies reflected the school's commitment to the promotion of cultural diversity. In some cases policies might be explicit (for example, those relating to racism and racist incidents), but more common was the intangible and almost unspoken way in which policies were formed and operationalized.

Policy implementation – from organizational principles to operational practice

Creating inclusive cultures

Booth *et al.* (2000: 9) identify the need to create 'a secure, accepting, collaborating, stimulating community in which everyone is valued, as the foundation for the highest achievements of all students'. School leaders

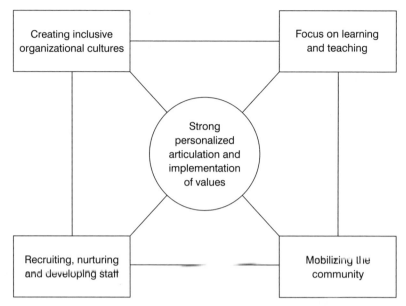

Figure 8.1 Leadership in multi-ethnic schools (After Dimmock *et al.* 2005: 7. Copyright NCSL. Reproduced from *Leading the Multi-Ethnic School* by kind permission of NCSL.)

in the case-study schools placed strong emphasis on the creation of such cultures and clearly prioritized these over systems and structures. Within the context of an ethnically diverse school the creation of an inclusive culture is more challenging. Arguably, inclusion is less problematic the more homogenous the population. Within a multi-ethnic school, diverse approaches are required to secure equitable outcomes and differential responses for different ethnic groups may be required to demonstrate that all are valued equally. School leaders in the case-study schools were committed to ensuring that all aspects of school life reflected the ethnic diversity of the school's local community and particular attention was paid to ensuring that collective bodies encouraged the representation and participation of all ethnic groups within the school. This points to the importance, and the confluence, of associational, distributive and cultural forms of social justice, manifest in a strong commitment to ensure the full participation of members of all communities in policy development in the school, and a willingness to devote additional resources where needed to achieve this.

The emphasis on associational justice was illustrated by the high level of student engagement in the life of the school, often exemplified by the active role played by students in Student Councils. However, inclusion was often about much more than this. For example, in one school a teacher training day for the staff was devoted to issues of raising academic achievement amongst ethnic groups identified through monitoring as 'underachieving' (the commitment of these resources itself illustrating how policy priorities need to be supported by appropriate resource allocations). Students from a range of ethnic backgrounds represented in the school were then involved in both planning and delivering the in-service training. Where this was the case students enthused that their contribution was both sought and valued.

A feature of inclusive cultures is the often intangible way that policy is operationalized. However, one key area of policy identified by student interviews as very explicit was in relation to racism and the handling of racist incidents. The experience of Burnage High School exposed the problems that arise when policies may be explicit, but when those working in the institution have no confidence in their application (Macdonald *et al.* 1989). In the case-study schools, student interviewees reported a high degree of confidence in the operation of anti-racist policies. Racism was not tolerated, and if it was reported, students were confident that appropriate action would be taken – 'Everybody knows the policy and everybody knows what will happen to them if they are involved in anything racist – there is hardly any racism here' (S5, student). There was no evidence of complacency about racism – either within the schools, or their immediate environs. However, there was a confidence that the schools took racism seriously and took appropriate action in response.

Nurturing and developing staff

A high priority for all the headteachers in the case-study schools was to recruit and develop the staff body. Whilst there was a clear focus on teaching staff, this was by no means exclusive, and there was a clear commitment to the development of inclusive staff cultures that valued everyone within the organization. Indeed, in many cases, hierarchies were deliberately challenged by the inclusion of non-teaching staff in policy forming processes that had previously been the preserve of teaching staff alone. Headteacher commitment to promoting their schools as successful multi-ethnic institutions was in part achieved by ensuring that new appointments were committed to the school's mission of racial equality. In this sense school leaders sometimes deliberately sought to shift the culture within the institution through the use of new appointments, hence creating the potential for new alliances and making it easier to secure organizational objectives. In some cases staff commitment to working in a multi-ethnic environment was assessed within formal recruitment and selection policies that ensured specific interview questions focused on ethnicity issues, whilst in other cases policy objectives were secured much more informally, described in the following terms by one headteacher – 'we're absolutely up front about what sort of school we are and what we stand for – if you don't like it, you won't want to come here. But everyone knows what they are coming to' (S4, Headteacher). Conversely, one headteacher described how they had exploited a redundancy situation to remove staff he had identified as resistant to the school's aspirations as a multi-ethnic institution.

The research also showed a strong commitment to the recruitment and development of staff from minority ethnic backgrounds. In all the case-study schools staff from minority ethnic backgrounds were under-represented, a disparity that worsened at managerial and leadership level. In many cases school leaders went to considerable efforts to both appoint and develop such staff. They worked within the law on equal opportunities, but were prepared to push at its boundaries in order to support individuals whose career progress had received few advantages, and many disadvantages. In one case a headteacher actively challenged the Home Office that was seeking to deport a member of the school's staff. The school had a significant number of Albanian/Kosovan refugees and the individual in question was the only person on the staff who could speak to these students in their first language (and in some cases their only language). Significantly, students involved in the research project indicated they valued an ethnically diverse staff, not simply because of the role models provided, but principally because this signalled the school's commitment to ethnic diversity.

Mobilizing the community

A strong feature of the case-study schools was to not see the educational process as being confined to the school boundaries, or the restrictions of the school day. Strong links with the school's wider community, not just the parent body, was clearly a high policy priority for all the schools and was effectively seen as essential. Community links were not a nicety that could be developed if resources were available. Rather such links were perceived as central, and were resourced accordingly. Working *beyond* the school was seen as pre-requisite if the students were to achieve *within* the school. This support for students, and the willingness to work inside and outside the school, derived from the personal values and convictions of the school leader and other staff:

> I think the staff realise we're an extended school. School is such a limited part of students' life. I could never take the view ... and there are Heads in the town I know, and there are Heads who have been here who take the view .... that once the child leaves the boundary of the school they're not their responsibility. I can never subscribe to that. I cannot take that attitude.
>
> (S2, Headteacher)

In the case-study schools, it was possible to identify a number of different ways in which links from within and beyond the school were developed. For example, in the case-study schools, staff had a detailed knowledge of individual students and their personal circumstances outside the school. Understanding what is happening, or has happened, in the lives of individual children, was seen as crucial to promoting achievement within school. This extended to strong links with parents. However, developing such links was often difficult as a number of barriers might exist to such contact (language issues, local working patterns, lack of safe transport options). In the case-study schools, in different ways, considerable efforts were undertaken to overcome these difficulties. An obvious example is the high level of outreach work in which school staff visited parents in places, and at times, that minimized the problems identified above. Finally, links with community groups were frequently prioritized, with substantial resources, often in terms of the personal time of individuals, invested in developing such links. Arguably, a feature of many minority ethnic groups in the UK is that they retain a more collectivist culture with a strong network of local organizations and self-help groups. These groups were seen as a powerful resource in terms of articulating community concerns and aspirations and working with such groups was a common strategy adopted by school leaders. In turn, those community representatives who were interviewed spoke warmly of the steps taken by school leaders to engage with their communities beyond the school gates. Communities felt valued and included where such steps were taken.

Focusing on teaching and learning

The improvement of classroom practice was understandably a common priority in the case-study schools. However, within the context of a multi-ethnic school this sometimes took quite distinctive approaches. For example, there was a clear recognition of the need for curriculum provision to reflect cultural diversity, and in some cases to draw on the cultural experience of students. In all case schools examples were provided of how some curriculum subjects reflected and valued cultural diversity, and how the curriculum was used to tackle racism and develop an active citizenship. Often these examples were to be found in similar curriculum areas – religious education, art and drama for example. In some further examples it was clear that subjects such as English or Humanities might be used to explore relevant issues, but in subjects such as Maths and the natural sciences cultural diversity was rarely on the map. The research pointed to a number of reasons why this might be the case. The emphasis on student achievement in national tests appeared to generate a functional and utilitarian approach to the curriculum amongst staff and students alike, particularly amongst older students (Beresford 2003). Published league tables of performance data, high stakes inspection arrangements and a euro-centric national curriculum seemed to conspire to create a situation in which not only was the focus almost exclusively on test performance, but developing a curriculum that exploited opportunities for inter-cultural learning was seen as something that would detract from, not improve, student performance.

Such evidence perhaps points once again to the difficulty of whole school policy penetrating classrooms, and the more pervasive influence of external control mechanisms such as inspections and published league tables. Many of the case-study schools were vibrant examples of their multi-cultural communities, at a whole school level. Cultural diversity was acknowledged and celebrated. Within the classroom, this rarely reflected the reality. Staff reported that external pressures often reduced professional self-confidence and the willingness to take risks, whilst expectations were lowered by the functional nature of the curriculum and its associated testing regime. One teacher interviewed, of South Asian heritage, commented that students from minority ethnic backgrounds 'expect to leave their culture outside the door when they enter the classroom'. Whilst they may want lessons to reflect their ethnic identity, they did not expect that they actually would. This evidence clearly supports Gillborn's (2002) assertion that 'permeation' – the inclusion of multi-cultural content across the curriculum – has failed to be effective. Many individual teachers, schools and whole curricular areas have remained unaltered behind a façade of supposed permeation (Gillborn 2002: 60).

Although there was limited evidence that students' classroom experience reflected cultural diversity there was considerable evidence of how students' classroom performance was monitored by ethnicity. This was in part due to the requirements of the updated Race Relations Act; however, school practice often exceeded basic legislative requirements. In one school there was evidence of highly sophisticated monitoring by ethnicity that sought to monitor achievement in relation to the linguistic background of students. In this school the headteacher invested considerable resourcing, supported by personal commitment, to monitoring student performance and to identify appropriate strategies for intervention. Where patterns of underachievement were identified the school was prepared to follow through with specific strategies focused on particular ethnic groups. In this case the school provided examples of specific initiatives it had developed to support Bangladeshi and African-Caribbean students in response to data generated by their own monitoring by ethnicity. However, the fact that curriculum monitoring by ethnicity appeared to be more sophisticated than the provision of curriculum content arguably points to the relative importance of an external agenda driven by accountability and performativity concerns, rather than a broader concern to develop a genuinely multi-cultural curriculum.

The outline presented here of how leaders at an institutional level developed their own policy agendas within the context of their multi-ethnic schools demonstrates the complexity of the policy process. Figure 8.2 presents the model developed throughout this volume related to the formulation and implementation of policy relating to issues of race and ethnicity.

The model here highlights the very significant tensions at all levels of the process. For example, within the socio-political environment there are tensions between the promotion of an equality agenda based on diversity and one that seeks to create a sense of social cohesion through the promotion of traditional notions of national identity. As strategic direction develops, tensions emerge between the promotion of a National (or nationalist?) Curriculum (mac an Ghaill 1993) and a more flexible curriculum that is

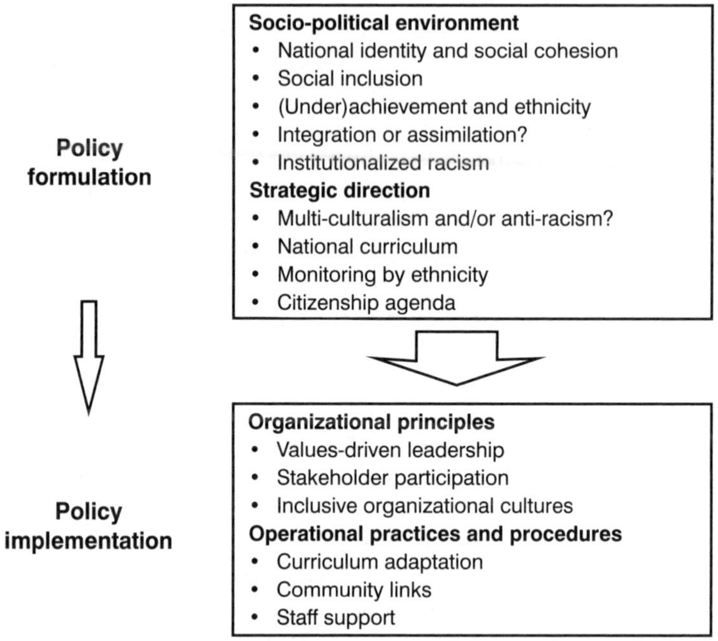

Socio-political environment
- National identity and social cohesion
- Social inclusion
- (Under)achievement and ethnicity
- Integration or assimilation?
- Institutionalized racism

Strategic direction
- Multi-culturalism and/or anti-racism?
- National curriculum
- Monitoring by ethnicity
- Citizenship agenda

Policy formulation

Organizational principles
- Values-driven leadership
- Stakeholder participation
- Inclusive organizational cultures

Operational practices and procedures
- Curriculum adaptation
- Community links
- Staff support

Policy implementation

Figure 8.2   Policy into practice: leading multi-ethnic schools

capable of reflecting the full range of cultures that may be present within a school. As the example of Burnage High School illustrated, confusion and contradiction at state level will inevitably be mirrored and reflected at institutional level as organizational principles and operational practices and procedures are developed. This raises important questions therefore about the extent to which leaders at an institutional level can work within, or against, the grain of state policy – to what extent is there a space for 'values-driven leadership'?

Policy development in the institution – creating a space for 'values-driven leadership'?

Gold *et al.*'s (2003) study of ten 'outstanding' Principals argued that these Principals placed their values at the centre of their leadership – referred to as 'values-driven leadership'. In these cases, Principals' values provided the 'moral compass' (Fullan 2003) that allowed them to navigate the murky waters, and in some cases the hazardous swamps, of external policy imperatives. Hence Gold *et al.* (2003), and others (Day *et al.* 2000; Moore *et al.* 2002), have argued that such school leaders were able to retain, and advance, their personal agendas based on principles informed by social justice, even in the face of a hostile wider environment (discussed further in Chapter 2). In these cases Principals seized on initiatives and external projects that might in any way support them in achieving their personal priorities and objectives. This was certainly a feature of the headteachers in the multi-ethnic schools in this research project. School leaders demonstrated considerable flair in identifying wider policy initiatives that would further their personal agendas in their own schools. In terms of multi-ethnic schools, and the commitment to promoting a form of cultural justice, it was clear that the policies of the Labour government had opened up new opportunities that could be exploited. However, a feature of several of these headteachers was the way in which they were able to make use of sometimes modest national developments, and utilize them to promote radical changes and improvement in practice at an institutional level.

For example, whilst it has been argued that 'social inclusion' has often been a vague and ill-defined concept (see Chapter 7) lacking any real policy coherence, school leaders in the case-study schools were able to use the language of inclusion to promote a particular set of values within their institution that articulated a more radical vision of social justice, often with a particular emphasis on issues of cultural diversity and cultural justice. In these schools, how inclusion was articulated, the language used to describe it, and the specific policies used to operationalize it, were central to the schools' ethos and mission. This allowed specific resource allocation decisions to be made that supported the school's commitment to social inclusion, but which may have been contentious in the extent to which they distributed resources in favour of marginalized or disadvantaged groups within, and indeed beyond, the school. Other opportunities to promote this agenda were provided by incidents and initiatives that school leaders were able to exploit in a similar way.

Publication of the Macpherson Report (1999) following the murder of Stephen Lawrence was used by one headteacher to significantly raise the profile of issues relating to ethnicity in the school, and for the school to look self-critically at the issue of institutional racism. Whilst using the report in this way appeared to be the exception rather than the rule, it illustrates how some school leaders were able to seize on national issues and to use these to shape debates, and ultimately policy, within their own institutions. A similar example is provided by the introduction of the Race Relations (Amendment) Act (2000). Examples were provided as to how this was used as a vehicle to highlight issues relating to ethnicity, to question existing practices and to develop new institutional policies. For example, the instance of monitoring by ethnicity referred to previously was developed partly in response to requirements contained in the RRA 2000.

These examples therefore illustrate how school leaders in multi-ethnic schools were able, if they chose, to utilize external policy initiatives to build support for their own objectives, in their own institutions. However, leadership was never about solely seizing opportunities from the positive, it was also about defending values in the face of less benign pressures. This was most starkly illustrated by the impact of the specific local context of individual schools and the way in which this shaped, and often challenged, the values of those working in the institution. In analysing a school's individual local context it was possible to distinguish between a number of distinct, but interdependent factors relating to the institution's community context, and a similar range of factors relating to the school's market context. This interplay of community and market factors that are unique to each school may be described as the school's micro-context. In the following sections the concept of micro-context is developed and one of the case-study schools is analysed in detail to explore the way in which policy development in an individual school is shaped by its micro-context, and how the ethical dilemmas generated by micro-context can shape and constrain leadership.

Multi-ethnic schools and the importance of micro-context

Micro-context and the community

The community features of a school's micro-context may be considered to refer to the largely demographic issues that shape local conditions. In identifying appropriate factors it is important to reflect the extent to which these are dynamic and change over time. Community factors may be considered to reflect the following:

Ethnic profile

An obvious community feature of multi-ethnic schools is the specific ethnic profile in the local community. What ethnic groups are represented in the community and what is the balance between them? Such differences could make a significant difference to the type of policies that could be developed in some schools. For example, in schools S1 and S3 where a significant proportion of the school population were from a specific ethnic group, it was relatively easy for the school to re-organize its school day during the holy month of Ramadan to allow pupils to more easily fulfil their religious obligations. This practice was greatly valued by students and the wider community who saw it as a tangible signal of the school's commitment to its students. However, in S4, the headteacher felt unable to make such changes and believed that to do so would generate resentment from other students. Here the ethnic profile within the community and the school shaped the types of policies leaders felt they could, or could not, pursue.

Socio-economic profile

A second key feature of a school's community profile was the socio-economic profile of its population. Throughout the research project interview, respondents frequently described white students in terms of class – 'white working class', or 'middle-class whites' were common descriptors. However, using class as a descriptor was as uncommon for black and minority ethnic students as it was common for white students. This seemed to understate the complex nature of issues in multi-ethnic schools where issues of ethnicity and class overlap. Many multi-ethnic schools are located in urban areas, and exhibit the classic symptoms of communities experiencing economic disadvantage. School contexts therefore are shaped not solely by their ethnic profile, but also by the socio-economic status of the school and its community and this may have a significant bearing on school leadership and policy (Harris 2002). These issues become particularly important when assessing patterns of academic achievement in schools. It is becoming increasingly clear that it is not possible to explain the significant variations in achievement between different ethnic groups in the UK, without taking into account the influence of social class, and indeed gender.

Social cohesion and stability

A final feature of the community dimension relates to the extent to which communities can be considered stable and cohesive over time. Economically disadvantaged communities have a tendency to experience more significant population movements within them. Inhabitants are more likely to be transient. A degree of turbulence in the local population, and hence in its schools, can present significant challenges for those leading such schools. In some of the case-study schools the degree of turbulence in recent years had increased substantially as economic and geo-political pressures have driven increased numbers to seek work or refuge and asylum in apparently more prosperous and secure environments. This often presented affected schools with significant issues to deal with as they tried to meet complex, diverse and constantly changing needs.

Micro-context and the market

The community factors identified above have an importance in shaping the local context of individual schools, and hence the context within which policy at an institutional level is developed. However, these factors provide only part of the picture. As important, if not more important within many contexts, is the inter-play of a number of market-based factors by which parents and pupils are encouraged to act as consumers and make choices between competing schools. As with the community dimension, it is possible to identify a number of interlocking elements that shape the local market context of individual schools:

Open or closed markets?

The research suggested that whether or not a local market for schools could be described as open or closed was highly significant. A closed market was one in which there was very limited surplus capacity. In such instances potential students were allocated to their local school and there were very restricted opportunities to choose an alternative. Hence opportunities for parental choice were heavily circumscribed. Movements of pupils across catchment areas, and between schools, was limited. In open markets the reverse was true. Parents were able to exercise choice because there was available capacity, and they often did so. Open markets were therefore characterized by much more movement. Competition, and the corresponding need to 'market' the school, was often much more explicit. Some of the case-study schools functioned in markets where issues of ethnicity decisively shaped parental preferences (Gewirtz *et al.* 1995). In such instances schools could become sharply polarized between ethnic groups despite the best efforts of those working in the schools to avoid this.

'League table' position

When market policies were first seriously introduced in England in 1988 they were accompanied by the publishing of examination performance data to allow 'consumers' to make informed choices based on apparently full information. Notwithstanding the efficacy of this policy there can be little doubt that the influence of 'league table' data has proved to be highly significant in shaping individual schools' contexts – a significance that is likely to be more substantial where markets are more open. Superior positioning in the league table, which may be based largely on the influence of community factors identified above, confers a degree of market power. Those schools with market power have more room for policy manoeuvre. Those lacking such power face a more limited range of policy options – compelled to focus on organizational survival. It is often at this point that ethical dilemmas are at their most acute.

Parental and community perception

Whilst a school's positioning in the local league table can be highly significant in framing a school's local context, it is clearly not the only consideration parents make when choosing schools. More nebulous, but equally important, is what might be generally referred to as parental and community perception of the school. Whilst inevitably this will be informed by academic performance, it is not the only consideration. More general perceptions of the school are likely to be based on perceptions of its culture, its ethos – in short, what type of school is it? What does it stand for, what are its values and who is it for?

Taken together these community and market factors provide a school's micro-context within which policy at an institutional level is developed. Such contexts are by definition unique to each school, but they are not detached from broader macro-factors. Markets, for example, are the product of macro-policies that encourage or discourage competition between schools.

However, the micro-context of each school exerts significant influence on policy development within the institution, at times presenting ethical dilemmas that threaten to compromise educational objectives and values.

Values under pressure: how micro-context shapes policy

This case-study focuses on one of the five schools in the research project, school S2. All the following quotations are from the recently appointed headteacher of S2. It is located near to the centre of a large English town and has had a troubled recent history having been identified as 'failing' by the Inspectorate. During the research project it was undergoing a major re-organization following merger with another local school, also considered to be 'failing'. The school's catchment area reflected a diverse community in which many different ethnic groups were represented. The most significant group was students of Bangladeshi heritage (14 per cent), but more recently there had been a significant movement of Eastern Europeans into the area, many seeking refuge and asylum. This ensured that not only was the school's catchment area ethnically diverse, with a wide range of different ethnic groups within the school, but that the community, at the time of the research project, lacked stability and cohesion. The inward movement of new ethnic groups had created new tensions and significant inter-ethnic rivalries. At one point this had escalated into a major confrontation near to the school, which was then sensationalized by the local media and exploited by far-right political parties. In addition to the above, the school was located in a socio-economically mixed local area with some areas of significant disadvantage.

Working within this context the headteacher had a passionate commitment to education as a means of improving life chances. Specifically there was an explicit commitment to a form of social justice that actively sought to achieve most, for those who had least. In this instance the headteacher articulated explicit egalitarian commitments based in particular on social class and ethnicity. Working in a community where a significant proportion of the population experience economic disadvantage, and the consequences of both localized and systemic racism, this headteacher was driven by the desire to make a difference, and to make most difference to those most disadvantaged. Throughout the school there was evidence of how the values of the headteacher were reflected in both her own personal practice and conduct, and in the policies, systems and procedures of the organization. This commitment was central to the school's reputation as a successful inclusive school in which all pupils are valued and encouraged to achieve. However, despite this success, the re-organization was seen as a last ditch attempt at survival. Prior to this the school had faced closure. The reasons for this lie in the complex interplay of community and market factors that shaped the school's micro-context.

Given the school's location, it was never likely to achieve the academic success of schools in more affluent parts of the town. In recent years this situation had deteriorated and the school was academically the lowest performing school in the town. However, the problems of improving league table performance could not be tackled by traditional school improvement strategies alone without seriously challenging the underlying values of the institution. Within the local area, over the years there had been significant over-capacity. Parents were able, and did, choose to send their children to schools outside their own catchment area. The largest school in the town had made a deliberate pitch at presenting itself as a model of traditional academic excellence. It adopted the badges of traditionalism (uniform, prize givings) and was intolerant of non-conformists (Gewirtz 2002). It had been quick to exclude those students who were considered not to fit. The consequence was to both draw in aspiring and able children from outside its catchment area, and to disperse those considered difficult to schools elsewhere. The impact on S2 was that it simultaneously found its more able students exiting its catchment, whilst it was being asked to receive increasing numbers of children with behavioural problems from elsewhere:

> We co-operate as local heads very well, but actually when it comes to the survival of your school would you rather have my agenda or would you rather have a settled, over-subscribed totally supportive population of white parents. I get rung up by other heads in the town who say 'we're having difficulty with this difficult child, you've got some spare places – would you like them?' If it went to appeal we couldn't say no anyway.
>
> (S2, Headteacher)

Inevitably, this depressed results further. However, it was also clear that parental preferences were not based purely on academic achievement, but about parental perceptions of the type of school S2 was, and the type of student who went there. Hence parental preferences reflected the racist assumptions of some parents of white children who opted out of S2 because of its ethnic profile (Gewirtz *et al.* 1995). In this instance the more committed the headteacher was to an agenda based on social inclusion, the more likely it was that this wider perception of the school would result in many of the more able, and the more affluent, opting to seek their education in schools elsewhere. Moreover, the imbalanced movement of pupils between catchment areas ensured that the apparently successful school was over-subscribed, whilst S2 had surplus places. The further consequence of this was that when a significant number of refugee and asylum-seeker youngsters were placed in the town then the overwhelming majority of these were allocated to S2, where there were places. Over-subscribed schools could wring their hands and express a willingness to accept these youngsters, but claim that in practical terms this was not possible because of lack of space.

The headteacher at S2 was determined to do the best by the children placed in her school. However, with inadequate resources the creation of additional need was likely to cause problems. In this instance tensions arose because supporting refugee and asylum seeker children could be seen as a deflection from the goal of raising examination results for other students. The problem for the school was that its examination results were seen as poor and the school had faced possible closure as a consequence:

> People despair – that is not too hard a word for it. You get someone in your lesson who hardly speaks any English, and you have to meet those needs. You get support, but it is yet another need in an already needy situation. A lot of staff will run with it, but convincing them is a permanent sales job – 'this is good for the children .... this is what we are here to do'.
>
> (S2, Headteacher)

> That is a hard thing to manage when you are being hammered for your exam results and you're telling staff to get the results, get the value-added, but if you get refugee children then they often come for six weeks then they're re-located. You put all that effort in, you get to know a child – and then they are moved on.
>
> (S2, Headteacher)

This was not about a lack of commitment to refugee and asylum-seeker children, on the contrary. But what is illustrated is the way in which values are challenged by the tension between a personal commitment to support individual children and the need to satisfy market-driven league table criteria. In this context these were not compatible objectives. Once again, as illustrated in the previous chapter, this case highlights the tensions between the demands of a market-driven system and the imperatives of an institutional agenda based on a commitment to social inclusion. In this instance the school leader's strong personal commitment to social justice was being challenged and undermined by the dominance of a standards and accountability agenda that effectively penalized those who prioritized educational and egalitarian values over market priorities:

> *Interviewer*: Would you say you have you been penalized for being a good inclusive school?

> *Headteacher*: I have no doubt in my mind, having gone down the inclusion route, that affects people's perceptions of what we are, and we do not get the 'top end' [based on academic ability]. I do think that is a consequence of how the school has chosen to present itself, to welcome everybody.
>
> (S2, Headteacher)

In this case the headteacher's commitment to being an open and inclusive school created an environment which was highly regarded by many in the local community. Students and parents who participated in the research project valued what the school sought to provide – a supportive environment in which all were valued and encouraged to succeed. However, a market mechanism that places a premium on the promotion of self-interest was resulting in a school community skewed along class and ethnic lines. As a result, community divisions were compounded by the local system of schooling, rather than challenged. Furthermore, the commitment to the school's core values potentially threatened its very existence. The ethical dilemma facing the headteacher was how to hold on to educational values that seemed incompatible with institutional survival.

Conclusion

The schools referred to in this chapter illustrate the importance of institutional policy agendas and the extent to which these are shaped by the values and commitment of school leaders. In such cases school leaders are able to shape policy agendas significantly to reflect values based on a commitment to social justice, and explicitly a form of cultural justice based on ethnicity and equality. In some cases there had been resistance within the institution. In other cases staff were sympathetic to the drift of policy, but saw school priorities as being in tension with the institution's ability to survive in a market environment where commitments to ethnic diversity were at odds with market success. In these instances school leaders were constantly negotiating the development of institutional policies, sometimes confronting resistance, at other times constructing alliances in order to circumvent it. In yet other cases they were proselytizing for policies to win the support and commitment of a sympathetic staff concerned about issues of job security and personal livelihoods.

A feature of these school leaders was the extent to which they were able to exploit opportunities provided by external policy agendas, and to seek some alignment between internal and external policies. In the case-study schools, this had become easier following the election of a Labour government. In this instance clear statutory initiatives, such as the Race Relations (Amendment) Act 2000, coupled with a broader commitment to a policy of 'social inclusion' had opened up spaces and opportunities that creative leaders were able to exploit. However, capitalizing on the national policy agenda was never about a simple case of implementation of external initiatives, but about using these to go beyond what was often envisaged in national policy. In these cases school leaders were using national initiatives to strengthen their own arguments for change within their own institutions, but using this as a base, as a departure point, from which to promote bolder and more radical initiatives.

However, just as school leaders in these schools were keen to make use of external policy agendas that supported their personal objectives for their institution, so too they had to respond to external policy initiatives that cut across and undermined these objectives. In Chapter 7 it was seen how external policy agendas that sought to promote equality objectives could be challenged and undermined by initiatives that promoted 'standards' and market-driven accountability. In very similar ways this was illustrated by some of the schools in these case-studies. Accountability mechanisms based on quasi-markets and parental preferences generated two identifiable outcomes – both inconsistent with the rhetoric of central government policy. First was the experience evident in some schools, of parental preferences being driven almost exclusively by ethnic differences. In these instances schools failed to be the multi-ethnic reflections of their local communities but instead became homogenously representative of particular ethnic groups. In such cases the efforts of those working in schools to promote genuinely diverse, multi-ethnically communities was undermined by parental preferences that created ethnic segregated institutions. Secondly was the experience of parents deliberately opting away from a school that may be seen as 'inclusive', particularly in terms of ethnicity. Here leaders' values are put starkly to the test – retain educational principles and risk school closure, or adopt policies and practices which ensure institutional survival but compromise educational values?

9 Conclusion

In this book, policy has been presented here as the capacity to operationalize values derived from discourses within the socio-political environment. This highlights the dual nature of policy as both product (a textual statement of values and principles) and process (the power to formulate textual statements into operational practices). It has been demonstrated that policy is a dialectic process in which all those affected by the policy may be involved in shaping its development. The policy process passes through a variety of stages and can take place at a number of different levels. To understand the policy process requires more than an understanding of the priorities of governments or of individual school leaders. It is both a continuous and a contested process in which those with competing values and differential access to power seek to form and shape policy in their own interests. The model that has been developed to illustrate the complexities of the policy process by examining how strategic direction is derived from the wider political agenda is formulated into organizational principles and operational practices.

It has been argued that the educational policy is extremely complex. This is reflected in Table 9.1 which provides a summary of the discourses that can be identified in the socio-political environment and which shape the strategic directions of the educational policy considered in this volume. These discourses cover a wide range of issues from economic utilitarianism and urban regeneration to social inclusions and integration or assimilation. The text of educational policy frequently reflects a variety of discourses that compete within the socio-political environment, an arena within which, by definition, a range of ideologies are struggling for supremacy. Such discourses will not only reflect differing values perspectives, but also the differential access to power since those with the power resources to mobilize can more readily shape policy debates. These discourses are therefore contested and often generate sets of expectations that cannot all be met and problems that cannot all be resolved, not least because resources are limited and some alternatives are mutually exclusive. Market accountability, public-private partnerships, multi-culturalism and a citizenship agenda are just a few examples considered in this book. Thus, although strategic direction of policy is largely a product of the dominant discourse within the socio-political environment, it is often subject to different interpretations that, in turn, produce alternative organizational principles which might include competition in the market place, cross-school collaboration and values-driven leadership and a battery of institutional practices and procedures such as performance management, teacher engagement and community participation.

This is a complex set of policy agendas, yet it is argued here that one of the effects of globalization was to create a situation in which national governments increasingly viewed education as an important adjunct to the economic development of the nation state. The result of this has been to elevate economic values to a primordial position in shaping the socio-political discourse within which policy is developed. This is not to adopt the view of world systems theory that emphasizes the dynamics operating at trans-national level while casting the nation state in the role of hapless pawn being sacrificed to the pressures of globalization (Gallagher 2005). It has been emphasized throughout that at state and institutional levels there are significant variations within a similar policy context. Nevertheless, similar discourses emerging from their socio-political environments can be identified in many countries. The discourse based on economics and national competitiveness is perhaps the most powerful of these. These economic rationalist discourses take at least two forms. The first, and perhaps the most dominant, is that based on human capital theory that directly associates education with economic survival, competitiveness, growth and prosperity. Educational institutions are now, more than ever before, required to produce students with the appropriate skills and capabilities to match national priorities. The second, closely linked with human capital, seeks to maximize output whilst simultaneously controlling the cost of inputs. The consequence of this has been a tendency to shift the resourcing of education provision to the private sector – both to commercial providers and individual consumers. These discourses are often linked to notions of citizenship such that educational institutions are tasked with inculcating their students with those values that enable them to become productive members of the nation state. The challenge in many contexts is to promote a sense of citizenship, and corresponding sense of national identity, in societies which are increasingly fragmented, not least in terms of cultural and ethnic diversity (Crouch *et al.* 2001).

Here, of course, there is a tension between the efficient use of resources and the provision of a fair and equitable education system that offers opportunities for the disadvantaged. In these circumstances the pursuit of economic growth resulting from, for example, liberalized markets may stand at odds with aspirations for a more equitable and just society. Such tensions highlight the conflicts between values that shape education policy and the way in which policy is forged between the pressures of competing values and the differing interest groups that advance them. Where economic

Table 9.1 Policy into practice – a composite model

Level	Features	Human capital	Social justice and citizenship	Accountability	Strategic planning	Education Action Zones	Multi-cultural school
Socio-political environment	• Contested discourses • Dominant language of legitimation • First order values shape policy	• Economic functionality • Labour market demands • Maximizing economic growth	• Definitions of citizenship rights • Social cohesion and national identity • Social justice and fairness	• Economic utility • Value for money	• Economic utility • National Identity • Value for money	• Standards' agenda /underachievement • Social inclusion • Urban regeneration • Economic modernization	• National identity and social cohesion • Social inclusion • (Under) achievement and ethnicity • Integration or assimilation • Institutionalized racism
Strategic direction	• Policy trends emerge • Broad policy established • Applied to policy domains	• Quasi-markets • Direct or indirect control • Skills and knowledge requirements	• Access and entitlement to services • Participation in service provision • Curriculum construction	• Market accountability • Direct and indirect control • Autonomous schools	• Accountability • Direct and indirect control • Autonomous school • Long-term planning	• Public/private partnerships • Multi-agency working • Curriculum vocationalization • Flexible employment contracts for teachers	• Multi-culturalism and/or anti-racism? • National curriculum • Monitoring by ethnicity • Citizenship agenda
Organizational principles	• Targets set • Success criteria defined • Patterns of control established	• Patterns of accountability • Outputs clearly defined • Control mechanisms	• Stakeholder participation • Institutional articulation of citizenship • Formal assessment of 'citizenship curriculum'	• Choice and autonomy • Individual and institutional performance • Targets/outputs clearly defined	• Values and beliefs • Head teacher leadership and responsibility • School-based resourcing	• Institutional agendas • Professional values • External targets • Cross-school collaboration	• Values-driven leadership • Stakeholder participation
Operational practices and procedures	• Organizational procedures determined • Monitoring mechanisms established • Second order values mediate policy	• Outcomes drive curriculum and assessment • Leadership and management • Parents as partners	• Curriculum re-organization • Resource allocation • Parent and student 'voice'	• Performance measures identified • Monitoring mechanisms established • Monitoring information published	• Short-term problem solving • Consultation and collaboration • Professional development	• Professional development • Teacher engagement • Community participation	• Curriculum adaptation • Community links • Staff support

values increasingly prevail over social values, education policy debate will take place within increasingly narrow parameters. The resulting discourse provides legitimation for operational practices linked to the curriculum assessment of students, performance management of teachers and institutional accountability mechanisms. As Lam (2001) points out:

> Conceptually, economic rationalism as the hegemonic cornerstone of educational changes aspires to guarantee the quality of human resources in preparation for the new economic world order. Harbored in the economic rationalistic perspective is a set of principles ... These include efficiency of operation, visible evidence of increased productivity, and an unambiguous system of accountability.
>
> (Lam 2001: 351)

There are, however, other possible discourses that might legitimate an alternative approach to educational provision. Grace (1988) moves some way towards this by drawing on an alternative socio-political discourse leading to a different conceptualization of the educative process:

> Might not education be regarded as a public good because one of its fundamental aims is to facilitate the development of the ... artistic, creative and intellectual abilities of all citizens, regardless of their class, race or gender status and regardless of their regional location.
>
> (Grace 1988: 214)

Grace uses the term 'public good' to designate a publically provided service intended to enhance the life of all citizens through the acquisition of moral, intellectual, creative, economic and political competencies. He argues that moral accountability ought to be to the community, not to the market place or to the economics of business:

> The responsibilities of educational leadership are to build educational institutions around central values ... values ... of democratic culture.
>
> (Grace 1995: 212)

This rationale for educational provision provides a much more powerful argument for the nurturing of ethical responsibilities and moral values than does a human capital approach (Grace 1994). It actually re-asserts the central role of public education, removed from the market and the motivations of private providers, as a citizenship entitlement (Freedland 2001). The attempt by Grace (1994) to develop an alternative discourse for the legitimation of educational policy re-focuses attention on some important questions. How far does the education offered foster a rationality that sees criticism as an essential part of the educative process? To what extent are people treated as means to an end or an end in themselves? Are people regarded solely as resources to be manipulated or are they developed as resourceful human beings? Is an ethos created within the institution in which personal growth is promoted and democratic processes prevail that can be replicated in society at large (Bottery 1992)? By considering these questions it becomes possible to conceive of education *for* democracy in which choice and diversity are linked to the aspirations of all, rather than as the means by which inequalities are reinforced (Lauder 1997).

Such an alternative socio-political discourse will need to generate organizational principles that recognize that the world is unpredictable and the only certainty is uncertainty. Such an environment requires an approach to education that is not based on a set of immutable, externally imposed targets or on the development of a set of predetermined skills designed to facilitate entry to the labour market. It has to recognize also, that in coping with the new future, important information may not be available, important alternatives may be ignored and important possible outcomes neglected. At the strategic level, the capacity to retain a distinct separation between means and ends, between outcomes and benefits and to rely on the linear relationship between them is greatly reduced in this new environment.

In schools and colleges, there must be an agreement about basic values and broadly acceptable means, which are not rooted in the traditional hierarchical management model with its rule-bound inflexibilities and emphasis on the separation of functions. These values must inform both the educative process and the leadership of it in such a way that leadership and management move away from the target driven and instrumental and take on a wider ethical dimension. This ethical dimension of educational leadership, as Starratt (2005) points out, might operate at five levels. At its most basic, ethical educational leadership involves acting humanely towards others. Educational leaders should also carry out their responsibilities as citizens and as public servants. The educational dimension of ethical leadership is grounded in the realization that education is more than simply a public service. It is about developing and supporting individuals. As an administrator and manager the ethical educational leader will treat everyone in the school with compassion, engaging them in the ethical exercise of the common, core work of the school. This requires the administrator to orchestrate the resources, structures and processes of the school within negotiated agreements about what the nature of the work is and what is expected from the various members of the school community (Starratt 2005). At what Starratt (2005) calls the ethical enactment of educational leadership:

> The leader has to be humane, ... even while appealing to the more altruistic motives of teachers and students ... The leader has to affirm their dignity and rights as autonomous citizens, even while appealing to their higher ... democratic ideals. The leader has to acknowledge the demanding nature of teaching and learning, ... even while appealing to the transformational possibilities of authentic learning ... Finally

the educational leader has to acknowledge the ethics of organisational life, the fact that every organisation imposes limits of the freedom ... of individuals ... and ... to see that ... basic contracts are honoured out of fairness and justice.

(Starratt 2005: 67)

Mechanisms for holding schools to account might become less of a process by which schools, colleges and their staff are: 'liable to review and [to] the application of sanctions if their actions fail to satisfy those with whom they are in an accountability relationship' (Kogan 1986: 25).

New forms of accountability could be based on responsiveness: 'the willingness of an institution – or, indeed, an individual – to respond on its own or their own initiative, i.e. the capacity to be open to outside impulses and new ideas' (Scott 1989: 17).

Scott suggests that the responsiveness of teachers and educational institutions may be best captured within a web of professional responsibilities that embody codes of practice and sets of values that are all the more influential because they are self-imposed. Thus, accountability would come from the adherence to what Sockett (1980) termed principles of practice rather than from the evaluation of results based on student performance. Here the teacher would be regarded as an autonomous professional, not as a social technician (Carter and Stevenson 2005), within the bureaucratic framework of a school. Scott (1989) also recognizes the centrality of the teacher in this approach to accountability. He claims that professionals know that it is the student who is at the centre of the accountability process:

> In discussing ways to improve the responsiveness of institutions and individuals in education it would be wrong to ignore the professional model. It does allow accountability to be exercised outside the immediate context of politics and the market.
>
> (Scott 1989: 20)

This commitment to students and other stakeholders is widely recognized as a central tenet of professional accountability:

> Professionals are judged by other professionals: they are accountable to their peers ... Legal and contractual accountabilities exist and can be used in extremes, but they are not what secures proper performance. Commitment to pupils and their parents, to the outcomes of training, to best practice and to ethical standards are more effective here.
>
> (Burgess 1992: 7)

In the climate of the twenty-first century, however, the concept of professional accountability may be criticized by some for its inward, provider-dominated focus in contrast to the consumer stance espoused by governments in many countries. 'The time has long gone when isolated, unaccountable professionals made curriculum and pedagogical decisions alone, without reference to the outside world ... Teachers in the modern world need to accept accountability' (DfEE 1998a: para. 13).

Nevertheless, professional accountability may be regarded as legitimate if considered alongside that of other forms:

> The difficulty here is the extent to which accountability and responsiveness with its implied levels of autonomy are applied in relation to each other. The ideas of 'autonomy' and 'responsibility' [are] conceptually linked with that of 'a profession'. An association of people would not be entitled to the status of a profession if it was not in a position to accept 'responsibility' for its activities. People can be held responsible for their activities only if they are free to decide between alternatives. In other words, 'responsibility' can only be ascribed to those who are free to act 'autonomously'.
>
> (Elliott *et al.* 1979: 8)

The organizational principles that shape the mechanisms for holding educational institutions to account might, therefore, produce different operational practices. These operational practices within schools must establish work relationships that are more multi-functional and holistic, based on a wider distribution of power within the organization. Co-operation, responsiveness, flexibility and partnership must replace our present inflexible structures. Planning needs to be based on a process of reaching agreement on a series of short-term objectives derived from negotiated and shared common values that take into account the wider aesthetic, ethical and social benefits that education can provide. Such an approach to leadership and management will be predicated on openness and collaboration where flexibility, creativity, imagination and responsiveness can flourish. It will require a new form of leadership that embraces a wide range of cultures and practices (Singapore Teachers' Union 2000). This agenda is not a clarion call to return to an age of unaccountable autonomy, but rather a plea to reconceptualize accountability in terms that recognize the central role of professionalism and partnership.

This language of legitimation locates the capacity to respond rapidly to changing situations with an agreed view of what might be possible based on a series of incremental responses to external change. In order to succeed it requires both a coherent sense of purpose and an enlightened approach to education that recognizes that the world for which we all seek to plan is neither predictable nor controllable. It is from this starting point that planning must evolve. Such planning should be based on a collaborative process of looking for what is right through sharing rather than competing and by accepting the validity of a range of different perspectives. Meanings will be founded on a commonality of experience, not on the defence of differences and constructed through reasoning with others and through narratives rather than analysis. Such processes must provide a foundation on which flexible yet inclusive policy formulation based on holistic relationships and focusing on integration rather than fragmentation can evolve. These policies will recognize that the sum is greater than the parts and celebrate the imaginative and the experimental not the narrow and restrictive.

The capacity to be flexible and the capability to be creative cannot be taught as an adjunct to literacy and numeracy, compartmentalized as a curriculum subject or measured and assessed. Nor can bureaucratic fiat address the crucial issue of how to teach children to be independent and creative (Lauder *et al.* 1998). The immediate human capital requirements of the economy and the concomitant managerialism, as perceived by politicians and industrialists, should not be the only factors that shape the context and text of policies that determine educational provision in any society. It should be remembered that, as schools and colleges struggle to cope with the needs of a changing, global world, the problems that confront them may be larger than any one group can solve:

> finding solutions will require cooperation and collaboration. Collaboration holds the possibility of higher quality decisions. As principals collaborate with teachers, they make use of the knowledge and expertise of those organizational participants most often in touch with the primary constituents of the school ... Collaboration can generate the social capital necessary for excellent schools as both parents and teachers participate in the problem-solving processes ... Collaboration in an atmosphere of trust holds promise for transforming schools into vibrant learning communities.
>
> (Tschannen-Moran 2001: 327–8)

Bibliography

Advisory Committee on School-based Management (2000) *Transforming Schools into Dynamic and Accountable Professional Learning Communities – School-based Management Consultation Paper*, Hong Kong Special Administrative Region of China: Education Department.

Agbo, S. (2004) 'The dialectics of education for modernization and the African diaspora: unstated features of the University as a vehicle for national development in Africa', paper presented at Athens Institute for Education and Research 6th International Conference on Education. 21–23 May, Athens.

Amsler, M. (1992) *Charter Schools: Policy Briefs 19*, San Francisco: Far West Laboratory for Educational Research and Development.

Andrews, G. (ed.) (1991) *Citizenship*, London: Lawrence and Wishart.

Angelo Corlett, J. (1991) (ed.) *Quality and Liberty: Analyzing Rawls and Nozick*, London: Macmillan.

Apple, M. (2000) 'Between neoliberalism and neoconservatism: education and conservatism in a global context', in N. Burbules and C. Torres (eds) *Globalization and Education: Critical Perspectives*, New York: Routledge.

—— (2003) *The State and the Politics of Knowledge*, London: RoutledgeFalmer.

—— (2004) *Ideology and Curriculum*, 3rd edn, London, RoutledgeFalmer.

Apple, M. and Beane, J. (1999) *Democratic Schools: Lessons from the Chalk Face*, Buckingham: Open University Press.

Ashton, D. and Sung, J. (1997) 'The Singaporean approach', in A. Halsey, H. Lauder, P. Brown and A. Wells (eds) *Education: Culture, Economy and Society*, London: Oxford University Press.

Audit Commission/OfSTED (2003) *School place planning: the influence of school place planning on school standards and social inclusion*, HMI 587, Online. Available HTTP: <http://www.ofsted.gov.uk> (accessed 29 March 2005).

Bacharach, S. and Lawler, E. (1980) *Power and Politics in Organizations: The Social Psychology of Conflict, Coalitions and Bargaining*, San Francisco: Jossey-Bass Publishers.

Bachrach, P. and Baratz, S. (1962) 'Two faces of power', in M. Haugaard (ed.) (2002) *Power: A Reader*, Manchester: Manchester University Press.

Bacon, R. and Eltis, W. (1976) *Britain's Economic Problem: Too Few Producers*, London: Macmillan.

Ball, S. (1987) *The Micro-Politics of the School: Towards a Theory of School Organization*, London: Routledge.

—— (1993) 'Education policy, power relations and teachers' work', *British Journal of Educational Studies*, 41(2): 106–21.

—— (1994) *Education Reform: A Critical and Post-structural Approach*, Buckingham: Open University Press.

—— (1999) 'Labour, learning and the economy: a "policy sociology" perspective', *Cambridge Journal of Education*, 29(2): 195–206.

Barber, M. (1992) *Education and the Teacher Unions*, London: Cassell.

—— (1993) 'Equality with diversity', *Education Today and Tomorrow*, 44(3): 16–17.

Barr, R. and Dreeben, R. (1983) 'Subject departments and the implementation of National Curriculum policy: an overview of the issues', *Journal of Curriculum Studies*, 24(2): 7–115.

Bartlett, W. (1992) *Quasi-markets and Educational Reforms: A Case Study*, Bristol: University of Bristol: School for Advanced Urban Studies.

Bassey, M. (2001) 'The folly of the global phenomenon of economic competitiveness as the rationale for educational development', *Research Intelligence*, 76, July: 30–6.

Beare, H. (2001) *Creating the Future School: Student Outcomes and the Reform of Education*, London: RoutledgeFalmer.

Beck, U. (1992) *Risk Society: Towards a New Modernity*, London: Sage.

Begley, P.T. (2004) 'Understanding valuation processes: exploring the linkage between motivation and action', *International Studies in Educational Administration*, 32(2): 4–17.

Bell, D. (2003) 'Urban schooling revisited', *Times Educational Supplement*, 21 November: 21.

Bell, L. (1989) *Control and Influence in a Teaching Union*, Sheffield: Sheffield City Polytechnic Papers in Educational Management, Number 88.

—— (1999a) 'Back to the future: the development of education policy in England', *Journal of Educational Administration*, 37(3) and (4): 200–28.

—— (1999b), 'The quality of markets is not strain'd. It droppeth as the gentle rain from Heaven upon the place beneath, Primary schools in the education market place', in T. Bush, L. Bell, R. Bolam, R. Glatter and P. Ribbins, (eds) *Educational Management, Redefining Theory, Policy and Practice,* London: Paul Chapman Publishers.

—— (2002a) 'Strategic planning and school management: full of sound and fury, signifying nothing?', *Journal of Educational Administration*, 40(5): 407–24.

—— (2002b) 'New Partnerships for Improvement' in T. Bush and L. Bell (eds) *The Principles and Practice of Educational Management*, London: Paul Chapman Publishers.

—— (2004a), 'Throw physic to the dogs. I'll none of it! Human capital and educational policy – an analysis', paper presented at Athens Institute for Education and Research 6th International Conference on Education. 21–23 May, Athens.

—— (2004b) 'Visions of Primary Headship', *Primary Headship*, 17: 8–9.

Bell, L. and Chan, D. (2004) 'Principals, leadership and strategic planning in primary schools in Hong Kong and England: a comparison', paper presented at the Commonwealth Council for Educational Administration and Management Regional Conference, *Educational Leadership in Pluralistic Societies*, Hong Kong. October.

Beresford, J. (2003) 'Developing students as effective learners: the student conditions for school improvement', *School Effectiveness and School Improvement*, 14(2): 121–58.

Blair, T. (2001) 'A second term to put secondary schools first', *The Times*, 5 September.

Blakemore, K. (2003) *Social Policy: An Introduction*, Buckingham: Open University Press.

Boisot, M. (1995) 'Preparing for turbulence: the changing relationship between strategy and management development in the learning organisation', in B. Garret (ed.) *Developing Strategic Thought: Rediscovering the Art of Direction-Giving*, London: McGraw-Hill.

Bolam, R. (2002) 'Professional development and professionalism', in T. Bush and L. Bell (eds) *The Principles and Practice of Educational Management*, London: Paul Chapman Publishers.

Booth, T., Ainscow, M., Black-Hawkins, K., Vaughan, M. and Shaw, L. (2000) *Index for Inclusion: Developing Learning and Participation in Schools*, Bristol: Centre for Studies on Inclusive Education (CSIE).

Bottery, M. (1992) *The Ethics of Educational Management: Personal, Social and Political Perspectives on School Organisation*, London: Cassell.

—— (1998) ' "Rowing the Boat" and "Riding the Bicycle" – metaphors for school management and policy on the late 1990s', paper presented to the 3rd ESRC Seminar, *Redefining Education Management*, Milton Keynes: Open University.

—— (2000) *Education, Policy and Ethics*, London: Continuum.

—— (2004a) *The Challenges of Educational Leadership*, London: Paul Chapman Publishing.

—— (2004b) 'Education and globalisation: redefining the role of the educational professional', University of Hull: Inaugural Professorial Lecture.

Bourdieu, P. (1997) 'The forms of capital', in A.H. Halsey, H. Lauder, P. Brown and A. Stuart Wells (eds) *Education: Culture, Economy, Society*, Oxford: Oxford University Press.

Bowe, R. and Ball, S.J. with Gold, A. (1992) *Reforming Education and Changing Schools*, London: Routledge.

Bowles, S. and Gintis, H. (1976) *Schooling in Capitalist America*, London: Routledge and Kegan Paul.

Brown, P. and Lauder, H. (1997) 'Education, globalisation and economics', in A.H. Halsey, H. Lauder, P. Brown and A.S. Wells (eds) *Education: Culture, Economy, Society*, Oxford: Oxford University Press.

Brown, P., Halsey, A.H., Lauder, H. and Wells, A.S. (1997) 'The transformation of education and society: an introduction', in A.H. Halsey, H. Lauder, P. Brown and A.S. Wells (eds) *Education: Culture, Economy, Society*, Oxford: Oxford University Press.

Bullock, A. and Thomas, H. (1997) *Schools at the Centre? A Study of Decentralisation*, London: Routledge.

Bulmahn, E. (2000) *Address to the American Association for the Advancement of Science*. Online. Available HTTP: <http://www.bmbf.de/reden.htm> (accessed 12 January 2004).

Burbules, N. and Torres, C. (eds) (2000) *Globalization and Education: Critical Perspectives*, New York: Routledge.

Burgess, T. (1992) 'Accountability with confidence', in T. Burgess (ed.) *Accountability in Schools*, Harlow: Longman.

Bush, T. (2002) 'Educational management: theory and practice', in T. Bush and L. Bell (eds) *The Principles and Practice of Educational Management*, London: Paul Chapman Publishers.

Bush, T. and Chew, J. (1999) 'Developing human capital: training and mentoring for principals', *Compare*, 28(2): 41–53.

Busher, H. and Harris, A. (2000) *Leading Subject Areas; Improving Schools*, London: Paul Chapman.

Caldwell, B. (1992) 'The principal as leader of the self-managing school in Australia', *Journal of Educational Administration*, 30(3): 6–19.

—— (2002) 'Autonomy and self-management: concepts and evidence', in T. Bush and L. Bell (eds) *The Principles and Practice of Educational Management*, London: Paul Chapman Publishers.

Caldwell, B. and Spinks, J. (1998) *Beyond the Self-Managing School*, London: Falmer Press.

Carlson, R. (1989) *Restructuring Schools. Internal Memorandum*, Washington, DC: Washington DC Public Schools.

Carter, B. and Stevenson, H. (2005) 'Teachers, class and the changing labour process', paper presented at 23rd Annual International Labour Process Conference, 21–23 March, University of Strathclyde, Glasgow.

Carter, D. and O'Neil, M. (1995) *International Perspectives on Educational Reform and Policy Implementation*, London: Falmer Press.

Carter, K. (2002) 'Leadership in urban and challenging contexts: investigating EAZ policy in practice', *School Leadership and Management*, 22(1): 41–59.

Carter, T. (1986) *Shattering Illusions: West Indians in British Politics*, London: Lawrence and Wishart.

Chan, D. (2002) *How primary school principals manage strategic planning: a comparative study between Hong Kong and England. A report for the National College of School Leadership*, University of Leicester: Centre for Educational Leadership and Management.

Cheng, Y.C. (1999) 'Recent educational developments in south east Asia', *School Effectiveness and Improvement*, 1(1): 3–30.

—— (2000) 'Change and development in Hong Kong: effectiveness, quality and relevance', in T. Townsend and Y.C. Cheng (eds) *Educational Change and Development in the Asia-Pacific Region: Challenges for the Future*, Lisse: Swets and Zeitlinger.

—— (2002) 'The changing context of school leadership: implications for paradigm shift', in K. Leithwood and P. Hallinger (eds) *Second International Handbook of Educational Leadership and Administration*, Dordrecht: Kluwer Academic Publishers.

Cheng, Y. and Cheung, W. (1999) 'Towards school-based management: uncertainty, meaning, opportunity and development', *International Journal of Education Reform*, 8(1): 25–36.

Cheung, W., Cheng, Y. and Tam, W. (1995) 'Parental involvement in school education: concepts, practice and management', *Journal of Primary Education*, 5(2): 57–66.

Chew, J. (2001) 'Principal performance appraisal in Singapore', in D. Middlewood and C. Cardno (eds) *Managing Teacher Appraisal and Performance: A Comparative Approach*, London: RoutledgeFalmer.

Chitty, C. (ed.) (1993) *The National Curriculum: is it working?* Harlow: Longman.

Choi, P. K. (2003) ' "The best students will learn English": ultra-utilitarianism and linguistic imperialism in education in post-1997 Hong Kong', *Journal of Educational Policy*, 18(6): 673–94.

Chong, K. and Boon, Z. (1997) 'Lessons from a Singapore programme for school improvement', *School Effectiveness and Improvement*, 8(4): 463–70.

Chubb, J. and Moe, T. (1992) *A Lesson in School Reform from Great Britain*, Washington DC: Brooking Institution.

Coote, A. (ed.) (2000) *New Gender Agenda: Why Women Still Want More*, London: Institute for Public Policy Research.

Costello, N., Michie, J. and Milne, S. (1989) *Beyond the Casino Economy*, London: Verso.

Cowham, T. (1994) 'Strategic planning in the changing external context', in M. Crawford, L. Kydd and S. Parker (eds) *Educational Management in Action: A Collection of Case Studies*, London: Paul Chapman Publishers.

Cribb, A. and Gewirtz, S. (2003) 'Towards a sociology of just practices: an analysis of plural conceptions of justice', in C. Vincent (ed.) *Social Justice, Education and Identity*, London: RoutledgeFalmer.

Crouch. C. (2001) 'Citizenship and markets in recent British education policy', in C. Crouch, K. Eder and D. Tambini (eds) *Citizenship, Markets and the State*, Oxford: Oxford University Press.

—— (2003) *Commercialisation or Citizenship: Education Policy and the Future of Public Services*, London: Fabian Society.

Crouch, C., Eder, K. and Tambini, D. (eds) (2001) 'Introduction: dilemmas of citizenship' *Citizenship, Markets and the State*, Oxford: Oxford University Press.

Crozier, G. (1998) 'Parents and schools: partnership or surveillance?', *The Journal of Educational Policy*, 13(1): 125–36.

Dahl, R. (1957) 'The concept of power', in Haugaard, M. (2002) (ed.) *Power: A Reader*, Manchester: Manchester University Press.

—— (1982) *Dilemmas of Pluralist Democracy*, New Haven, CT: Yale University Press.

Dale, R. (1989) *The State and Education Policy*, Milton Keynes: Open University Press.

Davies B. (1997) 'Rethinking the educational context: a reengineering approach', in B. Davies and L. Ellison (eds) *School Leadership for the 21st Century*, London: Routledge.

Davies B. and Davies B.J. (2005) 'Strategic leadership', in B. Davies (ed.) *The Essentials of School Leadership*, London: Paul Chapman Publishing.

Davies, B. and Ellison. L. (1999) *Strategic Direction and Development of the School*, London: Routledge.

Davies, L. (1990) *Equity and Efficiency? School Management in an International Context*, London: The Falmer Press.

Day, C., Harris, A., Hadfield, M., Tolly, H. and Beresford, J. (2000) *Leading Schools in Times of Change*, Buckingham: Open University Press.

Dearing, R. (1997) *National Committee of Enquiry into Higher Education*, London: The Stationery Office.

Department for Education (1994) *Our Children's Future, the Updated Parents' Charter*, London: HMSO.

DES (1986) *The Education Act*, London: Her Majesty's Stationery Office (HMSO).

—— (1988) *Education Reform Act*, London: HMSO.

DfEE (1997) *Excellence in Schools*, London: The Stationery Office.

—— (1998a) *Teachers Meeting the Challenge of Change*, London: The Stationery Office.

—— (1998b) *Home School Agreements: Guidance for Schools*, London: The Stationery Office.

—— (1998c) *The National Literacy Strategy*, London: The Stationery Office.

—— (1998d) *The Implementation of the National Numeracy Strategy*, London: The Stationery Office.

—— (2000) *Performance Management in Schools Performance Management Framework*, London: The Stationery Office.

—— 0051/2000, Online. Available HTTP: <http://teachernet.gov.uk/management/payand performance/performancemanagement/keydocuments (accessed 3 August 2004).

DfES (1992) *Choice and Diversity a new framework for schools*, Cardiff: The Welsh Office.

—— (2001a) *Schools Achieving Success*, London: The Stationery Office.

—— (2001b) *The Teaching Standards Framework*, London: Stationery Office.

—— (2001c) *Statutory Instrument 2001 No. 2855 Education (School Teacher Appraisal) (England) Regulations 2001*. Online. Available HTTP: <http://www.legislation.hmso.gov.uk/si2001/20012855.htm#5> (accessed 3 August 2004).

—— (2003) *Excellence and Enjoyment: A Strategy for Primary Schools*, London: DfES Publications.

—— (2004) *Five Year Strategy for Children and Learners*, London: HMSO.

Dickson, M. and Power, S. (2001) 'Education Action Zones: a new way of governing education?', *School Leadership and Management*, 21(2): 137–42.

Dickson, M., Power, S., Telford, D., Whitty, G. and Gewirtz, S. (2001) 'Education Action Zones and democratic participation', *School Leadership and Management*, 21(2): 169–82.

Dimmock, C. (1998) 'Restructuring Hong Kong's schools: the applicability of Western theories, policies and practices to an Asian culture', *Educational Management and Administration*, 26(4) 363–77.

—— (2000) *Designing the Learning-Centred School: A Cross-cultural Perspective*, London: Falmer Press.

Dimmock, C., Stevenson, H., Shah, S., Bignold, B. and Middlewood, D. (2005) *The Leadership of Schools with Substantial Numbers of Minority Ethnic Students: School Community Perspectives and their Leadership Implications*, Nottingham: National College for School Leadership.

Dye, T. (1992) *Understanding Public Policy*, Englewood Cliffs, NJ: Prentice Hall.

Education Commission (1984) *Education Commission Report No. 1*, Hong Kong: Government Printer.

—— (1990) *Education Commission Report No 4: Curriculum and student behavioural problems in schools*, Hong Kong: Government Printer.

—— (1992) *Education Commission Report No.5: The teaching profession*, Hong Kong: Government Printer.

—— (1996) *Education Commission Report No.6: Enhancing language proficiency: a comprehensive strategy*, Hong Kong Special Administrative Region of The People's Republic of China, Hong Kong: Government Printer.

—— (1997) *Quality School Education, Report No. 7*, Hong Kong: Government Printer.

—— (1999) *Education Blueprint for the 21st Century: Review of Academic System – Aims of Education (Consultation Document)*, Hong Kong Special Administrative Region of The People's Republic of China, Hong Kong: Government Printer.

Education Department (1997) *Medium of Instruction: Guidance for Hong Kong Secondary Schools*, Hong Kong: Government Printer.

—— (2000) *Transforming Schools into Dynamic and Accountable Professional Learning Communities: School-based Management Consultation Document*, Hong Kong: Advisory Committee on School-based Management.

Education and Manpower Bureau (1997) *Policy Program: The 1997 Policy Address*, Hong Kong Special Administrative Region of The People's Republic of China, Hong Kong: Government Printer.

Edwards, T. and Tomlinson, S. (2002) *Selection Isn't Working: Diversity, Standards and Inequality in Education*, Catalyst Working Paper, London: Central Books.

Edwards, W. (1991) 'Accountability and autonomy: dual strands for the administrator', in W. Walker, R. Farquhar and M. Hughes (eds) *Advancing Education: School Leadership in Action*, London: The Falmer Press.

Elliott, J., Bridges, D., Ebbutt, D., Gibson, R. and Nias, J. (1979) *School Accountability: Social Control or Dialogue*, Cambridge Accountability Project: Interim Report to the Social Science Research Council.

Elmore, R. (1988) *Early Experiences in Restructuring Schools: Voices From the Field*, Washington, DC: National Governors Association.

Entwistle, H. (1977) *Class, Culture and Education*, London: Methuen.

Farrell, C. and Morris, J. (2004) 'Resigned compliance: teacher attitudes towards performance-related pay in schools', *Educational Management, Administration and Leadership*, 32(1): 81–104.

Faulks, K. (1998) *Citizenship in Modern Britain*, Edinburgh: Edinburgh University Press.

Fergusson, R. (1994) 'Managerialism in education', in J. Clarke, A. Cochrane and E. McLaughlin (eds) *Managing Social Policy*, London: Sage.

—— (2000) 'Modernizing managerialism in education', in J. Clarke, S. Gewirtz and E. McLaughlin (eds) *New Managerialism New Welfare?*, London: Sage.

Ferlie, E., Ashburner, L., Fitzgerald, L. and Pettigrew, A. (1996) *The New Public Management in Action*, Oxford: Oxford University Press.

Fevre, R., Rees, G. and Gorard, S. (1999) 'Some sociological alternatives to human capital theory and their implications for research on post-compulsory education and training', *Journal of Education and Work*, 12(2): 117–40.

Fidler, B. (1996) *Strategic Planning for School Improvement*, London: Pitman.

—— (2002) 'Strategic leadership and cognition', in K. Leithwood and P. Hallenger (eds) *Second International Handbook of Educational Leadership and Administration*, London: Kluwer Academic Publishers.

Foskett, N. (2003) 'Market policies, management and leadership in schools', in B. Davies and J. West-Burnham (eds) *Handbook of Educational Leadership and Management*, London: Pearson Longman.

Foucault, M. (1977) *Discipline and Punish: The Birth of the Prison*, Harmondsworth: Penguin.

Freedland, M. (2001) 'The Marketization of Public Services', in C. Crouch, K. Eder and D. Tambini (eds) *Citizenship, Markets and the State*, Oxford: Oxford University Press.

Friedman, M. and Friedman, R. (1980) *Free to Choose*, New York: Harcourt, Brace and Jovanovich.

Fullan, M. (2003) *The Moral Imperative of School Leadership*, London: Sage.

Gahima, C. (2005) *Towards a Gender Balanced Education: An Investigation of the Role of Schools in the Retention of Girls in Rwandan Primary Schools*, unpublished MBA (Educational Management) thesis, University of Leicester.

Gallagher, T. (2005) 'Mediating globalisation: local responses to political and economic pressure', *British Educational Research Journal*, 31(1): 121–8.

Gaskell, J. (2002) 'School choice and educational leadership: rethinking the future of public schooling', in K. Leithwood and P. Hallinger (eds) *Second International Handbook of Educational Leadership and Administration*, Dordrecht: Kluwer Academic Publishers.

Gewirtz, S. (1998) 'Conceptualizing social justice: mapping the territory', *Journal of Education Policy*, 13(4): 469–84.

—— (2002) *The Managerial School: Post-welfarism and Social Justice in Education*, London: Routledge.

Gewirtz, S. and Ball, S.J. (2000) 'From "welfarism" to "new managerialism": shifting discourses of school headship in the education marketplace', *Discourse: Studies in the Cultural Politics of Education*, 21(3): 253–68.

Gewirtz, S., Ball, S.J. and Bowe, R. (1995) *Markets, Choice and Equity in Education*, Milton Keynes: Open University Press.

Gibton, D. (2004) 'The Empire Strikes Back! How (and Why) was England's Educational Policy Adopted by Israel's Whole System Reform', paper presented at the British Educational Management, Administration and Leadership Society annual conference, Yarnfield Park, Staffordshire, 8–10 October.

Giddens, A. (1984) *The Constitution of Society*, Cambridge: Polity Press.

—— (1998) *The Third Way: The Renewal of Social Democracy*, Cambridge: Polity Press.

—— (2001) (ed.) *The Global Third Way Debate*, Cambridge: Polity Press.

Gillborn, D. (2002) 'Inclusive schooling in multi-ethnic societies', in C. Campbell (ed.) *Developing Inclusive Schooling: Perspectives, Policies and Practices*, London: Institute of Education.

Giroux, H. (1992) 'Educational leadership and the crisis of democratic government', *Educational Researcher*, 21(4): 4–11.

Glatter, R. (2003) 'Collaboration, collaboration, collaboration: the origins and implications of a policy', *Management in Education*, 17(5): 16–20.

Gleeson, D. and Shain, F. (2003) Managing ambiguity in further education', in N. Bennett, M.Crawford and M. Cartwright (eds) *Effective Educational Leadership*, London: Paul Chapman Publishers.

Gold, A., Evans, J., Earley, P., Halpin, D. and Collarbone, P. (2003) 'Principled Principals? Values-driven leadership: evidence from ten case studies of "outstanding" school leaders', *Educational Management and Administration*, 31(2): 127–37.

Goldring, E. (1997) 'Parental involvement and school choice: Israel and the United States', in R. Glatter, P. Woods and C. Bagley (eds) *Choice and Diversity in Schooling: Perspectives and Prospects*, London: Routledge.

Gorard, S., Taylor, C. and Fitz, J. (2002) 'Does school choice lead to "spirals of decline"?', *Journal of Education Policy*, 17(3): 367–84.

Gordon, I., Lewis, J. and Young, K. (1997) 'Perspectives on policy analysis', in M. Hill (ed.) *The Policy Process*, London: Prentice Hall.

Gough, I. (1979) *The Political Economy of the Welfare State*, London: Macmillan.

Gould, B. (2001) *Times Higher Education Supplement*, 13 April.

Government of New Zealand (1998) *Tomorrow's Schools: The Reform of Educational Administration in New Zealand*, Wellington: Government Printer.

Grace, G. (1988) 'Education: commodity or public good?', *British Journal of Educational Studies*, 37(3): 207–21.

—— (1994) 'Education is a public good: on the need to resist the domination of economic science', in D. Bridges and T. McLaughlin (eds) *Education and the Market Place*, London: Falmer Press.

—— (1995) *School Leadership: Beyond Education Management – An Essay in Policy Scholarship*, London: Falmer Press.

—— (1997) 'Politics, markets, and democratic schools: on the transformation of school leadership', in A. Halsey, H. Lauder, P. Brown and A. Wells (eds) *Education: Culture, Economy, Society*, Oxford: Oxford University Press.

Gramsci, A. (1971) *Selections from the prison notebooks*, London: Lawrence and Wishart.

Green, A. (1997) *Education, Globalization and the Nation State*, Basingstoke: Macmillan Press.

—— (1999) 'Education and globalisation in Europe and East Asia: convergent and divergent trends', *Journal of Educational Policy*, 14(1): 55–72.

Green, F. (1989) 'Evaluating structural economic change: Britain in the 1980s', in F. Green (ed.) *The Restructuring of the UK Economy*, London: Harvester Wheatsheaf.

Hall, S. (2004) 'Rightwing think tank's school aims to teach traditional culture', *The Guardian*, 31 August.

Hallgarten, J. and Watling, R. (2001) 'Buying power: the role of the private sector in Education Action Zones', *School Leadership and Management*, 21(2): 143–58.

Halpin, D. and Troyna, B. (eds) (1994) *Researching Education Policy: Ethical and Methodological Issues*, London: Falmer.

Halsey, A., Heath, A. and Ridge, J. (1980) *Origins and Destinations: Family, Class and Education in Modern Britain*, Oxford: Clarendon Press.

Halsey, A., Lauder, H., Brown, P. and Stuart Wells, A. (1997) (eds) *Education: Culture, Economy, Society*, Oxford: Oxford University Press.

Hampden-Turner, C. (2003) 'Strategic dilemmas occasioned by using alternative scenarios of the future', in B. Garrett, (ed.) *Developing Strategic Thought: A Collection of the Best Thinking on Business Strategy*, London: Profile Books.

Hargreaves, A. (1995) *Changing Teachers, Changing Times: Teachers' Work and Culture in the Post-modern Age*, London: Cassell.

Hargreaves, D. and Hopkins, D. (1994) *The Empowered School: The Management and Practice of Development Planning*, London: Cassell.

Harman, G. (1984) 'Conceptual and theoretical issues', in J.R. Hough (ed.) *Educational Policy: An International Survey*, London: Croom Helm.

Harris, A. (2002) 'Effective leadership in schools facing challenging contexts', *School Leadership and Management*, 22(1): 15–26.

—— (2003) 'Teacher leadership as distributed leadership: heresy, fantasy or possibility?', *School Leadership and Management*, 23(3): 313–24.

Hatcher, R. (2005) 'The distribution of leadership and power in schools', *British Journal of Sociology of Education*, 26(2): 253–68.

Hay, D. (1985) 'International trade and development', in D. Morris (ed.) *The Economic System in the UK*, Oxford: Oxford University Press.

Hayek, F. (1973) *Law, Legislation and Liberty, Volume 1*, London: Routledge and Keagan Paul.

Hills, G. (2004) 'Making the grade', *Royal Society of Arts Journal*, April: 26–7.

Hirsch, D. (1997) 'Policies for school choice: what can Britain learn from abroad?', in R. Glatter, P. Woods and C. Bagley (eds) *Choice and Diversity in Schooling: Perspectives and Prospects*, London: Routledge.

Hodkinson, P., Sparkes, A. and Hodkinson, H. (1996) *Young People's Leisure and Lifestyles*, London: Routledge.

Hopkins, D., West, M. and Ainscow, M. (1996) *Improving the Quality of Education for All: Progress and Challenge*, London: David Fulton.

House, E. (1973) *School Evaluation: The Politics and Process*, San Francisco, CA: McCutchan Publishing Corporation.

Hoyle, E. (1982) 'Micropolitics of educational organisations', *Educational Management and Administration*, 10(2): 87–98.

Hoyle, E. and John, P. (1995) *Professional Knowledge and Professional Practice*, London: Cassell.

Ironside, M. and Seifert, R. (1995) *Industrial Relations in Schools*, London: Routledge.

Jacques, M. (2005) 'No monopoly on modernity', *The Guardian*, 5 February.

James, C. and Phillips, P. (1995), 'The practice of educational marketing in schools', in M. Preedy, R. Glatter and R. Levaçič (eds) *Educational Management: Strategy, Quality and Resources*, Buckingham: Open University Press.

Jayne, E. (1998) 'Effective school development planning', in D. Middlewood and J. Lumby (eds) *Strategic Management in Schools and Colleges*, London: Paul Chapman Publishing.

Jennings, R. (1977) *Education and Politics: Policy Making in Local Education Authorities*, London: Batsford.

Jenson, J. and Phillips, S.D. (2001) 'Redesigning the Canadian citizenship regime: remaking the institutions of representation', in C. Crouch, K. Eder and D. Tambini (eds) *Citizenship, Markets and the State*, Oxford: Oxford University Press.

Jessop, B. (1994) 'The transition to post-Fordism and the Schumpeterian workfare state', in R. Burrows and B. Loader (eds) *Towards a Post-Fordist Welfare State?* London: Routledge.

Jones, K. and Bird, K. (2000) ' "Partnership" as strategy: public-private relations on Education Action Zones', *British Education Research Journal*, 26(4):491–506.

Joseph, Sir K. (1976) *Stranded on the Middle Ground*, London: Centre for Policy Studies.

Kam, H.W.K. and Gopinathan, S. (1999) 'Recent developments in Singapore', *School Effectiveness and Improvement*, 10(1): 99–118.

Karstanje, P. (1999) 'Decentralisation and deregulation in Europe: towards a conceptual framework', in T. Bush, L. Bell, R. Bolam, R. Glatter and P. Ribbins (eds) *Educational Management: Redefining Theory, Policy and Practice*, London: Paul Chapman Publisher.

Kemp, D. (2000) Australian Ministry for Education Online. Available HTTP: <http://www.detya.gov.au/archieve/Ministers/kemp/may00> (accessed 12 January 2004).

Killeen, J., Turton, R., Diamond, W., Dosnon, O. and Wach, M. (1999) 'Educational policy and the labour market: subjective aspects of human capital', *Journal of Educational Policy*, 14(2): 99–116.

Kogan, M. (1975) *Educational Policy-Making: A Study on Interest Groups and Parliament*, London: George Allen and Unwin Ltd.

—— (1986) *Education Accountability*, London: Hutchinson.

Kreysing, M. (2002) 'Autonomy, accountability, and organizational complexity in higher education: the Goettingen model of university reform', *Journal of Educational Administration*, 40(6): 552–60.

Kwong, J.Y.S. (1997) 'Managing the present for the future', in J. Tan, S. Gopinathan and W.K. Ho (eds) *Education in Singapore*, Singapore: Prentice Hall.

Labour Party (2003) *The Best Education For All*, London: Labour Party.

Lam, Y. (2001) 'Economic rationalism and education reforms in developed countries', *Journal of Educational Administration*, 39(4): 346–58.

Latham, M. (2001) 'The Third Way: an outline', in A. Giddens (ed.) *The Global Third Way Debate*, Cambridge: Polity Press.

Lauder, H. (1997) 'Education, democracy and the economy', in A.H. Halsey, H. Lauder, P. Brown and A.S. Wells (eds) *Education: Culture, Economy, Society*, Oxford: Oxford University Press.

Lauder, H., Jamieson, I. and Wikeley, F. (1998) 'Models of effective schools: limits and capabilities', in R. Slee and G. Weiner with S. Tomlinson (eds) *School Effectiveness for Whom? Challenges to the School Effectiveness and Improvement Movements*, London: Falmer Press.

Lawton, D. (1980) *The Politics of the School Curriculum*, London: Routledge and Kegan Paul.

Lazaridou, A. (ed.) (2004) *Contemporary Issues on Educational Administration and Policy*, Athens: Athens Institute for Education and Research.

Leadbetter, C. (1999) *Living in Thin Air*, London: Viking.

LeGrand, J. (1982) *The Strategy of Equality: Redistribution and the Social Services*, London: George Allen and Unwin.

—— (1990) *Quasi-markets and Social Policy*, Bristol: University of Bristol, School for Advanced Urban Studies.

Leiberman, A. and Miller, L. (1999) *Teachers Transforming their World and their Work*, New York: Teachers College Press.

Leithwood, K. (2001) 'Criteria for appraising school leaders in an accountability policy context', in D. Middlewood and C. Cardno (eds) *Managing Teacher Appraisal and Performance*, London: Routledge Falmer.

Leithwood, K., Jantzi, D. and Steinbach, R. (2002) 'Leadership practices for accountable schools', in K. Leithwood and P. Hallinger (eds) *Second International Handbook of Educational Leadership and Administration*, Dordrecht: Kluwer Academic Publishers.

Levin, B. (2003) 'Education policy: commonalities and differences', in B. Davies and J. West-Burnham (eds) *Handbook of Educational Leadership and Management*, London: Pearson Longman.

Levin, H.M. and Kelly, C. (1997) 'Can education do it alone?', in A.H. Halsey, H. Lauder, P. Brown and A.S. Wells (eds) *Education: Culture, Economy, Society*, Oxford: Oxford University Press.

Lister, R. (2003) *Poverty*, London: Polity Press.

Llewellyn, J., Hancock, G., Kirst, M. and Roeloffs, K. (1982) *A Perspective on Education in Hong Kong: Report by a Visiting Panel*, Hong Kong: Government Printer (The Llewellyn Report).

Lloyd, C. and Payne, J. (2003) 'The political economy of skill and the limits of educational policy', *Journal of Educational Policy*, 18(1): 85–107.

Lodge, P. and Blackstone, T. (1982) *Educational Policy and Educational Inequality*, Oxford: Martin Robertson and Company.

Lukes, S. (1974) *Power: A Radical View*, London: Methuen.

Lumby, J. (2002) 'Vision and strategic planning', in T. Bush, and L. Bell, (eds) *The Principles and Practice of Educational Management*, London: Paul Chapman Publishers.

Lunt, I. and Norwich, B. (1999) *Can Effective Schools be Inclusive Schools?*, London: Institute of Education.

Mace, J. (1987) 'Education, the labour market and government policy', in H. Thomas and T. Simkins (eds) *Economics and the Management of Education: Emerging Themes*, London: The Falmer Press.

mac an Ghaill, M. (1993) 'The National Curriculum and equal opportunities', in C. Chitty (ed.) *The National Curriculum: Is it Working?*, Harlow: Longman.

McClenaghan, P.M. (2003) 'Response to "Social Capital: an analytical tool for exploring lifelong learning and community development"', *British Educational Research Journal*, 29(3): 435–9.

Macdonald, I., Bhavnani, R., Khan, L. and John, G. (1989) *Murder in the Playground: the Report of the Macdonald Inquiry into Racism and Racial Violence in Manchester Schools*, London: The Longsight Press.

Machlup, L. (1984) *Knowledge, its Creation, Distribution, and Economic Significance, Volume 1, Knowledge and Knowledge Production*, Princeton, NJ: Princeton University Press.

McNamara, O., Hustler, D., Stronach, I., Rodrigo, M., Beresford, E. and Botcherly, S. (2000) 'Room to manoeuvre: mobilising the "active partner" in home-school relations', *British Educational Research Journal*, 26(2): 473–90.

McNay, I. and Ozga, J. (1985) 'Introduction: perspectives on policy', in I. McNay and J. Ozga (eds) *Policy-Making in Education: The Breakdown of Consensus*, Buckingham: Open University Press.

Macpherson, W. (1999) *The Stephen Lawrence Inquiry*, CM 4262-I, London: The Stationery Office.

Mansell, W. (2000), 'Zones told to come up with results', *Times Educational Supplement*, 17 March.

Mansell, W. and Thornton, K. (2001) 'Blunkett faces up to zone failings', *Times Educational Supplement*, 9 March.

Marginson, S. (1993) *Education and Public Policy in Australia*, Cambridge: Cambridge University Press.

Marshall, T.H. (1950) *Citizenship and Social Class, and Other Essays*, London: Cambridge University Press.

—— (1981) *'The Right to Welfare' and Other Essays*, London: Heinemann.

Martin, Y. (1995) 'What do parents want?', *Management in Education*, 9(1): 9–11.

Middlewood, D. (2002) 'Appraisal and performance management', in T. Bush and L. Bell (eds) *The Principles and Practice of Educational Management*, London: Paul Chapman Publishers.

Middlewood, D. and Cardno, C. (2001) 'The significance of teacher performance and its appraisal', in D. Middlewood and C. Cardno (eds) *Managing Teacher Appraisal and Performance: A Comparative Approach*, London: RoutledgeFalmer.

Ministry of Trade and Industry (1991) *The Strategic Economic Plan: Towards a Developed Nation*, Singapore: Economic Planning Committee.

Mintzberg, H. (1995) 'Strategic thinking as seeing', in B. Garrett (ed.) *Developing Strategic Thought: Rediscovering the Art of Direction-giving*, London: McGraw-Hill.

Mok, K. (2003) 'Decentralization and marketization of education in Singapore: a case study of the school excellence model', *Journal of Educational Administration*, 41(4): 348–86.

Monteils, M. (2004) *The Analysis of the Relation between Education and Economic Growth*, paper presented at the 6th Annual Athens Institute for Education and Research Conference, Athens, 21–23 May.

Moore, A., George, R. and Halpin, D. (2002) 'The developing role of the headteacher in English schools: management, leadership and pragmatism', *Educational Management and Administration*, 30(2): 175–88.

Murch, I. (1997) 'Behind the headlines: learning lessons from The Ridings', *Education Today and Tomorrow*, 49(1): 9–10.

Murphy, J. (1991) *Restructuring Schools: Capturing and Assessing the Phenomena*, New York: Teachers College Press.

Nash, R. (1989) 'Tomorrow's schools: state power and parent participation', *New Zealand Journal of Educational Studies*, 24(2): 113–28.

New Zealand Treasury (1987) *Government Management: Brief to the Incoming Government (Vol. 2) Educational Issues*, Wellington: Government Printer.

NTFAE (2004) *Summary of the Report of the National Task Force for the Advancement of Education (Interim Report) (The Dovrat Report)*, National Program on Education, Tel Aviv: NTFAE.

NZSTA (1999) *Guidelines for Boards of Trustees: The Management of the Principals by the School Board of Trustees*, Wellington: Government Printer.

Ng, S-W. (1999) 'Home-school relations in Hong Kong: separation or partnership', *School Effectiveness and Improvement*, 9(4): 551–60.

Normore, A. (2004) 'The edge of chaos: school administrators and accountability', *Journal of Educational Administration*, 42(1): 55–77.

Nozick, R. (1974) *Anarchy, State and Utopia*, Oxford: Basil Blackwell.

O'Connor, J. (1973) *The Fiscal Crisis of the State*, New York: St Martins Press.

OECD (1996a) *Measuring What People Know: Human Capital: Accounting for the Knowledge Economy*, Paris: OECD.

—— (1996b) 'An overview of OECD work on teachers, their pay and conditions, teaching quality and the continued professional development of teachers', paper presented to the 45th International Conference on Education, 30 September – 5 October, Geneva: UNESCO.

OfSTED (1992) *Framework for the Inspection of Schools*, London: HMSO.

—— (2003) *Excellence in Cities and Education Action Zones: Management and Impact*, London: OfSTED/HMI.

Ogilvie, D. and Crowther, F. (1992) 'Politics in educational administration: the dominant paradigm and an ethical analysis' in F. Crowther and D. Ogilvie (eds) *The New World of Educational Administration*, Victoria: Australian Council for Educational Administration.

Olssen, M. (2004) 'From the Crick Report to the Parekh Report: multiculturalism, cultural difference and democracy – the revisioning of citizenship education', *British Journal of Sociology of Education*, 25(2): 179–92.

Olssen, M., Codd, J. and O'Neill, A. (2004) *Education Policy: Globalization, Citizenship and Democracy*, London: Sage.

O'Neill, J. (1997) 'Teach, learn, appraise; the impossible triangle', in J. O'Neill (ed.) *Teacher Appraisal in New Zealand*, Palmerston North: ERDC Press.

Osler, A. (2000) 'The Crick Report: difference, equality and racial justice', *The Curriculum Journal*, 11(1): 25–38.

Parekh, B. (1991) 'British citizenship and cultural difference', in G. Andrews (ed.) *Citizenship*, London: Lawrence and Wishart.

Picot, B. (1988) *Report of the Task Force to Review Educational Administration; Administration for Excellence; Effective Administration in Education (The Picot Report)*, Wellington: NZ Government Printer.

Plant, R. (1991) 'Social rights and the reconstruction of welfare' in G. Andrews (ed.) *Citizenship*, London: Lawrence and Wishart.

Pollitt, C. (1992) *Managerialism in the Public Services*, Oxford: Blackwell.

Polsby, N. (1963) *Community Power and Political Theory*, London: Yale University Press.

Preedy, M., Glatter, R. and Levaçi, R. (1997) (eds), *Educational Management: Strategy, Quality and Resources*, Buckingham: Open University Press.

PriceWaterhouse Coopers (undated) *Evaluation of the Role of Teachers in Education Action Zones*, London: NUT.

Psacharopoulos, G. (1986) 'Links between education and the labour market, a broader perspective', *European Journal of Education*, 21(4): 409–15, quoted in Killeen, J., Turton, R., Diamond, W., Dosnon, O. and Wach, M. (1999) (eds) 'Educational and the labour market: subjective aspects of human capital', *Journal of Educational Policy*, 14(2): 99–116.

Puffitt, R., Stoten, B. and Winkley, D. (1992), *Business Planning for Schools*, London: Longman.

QCA (1998) *Education for Citizenship and the Teaching of Democracy in Schools (the Crick Report)*, London: Qualifications and Curriculum Authority (QCA).

Rasmussen, P. (2002) 'Education for everyone: secondary education and social inclusion in Denmark', *Journal of Education Policy*, 17(6): 627–42.

Rawls, J. (1972) *A Theory of Justice*, Oxford: Clarendon Press.

Ream, R.K. (2003) 'Counterfeit social capital and Mexican-American under-achievement', *Educational Evaluation and Policy Analysis*, 25(3): 237–62.

Rees, G., Williamson, H. and Istance, D. (1996) 'Status zero: a study of jobless school-leavers in South Wales', *Research Papers in Education*, 11(2): 219–35.

Reyes, P., Scribner, J.D. and Scribner, A.P. (1999) *Lessons from High Performing Hispanic Schools: Creating Learning Communities*, New York: Teachers College Press.

Riley, K. (2004) 'Reforming for democratic schooling: learning from the future not yearning for the past', in J. MacBeath and L. Moos (eds) *Democratic Learning: The Challenge to School Effectiveness*, London: RoutledgeFalmer.

Rutayisire, J. (2004) 'Education for Social and Political Reconstruction: the Rwandan Experience from 1994 to 2004', paper presented at the British Association for Comparative and International Education Conference, University of Sussex, 3–5 September.

Rutayisire, J., Kabano, J. and Rubagiza, J. (2004) 'Redefining Rwanda's future: the role of curriculum in social reconstruction', in S. Tawil and A. Harley (eds) *Education, Conflict and Social Cohesion*, Geneva: UNESCO (International Bureau of Education).

Saiti, A. (2003) 'Evidence from Greek secondary education', *Management in Education*, 17(2): 34–8.

Saitis, C. (1999) 'Management in the Greek system of higher education', *Mediterranean Journal of Education Studies*, 4(2): 69–90.

Sallis, J. (1988) *Schools, Parents and Governors: A New Approach to Accountability*, London: Routledge.

Sautter, R. (1993) *Charter Schools: A New Breed of Public Schools. Policy Briefs 2*, Oak Brook: North Central Regional Educational Laboratory.

Schultz, T.W. (1997) 'Investment in human capital', in J. Karabel and A.H. Halsey (eds) *Power and Ideology in Education*, New York: Oxford University Press.

Scott, P. (1989) 'Accountability, responsiveness and responsibility', in R. Glatter (ed.) *Educational Institutions and their Environments: Managing the Boundaries*, Milton Keynes: Open University Press.

Scruton, R. (1984) *The Meaning of Conservatism*, London: Macmillan.

Shyaka, A. (2005) *The Rwandan Conflict: Origin, Development, Exit Strategies*, Kigali: National University of Rwanda.

Sibomana, A. (1997) *Hope for Rwanda*, London: Pluto Press.

Simkins, T. (1997) 'Autonomy and accountability', in B. Fidler, S. Russell and T. Simkins (eds) *Choices for Self-Managing Schools*, London: Paul Chapman Publishers.

Simon, B. (1988) *Bending the Rules: The Baker 'Reform' of Education*, London: Lawrence and Wishart.

—— (1991) *Education and the Social Order 1940–1990*, London: Lawrence and Wishart.

Singapore Teachers' Union (2000), *Towards a World Class Education System through Enlightened School Management/Leadership and Meaningful Educational Activities*, Singapore: STU.

Smyth, J. (ed.) (1993) *A Socially Critical View of the Self-Managing School*, London: The Falmer Press.

Sockett, H. (1980) 'Accountability: the contemporary issues', in H. Sockett (ed.) *Accountability in the English Educational System*, London: Hodder and Stoughton.

Southworth, G. (1998) *Leading Improving Primary Schools*, London: Falmer Press.

STA (1998) *Trojan Horses: Education Action Zones – The Case Against the Privatisation of Education*, London: Socialist Teachers' Alliance.

Starratt, R. (2005) 'Ethical leadership', in B. Davies (ed.) *The Essentials of School Leadership*, London: Paul Chapman Publishing.

Stevenson, H. (2003a) 'On the shop floor: exploring the impact of teacher trade unions on school-based industrial relations', *School Leadership and Management*, 23(3): 341–56.

—— (2003b) 'Evaluating the EAZ experiment: learning lessons from the zones', paper presented at British Educational Leadership Management and Administration Society Annual Conference, Kents Hill, Milton Keynes, 3–5 October.

—— (2004) 'Building Capacity through Leadership Development: Exploring the Contribution of Education Action Zones', paper presented at British Educational

Leadership and Management Association/Standing Conference for Research on Educational Leadership and Management Research Conference, St Catherine's College, University of Oxford, 8–10 July.

—— (2005) 'From "school correspondent" to workplace bargainer? The changing role of the school union representative', *British Journal of Sociology of Education*, 26(2): 219–34.

Stoll, L. and Fink, D. (1996) *Changing Our Schools*, Buckingham: Open University Press.

Stone, M. (1985) *The Education of the Black Child: The Myth of Multiracial Education*, London: Fontana.

Strange, S. (1997) *Casino Capitalism*, Manchester: Manchester University Press.

Sui-Chu, E. (2003) 'Teachers' views on educational decentralization towards parental involvement in an Asian educational system: the Hong Kong case', *International Studies in Educational Administration*, 31(3): 43–57.

Swann, Lord (1985) *Education For All: The Report of the Inquiry into the Education of Children from Ethnic Minority Groups*, London: HMSO.

Taylor, S., Rizvi, F., Lingard, B. and Henry, M. (1997) *Educational Policy and the Politics of Change*, London: Routledge.

Theakston, J., Robinson, K.D. and Bangs, J. (2001) 'Teachers talking: teacher involvement in Education Action Zones', *School Leadership and Management* 21(2): 183–98.

Third Outline Perspective, The (2001) *Plan 2001–10, Malaysia*, Kuala Lumpur: Ministry of Education.

Thrupp, M. (2004) 'Mediating Headteachers: A Case of Wishful Thinking?', paper presented at British Educational Leadership and Management Association/ Standing Conference for Research on Educational Leadership and Management Research Conference, St Catherine's College, University of Oxford, 8–10 July.

Thrupp, M. and Willmott, R. (2003) *Education Management in Managerialist Times: Beyond the Textual Apologists*, Maidenhead: Open University Press.

Tomlinson, H. (2000) 'Proposals for performance-related pay for teachers in English Schools', *School Leadership and Management*, 20: 281–98.

Tomlinson, S. (2001) *Education in a Post-Welfare Society*, Buckingham: Open University Press.

Tooley, J. (1994) 'In defence of markets in educational provision', in D. Bridge and T. McLaughlin (eds) *Education and the Market Place*, London: Falmer Press.

Torres, C. (2002) 'The state, privatization and educational policy: a critique of neo-liberalism in Latin America and some ethical and political implications', *Comparative Education*, 38(4): 365–85.

Tschannen-Moran, M. (2001) 'Collaboration and the need for trust', *Journal of Educational Administration*, 39(4): 308–31.

Van der Heijden, A. and Eden, C. (1998) 'The theory and praxis of reflective learning in strategy making', in C. Eden and J. Spender (eds) *Managerial and Organisational Cognition: Theory, Methods and Research*, London: Sage.

Vincent, C. (ed.) (2003) *Social Justice, Education and Identity*, London: Routledge Falmer.

Walford, G. (1994) 'Political commitment in the study of the City Technology College, Kingshurst', in D. Halpin and B. Troyna (eds) *Researching Education Policy: Ethical and Methodological Issues*, London: Falmer Press.

—— (1996) 'School choice and the quasi-market', *Oxford Studies in Comparative Education*, 6(1): 7–15.

—— (2001) *Doing Qualitative Educational Research*, London: Continuum.

Walker, A. and Dimmock, C. (eds) (2002) *School Leadership and Administration: Adopting a Cultural Perspective*, London: Routledge Falmer.

Wallace, M. and McMahon, A. (1994) *Planning for Change in Turbulent Times: The Case of Multi-Racial Primary Schools*, London: Cassell.

Weber, M. (1922) in R.A. Dahl 'Power', in M. Haugaard (ed.) (2002) *Power: A Reader*, Manchester: Manchester University Press.

Weeks, R. (1994) 'The deputy head and strategic planning', in M. Crawford, L. Kydd and S. Parker (eds) *Educational Management in Action: A Collection of Case Studies*, London: Paul Chapman Publishers.

Welle-Strand, A. and Tjeldvoll, A. (2002) 'The Norwegian unified school – a paradise lost?', *Journal of Education Policy*, 17(6): 673–86.

Whipp R. (1998) 'Creative deconstruction: strategy and organisations', paper presented to the ESRC Seminar Series, Redefining Educational Management, Cardiff.

Whitfield, D. (2000) 'The Third Way for education: privatisation and marketisation', *Forum: for Promoting 3–19 Comprehensive Education*, 42(2): 82–5.

Wilkins, R (2002) 'Hardy pioneers of the zone frontier', *Education Journal*, 63: 9.

Wilkinson, R. (1996) *Unhealthy Societies: The Afflictions of Inequality*, London: Routledge.

Wong, E., Sharpe, F. and McCormick, J. (1998) 'Factors affecting the perceived effectiveness of planning in Hong Kong self-managing schools', *Educational Management and Administration*, 26(1): 67–81.

Wong, K-C., (1995) 'Education accountability in Hong Kong: lessons from the School Management Initiative', *International Journal of Educational Research*, 23(6): 519–29.

Woodhall, M. (1997) 'Human capital concepts', in A.H. Halsey, H. Lauder, P. Brown and A.S. Wells (eds) *Education: Culture, Economy, Society*, Oxford: Oxford University Press.

Wragg, E., Haynes, G., Wragg, C. and Chamberlain, R. (2005) *Performance Pay for Teachers: The Experiences of Heads and Teachers*, London: RoutledgeFalmer.

Wright, N. (2001) 'Leadership, "bastard leadership" and managerialism: confronting twin paradoxes in the Blair education project', *Educational Management and Administration*, 29(3): 275–90.

—— (2003) 'Principled "bastard leadership"? A rejoinder to Gold, Evans, Earley, Halpin and Collarbone', *Educational Management and Administration*, 31(2): 139–44.

—— (2004) 'Intelligent Accountability: A Beginning of the End for "Bastard Leadership"?', paper presented at British Educational Leadership and Management Association /Standing Conference for Research on Educational Leadership and Management Research Conference, St Catherine's College, University of Oxford, 8–10 July.

Yancey, P. (2000) *Parents Foundation Charter Schools: Dilemmas of Empowerment and Decentralization*, Peter Lang: New York.

Yip, J., Eng, S.P. and Yap, J. (1997) '25 years of educational reform', in J. Tan, S. Gopinathan and W.K. Ho (eds) *Education in Singapore*, Singapore: Prentice Hall.

Zohar, D. (1997) *Rewiring the Corporate Brain: Using the New Science to Rethink How We Structure and Lead Organisations*, San Francisco, CA: Berrett-Koehler Publishers.

Index